Jason D. Mark

Grimm Fate

From France to Stalingrad:
The Life and Death of a
German Battalion

Grimm Fate

From France to Stalingrad: The Life and Death of a German Battalion

Written by: Jason D. Mark

Copyright © 2025 Leaping Horseman Books

Leaping Horseman Books
Sydney, NSW Australia
Email: info@leapinghorseman.com.au
Web: www.leapinghorseman.com

We are interested in hearing from authors with manuscripts or book ideas on related topics.

First published: 2025

National Library of Australia ISBN 978-0-9922749-8-6

CONTENTS

Bodensee

Ramifications of the Second World War are still felt today. It propelled the Soviet Union – a backward, feudal society before the war – into a leading position as a global superpower and set the scene for the tense East-West standoff that persisted until the 1990s and has since re-emerged. Despite this, the conflict on the Eastern Front is just a vague concept to most Westerners. Hollywood and popular media would have us believe that the war was won in the West in 1944-1945, yet by that stage, the Red Army had been battling Germany for three years, pushing the vaunted Wehrmacht back from the doors of Moscow and Stalingrad and out of Soviet Russia. However, thanks to those movies and books, we understand what the experience was like for the Allies fighting in Western Europe, Italy and the African desert. But what was it like on the Eastern Front?

The pitilessness and unbridled ferocity of the clash between Hitler's Germany and Stalin's Soviet Union cannot be conveyed in an overarching study of the entire conflict, it needs to be viewed at the micro level, from the perspective of a group of men bonded together by hardship, it needs to be viscerally felt through the experiences of the individual. This book was written with one objective in mind: to convey the German experience on the Eastern Front at the human level, to look at the greatest conflict in history through the eyes of ordinary male citizens, menfolk drawn from all walks of life. Why choose the 305th Pioneer Battalion, some may ask. In all honesty, the acquisition of an incredible hoard of documents prompted the creation of this book, so the battalion was selected by default. It was not a glamorous unit, it did not belong to one of the elite divisions, such as the SS-Leibstandarte or Grossdeutschland, not even to a panzer division, instead, it was part of a run-of-the-mill infantry division. However, the depth of the material and the battalion's ordinariness actually make it perfect for an examination of this type. Apolitical officers, reservists mainly, were in command, while its ranks were filled with slightly older recruits, very few of whom had seen action. By the end of its time in Russia, no member of the battalion would be decorated with high awards like the Knight's Cross or German Cross in Gold, though many did receive a posthumous wooden cross, while hundreds of others simply disappeared in the apocalyptic climax to Stalingrad and ensuing captivity.

Revisionists have attempted to portray Operation Barbarossa, the German invasion of Soviet Russia, as a pre-emptive strike, a bold attack to smash Red armies arrayed along the border, ready to surge westward through central Europe, but there is no denying that Germany was the aggressor. It was the Wehrmacht that crossed the border in June 1941, it was the Wehrmacht that pushed deep into the Soviet Union and waged all-out war with the goal of creating "living space" for the German people and relegating the local populace to serfdom and slavery. That being said, the vast majority of German soldiers were not evil, they were simply faithful, obedient members of society bearing arms at their government's behest, initially proud that perceived historical wrongs were being righted, but also content to finally face the growing Communist threat in the East. Many did not know they were the bad guys until told by the victors, until they saw the ghastly excesses of the Nazi regime, the foul deeds carried out in their name.

When the 305th Pioneer Battalion arrived on the Eastern Front in May 1942, it was a first time visit for the vast majority of its men, but they had a sense of what awaited them. Everyone knew someone who had been killed in Russia. Stories about the brutality of combat on the Eastern Front had travelled through the soldier's grapevine, so they were under no illusions, in contrast to those who participated in the opening stages of Barbarossa. They believed they would prevail, but that ultimate victory would come at a cost. How would the battalion compare with more experienced units? Were they less jaded and more inspired than units that had been at the front since Barbarossa began? Was the arrival of a combat-hardened commander enough to compensate for the almost complete lack of experience amongst both officers and men? How did the ongoing casualties affect the remaining soldiers and the battalion's combat performance? And what of their families? How did the war and loss of loved ones affect them? Finally, how did this war alter the soldiers themselves? Nothing makes unrelated humans bond closer than shared adversity, and nowhere is that felt more intensely than in a combat unit deployed in a war zone. Every soldier will agree that the blend of mateship, loyalty and cameraderie is what makes war bearable. How did these fire-forged friendships withstand the Stalingrad debacle and post-war turmoil? These are some of the questions that will be examined throughout the course of this book. It is hoped that the answers will serve as a warning to those called to arms by their government.

– – –

The acquisition of documents used in this book, most of them collected and preserved by former company commander Richard Grimm, is one of those rare culminations of perseverance and good fortune that stand out in a researcher's life.

Most searches lead to a dead-end or a contact that only produces a snippet of information, occasionally a small collection of photos or documents. Rarely, however, will a search end with the proverbial pot of gold. Please bear with me as I explain the process of contacting Grimm's daughter because it is a prime example of the detective work that goes into writing a factual book. Without this tenuous string of luck and circumstance, these documents would not have been found and you would not now be holding this book in your hands.

The name "Grimm" was almost unknown to me when I began the research phase of "Island of Fire" in early 2003. Over the next year or so, I had nothing else to add to his biography except for the fact that he was an Oberleutnant. His name did not appear in any records at my disposal. The breakthrough came in April 2004 when the son of Major Wilhelm Traub (commander of the 305th Pioneer Battalion) sent me his mother's "Suchakte" [search file], a heartbreaking collection of papers and letters documenting Frau Traub's quest to discover what happened to her missing husband. Within the folder was a copy of a letter dated 30 October 1949 from Grimm to the wife of Gefreiter Karl Welker, a soldier still listed as MIA. The letter was rich with facts and detail that I'd seen nowhere else. It was obvious that Grimm was a meticulous man interested in the history of his old battalion and keen to resolve the many MIA cases. Surely such a historically-minded person still possessed their archive? With that in mind, I set about tracking down this "Oberleutnant Grimm."

Further personal details proved frustratingly resilient to discovery. It wasn't until eight months later that two vital pieces of information emerged: Grimm's first name (Richard) and home address in 1949, both revealed in a letter written by Karl Binder, a veteran of 305th Infantry Division, survivor of Stalingrad and self-appointed historian who maintained a roster of "Stalingrad-Heimkehrer" (soldiers captured at Stalingrad who returned from captivity). This particular letter came from another line of research that had nothing to do with Grimm. With his home address, I now possessed a solid starting point for my search. An enquiry to the registry office of Stuttgart, Grimm's home town in 1949, was sent off on 17 January 2005. A week later, I received a reply: Richard Grimm had moved to Kirchheim unter Teck in 1950. I sent off an enquiry to Kirchheim that very same day and received an immediate reply, almost always a bad sign when dealing with bureaucratic agencies: all they could tell me was that Grimm had died in 1990. I dashed off a quick email asking for their help in establishing contact with Grimm's family. In particular, I wanted to know if Grimm's children still lived in Kirchheim. Silence. I sent my request again in April. Once again, no reply. I tried once more in July, with the same result. Being blatantly ignored like this is an uncommon occurrence when dealing with German officials, but that was of little consolation to me: I needed to contact Grimm's family. I sidestepped my contact person at Kirchheim by locating someone else in the same office. I lodged a tactfully worded complaint about my three unanswered enquiries and once again asked about contacting Grimm's family. Two days later I received a reply stating that yes, they had information about Grimm's children, but they could not divulge their addresses due to privacy laws… unless I was related. Those less scrupulous than I may have lied and said they were indeed next-of-kin, but such an untruth would inevitably rear its ugly head in the future, so I decided to lay all my cards on the table. In a lengthy email, I laid out all the information I had found about Grimm, described my contacts with other members of the battalion and stressed the importance of Grimm's archives – whose existence I only presumed – to my work. Abiding by the strict privacy laws, I asked Kirchheim to contact Familie Grimm on my behalf, explain my intentions and pass on my details. Two weeks later I received a positive reply from Kirchheim:; they agreed with my suggestion and contacted Grimm's family. Before I had time to thank them, I received another email, this time from Grimm's daughter. I was astounded. Her very first sentences were as follows: "It is to be deeply regretted that the city of Kirchheim took so long to inform me about you. I am the daughter of Richard Grimm and have the documents and diary entries of my father." In an email a short time later to Agnes Moosmann, my research assistant in Germany, I wrote: "I think this may be our biggest discovery yet!" Prophetic words.

The two-year search for Grimm's family was swiftly repaid when his daughter, Edda, began sending through scans of her father's 108-page manuscript. Within two weeks I had received all the pages, plus two separate 7-page accounts of the fighting in Stalingrad in October-November 1942. And this was only the beginning. The "Schatzkiste" [treasure chest], as we jokingly called it, contained stacks of other documents, letters and photos. (When my wife and I first visited Edda in 2007, I was admittedly a little disappointed to see that the Schatzkiste was just a cardboard box and not the brass-hinged wooden chest I'd always imagined.) Over the next few months, hundreds of those documents and letters were extracted and examined, and they generated countless leads to other veterans and families. The complete destruction of 6th Army ensured that very little documentary evidence left the pocket, but Grimm had gathered more than enough to record the rise and demise of his battalion and its men. A few words from Grimm himself, written in 1986, are the best explanation:

In the 76th year of my life, I want to bring together all available records and reports.[1] This is as the former commander of this company. Added are notes from memory, letters to comrades, photocopies of reports to Battalion together with reconnaissance reports. They are truthful reports and it can be recognised that the reports were written on the company typewriter kept at the baggage train.

The records exist because prior to the beginning of the campaign I had decided to write a detailed report about pioneer operations under my command after each action, for my comrades after the war. Company members suspected that because of my daily dictation – which were amended by the clerks – that I was going to write a book. I still heard this after the war. I was not so presumptuous that I wanted to write a book.

The final battle, the encirclement of Stalingrad, led to the destruction of all company documents. Why or how I managed to take my reports and records out of Stalingrad still remains unclear to me. I handed these things to my wife upon my arrival in our Degerloch residence. Bravely and cautiously, she tucked them away until war's end under some hay in the harnessing shed on my parents' piece of meadowland on the Notzinger Steige (in Kirchheim/Teck).

During the advance, I received a 1:100,000 map because I was a company commander, and there were often no other copies for the platoon leaders. Every day I marked our precise route with march performance, crossings, etc., on this map. At the baggage train it was joined together with glue and rolled up. At every opportunity, whenever I had to "go back to the baggage train" or if the baggage train caught up to us, I had a look at this approximately two-metre-long map. Who knows whether this map is located in some Russian archive. If I'd had this map with me when the Americans were around, after the war, I surely could have gained an advantage because upon returning home, all soldiers were immediately questioned

---

1   Grimm did not create his war-time reports with an eye to future publication. This is how he described them: "The combat reports were not Kriegsberichten [war communiqués written by a reporter]. I only wanted to record the mission of the company and its performance, as well as its casualties. I did this against regulations. Every day I dictated the tactical elements to the clerk with a message pad and the clerk himself went to the platoons to gain an outline about wounded and dead, number of prisoners, pioneer material used – mines of each type, barricading material like plain wire, Spanish riders, etc. The reconnaissance reports were dictated and typed up on a machine at the supply train in order to be forwarded. Special incidents had to be immediately reported to the commander, Major Beismann."

about their knowledge of Russia before they could even embrace their families.

It depends upon the state of my health as to whether I can attach photos to these records. I've got no time to lose by creating a preliminary draft. I'll type directly into the machine with the aid of my notes.

Grimm would spend the final four years of his life working on this draft. When cancer finally took him, Grimm had completed his draft and appended it with well-captioned photos. This book would have been impossible without it.

– – –

How do I even begin to thank Edda for her help? In addition to providing the initial bonanza of documentation that was the genesis of this book, she transcribed hundreds of pages of handwritten letters into typewritten text and fulfilled countless requests for further information, but most significantly, proactively sought out veterans and next-of-kin of deceased or MIA soldiers. Her status as Grimm's daughter opened doors, rekindled memories and produced further leads. We have become good friends and she welcomes me into her home whenever I visit Germany.

Edda was part of a formidable trio that pooled resources and assisted me throughout the course of this book. Agnes Moosmann (†) and Robin Diez have assisted me on previous books, and once again they navigated the often frustrating world of German bureaucracy to procure information and documents from dozens of local archives and registry offices. My deepest thanks and eternal friendship to them both.

Major contributors to this project were veterans Georg Zeller (†) and Josef Zrenner (†), the daughter of Oberleutnant Ludwig Beigel, the family of Hauptmann Paul Göhner, the son of Leutnant Hubert Homburger, the daughter of Hauptmann Hermann Klein, the daughter of Major Eugen Rettenmaier, the wife of Obergefreiter Heinz Rinck, the grandson of Oberleutnant Heinz Schaate, the son (†) of Major Wilhelm Traub and the nephew of Obergefreiter Friedrich Vorherr. As there are too many veterans and families to list here, I want to thank them collectively for supplying unique material and entrusting their memories to me.

Finally, my everlasting love to Belinda, my wife, for her unstinting support and devotion, but especially for nurturing our two children.

Jason D. Mark
Sydney, September 2025

"The Most Pleasant Period of the War"

February 1941. Leutnant Richard Grimm surveyed the scene, his compact stature attracting little attention, his cherubic face masking the disciplinarian inside. The town and surrounding landscape may have been snowy and bleak, but a feeling akin to the first day of school hung in the air. Small cliques, formed over many years and shared experiences, chatted and laughed. Cigarette smoke mixed with breath vapour. A steady influx of newcomers gravitated to the larger groups. Riedlingen's small town square soon filled with soldiers and their kit. Curious residents gazed out of their windows. This was just another posting to a foreign unit for Grimm, his third in seventeen months. Everybody was a stranger, apart from his company commander, the diminutive Oberleutnant Hans-Joachim Bloch, who stood barely 5'2". At the appointed hour, Bloch mounted an ersatz dais and welcomed the men to the 305th Pioneer Battalion. He then set his officers and senior NCOs to work billeting the men with local families. Grimm and fellow platoon leader Leutnant Siegfried Lutz were assigned a house in the heart of the town. Their hosts, a husband, wife and teenage boy, received them with open arms. As with all the soldiers, Grimm and Lutz were treated like sons. Gefreiter Josef Zrenner, a clerk on battalion staff, remembers this as a wonderful time for the unit. As a Bavarian, he was charmed by the Swabians and their dialect, and even sixty years later could name all the pubs in town. More memorable for a young man, however, were the many beautiful ("but decent!") girls. Several men eventually married Riedlingen lasses.

Circumstances denied Grimm the chance to settle in. After leading a training course in the snow-covered hills around the town, a telegram summoned him home: his 65-year-old father, a powerful and respected factory owner, had passed away unexpectedly. Grimm's superiors were understanding and granted immediate compassionate leave. Whilst his homecoming was under a black cloud, it was offset by conflicting emotions of happiness. Grimm had been called away for service a week after marrying Charlotte – close friends called her Lotte – in August 1939, so they had barely spent any time together as husband and wife. The longest period of marital union had begun in December 1939 after he was recalled to his civilian job to assist with a secret project for six weeks. Shortly after rejoining his unit, Grimm received news from his wife that his furlough had borne fruit; she was due in October. Hot on the heels of this good news came bad: his father's will and testament handed control of the family company, which manufactured electrical parts for many of Germany's famous auto companies now geared for war production, to Grimm's older brothers, Wilhelm and Fritz. Grimm and his sister missed out.

– – –

The first months of 1941 were tumultuous for Grimm. Not so for his country. Much of Western Europe had been conquered, Great Britain was cornered on its islands and the non-aggression pact with the Soviet Union had secured Germany's eastern frontier. Half a year had passed since German soldiers were welcomed home with floral garlands and kisses from euphoric Frauleins. Many had been doubtful about the Führer's decision to wage war on France and Britain but now, with most of continental Western Europe under German control and the war seemingly at an end, the future looked bright. Sure, Britain was still undefeated and putting up a fight in the distant North African theatre, but Rommel was taking care of things there.

Despite being in the military – on and off – for six and a half years and being called up a week before the Wehrmacht invaded Poland, Grimm had yet to see action. He had volunteered in October 1934. Due to his ten years of schooling, he was a so-called "yearling", only required to serve one year, while those with less education needed to serve two. Many enlisted in the army or labour service because of Germany's poor economic situation, but Grimm chose this path to improve his future prospects and to escape a difficult situation with his father and the family business. Upon completion of his service on 30 September 1935, he returned to his civilian career as a commercial clerk, simultaneously fulfilling his duties as a reserve soldier over the years. On 1 November 1938 he was promoted to Leutnant der Reserve. Shortly before the invasion of Poland he was recalled to active service, though he belonged to one of the divisions arrayed along Germany's Westwall, ready to repel the expected Anglo-French reaction. Participation in training courses at the pioneer schools in Pardubitz and Dessau-Rosslau kept him from the Western Campaign in May-June 1940, though he did arrive at the very end, driving through smouldering cities and passing POW columns en route to his new unit, the 25th Pioneer Battalion, in Bourges. Now in February 1941 he was with the pioneer battalion of the newly formed 305th Infantry Division, one of nine new divisions currently springing up over Germany.

After victory in France, Hitler decided to assign the bulk of equipment and manpower to the Kriegsmarine (navy) and the Luftwaffe and scale back the army from 156 to 120 divisions. During the autumn and winter of 1940, however, this decision was reversed when he decided to attack the Soviet Union in spring 1941, so rather than being downsized, the army was to be expanded to 180 division. In October 1940 the high command ordered the formation of new divisions for occupation duty so that combat units stationed in France could be freed for deployment in the east. These "bodenständige" divisions, static units, had fewer personnel, less firepower and reduced mobility compared to regular divisions, while some of their armaments came from captured stocks. Responsibility for their recruitment and formation fell to Germany's Wehrkreis (military district) system. Grimm was assigned to one of these new units (the 305th Infantry Division) because the division's formation and his posting were both controlled by Wehrkreis V (Military District V) in the south-western German state of Baden-Württemberg. Two-thirds of the new division's strength came from 78th and 296th Infantry Divisions, while new recruits fleshed out the framework.

This same "transfer and expand" method was used to create 305th Pioneer Battalion. The staff, light pioneer column (commonly abbreviated to le.Pi.Kol.) and two companies – the 1st and the 3rd – came from Wehrkreis V units and their men had the traditional Swabian traits of industriousness, conviviality, piety and a quaint sense of humour. The remaining company, the 2nd, had a slightly different character: while some men hailed from Swabia, most came from other southern areas like Würzburg, Nürnberg, Ingolstadt and even Austria. Nevertheless, as "southerners" with dialects and sing-song accents less easily understood by others, they shared a common bond.

An explanation of the origins of each pioneer company will be of benefit. The 1st Company was in fact just 2./178th Pioneer Battalion with a new name. Formed in January 1940, it did not participate in the Western Campaign, though it did march into Belgium and northern France with the rest of 78th Infantry Division as occupation troops. Leading the company was 41-year-old Oberleutnant Hermann Klein, a First World War veteran with almost 15 years of military service. One of his platoon leaders was Leutnant Kurt Birk, eldest son and heir apparent of a respected manufacturing family from Tuttlingen. In his luggage, as always, was his beloved accordion.

Grimm's 2nd Company was the rebranded 2./296th Pioneer Battalion, a unit whose garrison was in northern Bavaria but its origins were part-Austrian. Some of its members had accrued a little combat experience during the final weeks of the French campaign. Two of the company's platoon leaders, Leutnant Ludwig Beigel and Leutnant Peter Buchner, were best buddies. Buchner was the eldest son of a farming family from a tiny Bavarian village. Beigel and Buchner volunteered for service in the 10th Pioneer Battalion on the same day in November 1937. Since then, their careers had been identical: transferred to the same units, attended the same three-month officer training course and received parallel promotions. Barely a week after receiving their officer's commission, they arrived at 2./296th Pioneer Battalion but France had already fallen, so they had no chance to prove themselves in action. A short while later, Buchner was shifted to Klein's 1st Company. The new commander of 2nd Company, Oberleutnant Hans-Joachim Bloch, was brought in from the 25th Pioneer Battalion. Grimm came from the same unit. Bloch had gained a taste of combat during the second stage of the war against France but did not lead troops into battle. Grimm fulfilled the dual roles of platoon leader and Bloch's deputy. The last officer in the company was Leutnant Siegfried Lutz, a giant man whose wire-frame glasses and genial smile made him seem more like a pastor than a combat engineer.

The formation of 3rd Company was vastly different to the other pair. It was organised from elements of a fortress engineer battalion, a corps-level unit responsible for building and maintaining fortifications. The light pioneer column (le.Pi.Kol.) had the same origins. The men in both units tended to be older. At least five of 305th Pioneer Battalion's officers originated from these fortress units, including two that arrived with the new 3rd Company: Leutnant Friedrich Häntzka, a tough career NCO with the Iron Cross ribbon in his buttonhole recently promoted from the ranks, and Leutnant Heinz Schaate, a tall bespectacled Stuttgarter. Schaate looked so similar to Leutnant Lutz that many thought they were twins, so the pair jokingly called themselves "Zwillingsbrüder" (twin brothers). Transferred in from the 35th Pioneer Battalion to lead 3rd Company was Oberleutnant Hans Oster, a Great War veteran. Oster had led troops into battle in the Low Countries and France during the 1940 campaign, and in doing so, had earned the clasp to his World War One Iron Cross Second Class.

Appointed CO of 305th Pioneer Battalion was the commander of the aforementioned fortress battalion, Oberstleutnant Hans Hertel, a 55-year-old career officer and Great War veteran. When the armistice was signed in 1918, Hertel was a decorated Hauptmann and battalion commander with years of experience gained on the bloodiest sectors of the Western Front, from Flanders in the north to Verdun in the south. In 1919 he joined Stahlhelm, Bund der Frontsoldaten ("Steel Helmet, League of Frontline Soldiers"), a paramilitary organisation that arose after the German defeat. After demobilisation in 1920 he married and started working at the family factory, of which he took control after the death of his brother. After Hitler came to power in January 1933, Hertel became active in the SA (Sturmabteilung) later the same year, but never joined the Nazi Party. Hitler's program to expand the Wehrmacht prompted Hertel's recall to active service in 1935. Sometimes cantankerous, always controlling, he took his new battalion firmly in hand. Over the coming months he took great pleasure in educating his officers about the finer things in life, like food, wine, art and history. He also arranged day trips to local landmarks, museums, wineries and architectural wonders, such as the baroque cathedral in Zwiefalten.

The staff of Hertel's new battalion came from the 35th Pioneer Replacement Battalion, again, another Swabian unit. His adjutant was a 22-year-old reservist, Leutnant Max Fritz, who had received his officer shoulderboards just a few months earlier. Fritz was a reserved man who still lived with his parents, but quickly proved to be unsurpassed in his new role. The battalion doctor was university graduate Unterarzt Dr. Werner Dopfer, a jocular man whose charisma, perpetual good humour and impish smile quickly endeared him to all. Oberveterinär Dr. Erich Scheffel was in charge of the battalion's veterinary services. Although born a Saxon, he had set up practice in Stuttgart and years of running it provided a wealth of anecdotes with which he regaled his new comrades. Everyone agreed he was an outstanding storyteller. The paymaster, Oberzahlmeister Alfred Gabelmann, serious and introverted, was a long-time soldier who knew his responsibilities inside out. The commissary officer was Leutnant Dr. Reinhold Haan, a lanky man with protruding ears transferred in from the mountain troops. Finally, there was Inspektor Georg Zeller, the mercurial and well-liked technical inspector. Before conscription he was a civil engineer at Firma Hahn & Kolb in Stuttgart, manufacturers of industrial tools like lathes, drill presses and boring machines. Grimm had worked as a salesmen at this same factory before the war. Zeller's functions within the battalion included maintenance and bookkeeping of the battalion's motor vehicles and pioneering equipment, as well as overseeing driving tests and issuing licences, tasks that brought him into contact with most of the unit's men. As a result, he was one of the most visible officers, while his nature made him one of the most popular. The predominantly Swabian staff officers swiftly melded into an efficient team.

– – –

The 305th Infantry Division was officially founded on 15 December 1940. The senior commanders were the first to arrive, meeting in Ravensburg for a conference two days before Christmas to discuss the organisation of their division. The same participants, plus some new staff officers, met again on 18 January 1941 to discuss formation and training. Recruits aged 22 to 36 began arriving in late January to flesh out the division. The 305th Pioneer Battalion was based in Riedlingen, with the exception of 1st Company, which was housed in nearby Unlingen. All troops were billeted with local households. The other units of the division – almost immediately nicknamed the Bodensee Division because of its emblem – were similarly accommodated in towns throughout the area. The population was friendly, the weather mild and training exercises were carried out in the rolling fields of the Donau River valley. When the river broke its banks during the winter thaw, the pioneers assisted the townsfolk of Riedlingen by shoring up the broken levee with large props, ballast and earth. The troops enthusiastically contributed over 2000 Reichsmarks to an annual welfare drive and helped with local events. Overall, the atmosphere was relatively relaxed. Spouses and girlfriends visited on selected weekends. In this way, the officers' wives quickly became friends and established a support network.

Leutnant Grimm returned to the battalion shortly after an advanced order arrived for the division's transfer to France. Preparations were completed and the entire unit was ready to move in the first week of April. The pioneer battalion departed in two consignments, two days apart. The destination was Bordeaux in south-western France. As part of 7th Army, the division's main task was to protect part of the demarcation line with Vichy France. The pioneer battalion was designated the division's operational reserve and housed in the Libourne barracks. The lodgings pleased Grimm. His room had running water and the addition of his personal radio together with glassware and a small electric cooker purchased from a local shop made it quite comfortable. The discovery of cheap but excellent wine was also a thrill. The soldiers would stroll along the town's tree-lined avenues and riverside esplanade in the evenings. Theatres playing French movies were popular, as was a Wehrmacht variety show featuring a talented magician. The pace of life during the first weeks in France is evident from Grimm's chagrin when a much anticipated siesta was thwarted by a call from Hertel to get some work done. The casual atmosphere was broken for good on 22 April when the division commander, General Pflugradt, inspected the battalion's training area and barracks. The adjacent Boisbretrau grounds were to become the division's main training base and it was up to the pioneers to get everything ship-shape. General Pflugradt's visit had implications, as Grimm knew full well when he wrote to his wife: "Not as many letters will come [from me] next week [because] we had senior visitors yesterday."

The division's units began a fortnight of intensive training on 28 April but the pioneer battalion was excused this drill. A large-scale manoeuvre was carried out on 15 May with opposing sides split into red and blue teams. The pioneer battalion did not participate but Oberstleutnant Hertel was invited along as a spectator.

In mid-May the men of 2nd Company experienced "an unexpected change", as Grimm called it. "My dear good commander Bloch has been transferred [...] so now I lead the company until an old Hauptmann arrives. I would have much preferred Bloch to stay because all of us have become used to him and he really is a fine commander. Still, we hope for the best, come what may." Nobody, including Bloch himself, knew that the true purpose of his transfer was to take command of a pioneer company for Operation Barbarossa, the German invasion of the Soviet Union. The "old Hauptmann" arrived on 23 May: Grimm handed over the reins of 2nd Company to Hauptmann Wilhelm Traub. His glasses and paunch gave him a benign, headmasterly appearance that contrasted sharply with his resplendent medals and ribbon bar. Traub was yet another old soldier who had fought in the trenches of the Western Front and Russia during World War One. Demobilised in 1919, he began teaching at the naval school in Wesermünde in 1925, a role he fulfilled for the next 14 years. Recalled for service in early 1940, he commanded a pioneer company with great success. For a seemingly suicidal amphibious assault across the fire-lashed Meuse and the subsequent pursuit, Traub received clasps to both his Great War Iron Crosses, making him one of two officers in the 305th to have earned the First Class grade during the current war (the other was Hertel). A broken kneecap was cited as the reason for his removal as commander of 2./269th Pioneer Battalion just a month before the invasion of Russia. Traub and his military experience were warmly welcomed by the battalion.

With a new company commander in place, Grimm was given a fresh assignment that brought him into closer contact with Oberstleutnant Hertel, a prospect that did not entirely thrill him because as time went by, he saw traits in Hertel's character and leadership style that were not to his liking:

Every day, or at least every second day, we ride from 11 to 12 o'clock. In this way we learn to stay in the saddle without using stirrups. It's lucky the course is covered with deep sawdust, so that somersaults [off the horse] cannot be felt. The 'old man' stands on the course with his whip and yells. If he thinks he can educate his gentlemen in this way, then he's sadly mistaken.

Regardless, Grimm would have to adjust to Hertel's brusque manner because his new role was on the battalion staff as commissary officer:

Tomorrow I set up my commissary section on the staff. This is a position equal in responsibility to a company leader. The 'old man' spoke very highly of my special mission and he found the trust to charge me with this task.

The division was ordered to hand over its positions along the demarcation line and take over a section of coastline west of Bordeaux in the last days of May 1941. The transfer was to be implemented by 12 June at the latest. The division received strict instructions to leave billets in the condition they would wish to receive them and those who took along non-Wehrmacht material would be prosecuted. The pioneer battalion's new billets were in St. Georges de Didonne, a small town on the north shore of the Gironde estuary. With its broad sandy beaches, turquoise waters and balmy climate, this small coastal enclave felt like paradise. The officers of battalion staff had prime accommodation, as Grimm excitedly related in a letter home:

My commander, the adjutant, the paymaster and I live alone in a beach house. First, there was obviously a struggle against dirt and dust, but now I can say that we live like princes. If you were ever to see my four-poster bed! Around it, hanging down from the ceiling, is a rich fabric. I think it's some kind of mosquito protection. I can close it up nicely and even have light inside. Of course, running water is not missing here. Beautiful lounges with a marvellous billiards table. I am beginning to become a handy player. The garden is very nice, probably a bit overgrown, but there are masses of beautiful roses. In the evening, when the cool sea breeze comes in, we light the marble fireplace just for fun, it's such a little pleasure. I can even swim because all I need do is step out of the garden gate and into the water. I have also found a small restaurant, so I always have some bone marrow, eat fabulous plaice, etc. I love the fish here, it's like our freshwater fish… One might almost think that I'm living in paradise.

A few days working with Hertel, however, spoiled Grimm's Eden, as he explained to his wife:

I've already told you a lot about the quirks of my commander. Working with him is not easy. On top of that, it's like I'm chained to my desk. He monopolises all of our free time until he goes to bed. We have to figure out how to get away from him. My Sundays and Saturday afternoons are no longer free. So, not much is different here. However, a lot will change when it gets hot, so I'm looking on the bright side.

His theory was put to the test when sweltering weather set in. Despite Grimm's optimism, Hertel kept him and the other staff officers on a short leash: "The water was nice and warm. At 7 o'clock in the evening it was still 32° C in the shade. We had to reschedule our working hours because of the heat. The only thing is that we have to stay seated for at least two hours during lunch with our old boss and no-one can get away…" One mellow afternoon, fed up with their gilded prison, Grimm and other staff members snuck away and strolled down to the beach. After a delightful dip, they ambled back, only to realise they had fallen afoul of one of Hertel's quirky habits: he liked to observe the beach through a scissors telescope and had spotted the truants. Face aglow with rage, he scolded his officers as if they were naughty schoolboys.

Life was easier for the enlisted men. As jacks-of-all-trades, the pioneers were used for various purposes and not tied down to defending the coast or engaging in manoeuvres. Oberpionier Willi Füssinger from 1st Company explained to his family how his talents were used:

A few days ago, at low tide, our company commander discovered a relatively large fishing boat embedded in the mud. One evening, with a good deal of trouble, we dragged it out. I saw that only a few boards in the bottom were broken and the panel at the stern was caved in. Now I, a wainwright, have become a boatbuilder. I had to repair the boat the next day and repaint it and at noon yesterday it was once again seaworthy. The first test run went well and we were out on the water for over three hours. Of course our Oberleutnant, two Leutnants and the Spieß came along. In the next few days I will make five rudders but I'll have to shift my work into the shade. At the moment it is 41° C, over 50° in the sun. It could get hotter still. We don't do any work from midday to 4 o'clock, we just swim. I have become a good swimmer.

One Saturday afternoon, during a rare break from deskwork, Grimm was involved in an adventure on the high seas. A strong wind had whipped the sea into white caps. Against all advice – and orders – a pair of men had gone out too far in a paddle boat and found themselves being pushed towards a wave-battered headland. Fortunately for the hapless pair, Grimm and a few friends, who were riding the wind in the boat of a friendly local sailor, steered towards them. "You're bold lads!" Grimm called out, unaware of the seriousness of their plight. At that moment, he realised they were rudderless and heading for the breakers. Too late. The waves crashed upon the pleasure craft and pulverised it. The two men floundered in the churning water. The risk of drowning emphasised the ludicrous ignorance of the two men: heading out into crashing seas in a gaily coloured paddle boat! "It was funny but exhausting work for us," Grimm recalls "because we had to drag those guys from the water." After dropping off the soggy duo at the next small harbour, Grimm's sailboat headed back out for another two hours. He was not going to waste a precious day off.

Despite ongoing duties, plenty of opportunities existed to get to know an interesting foreign country and its residents. The sun shone constantly, restaurants offered excellent cuisine and wine, and the locals, while not especially friendly, tolerated the occupiers with Gallic aplomb. The partisan threat was virtually nil. Men wandered into the countryside, filling their mess-tins with blackberries or bartering with farmers for eggs and milk. Whoever experienced spring and summer 1941 in south-west France remembered it as the most pleasant of the war.

Any complaints about training, heat and surly commanders disappeared when news of the commencement of Operation Barbarossa on 22 June arrived "like a bombshell." "The immense battle of decision has started in the east," Grimm wrote to Lotte. "Much will be seen in eight weeks. My best wishes accompany both of your brothers into this tough struggle. To be a soldier, you must be hard. Now we have proof why the war with England could not be finished off over the past year. We hope for the best in this battle." Oberpionier Füssinger summarised the common beliefs drummed into most Wehrmacht soldiers by years of nationalistic indoctrination:

What has been thought about for a long time has started in the east. I would be in the attack there if I was still with my old division... How long this campaign will last, no-one knows, but I believe the Führer has made preparations for a swift victory. My opinion is that this conflict would have come about sooner or later. It's good that the Russians, not us, are at the wrong end of the stick. So far these false brothers have been profiting from the war and now they are up against us. Hopefully Bolshevism will be beaten this time. Perhaps this campaign will bring the decision closer. We soldiers hope that the Lord is again with our weapons...

It was soon realised that victory in Russia would demand a heavy price after news trickled in about friends and comrades who had been killed. Oberpionier Füssinger learned that two friends had already been severely wounded, one missing and another killed. Grimm was stunned when he received a letter from an old comrade: "My good-hearted former commander, Oberstleutnant Klotz from the 35th Pioneer Battalion, has been killed in the east. This has really bowled me over." Weeks later, after Hertel had got deeper under his skin, Grimm was still mulling over Klotz's demise:

If I had to draw a comparison between my first commander in the war, Oberstleutnant Klotz, and now, I keep asking why, of all people, did such a magnificent man, leader and superior have to die. This example faces me every day... a few days ago I handed in my resignation and requested a transfer. "Impossible!" was the answer. Thus I comfort myself and think, well, that's that. This evening the Oberzahlmeister – his favourite, of all people – tendered his resignation too. I'm anxious to see what comes of it.

The leisurely pace in France allowed some officers to be sent on training courses. Leutnant Beigel attended a three-week course for platoon leaders at the Army Engineers School in Angers, midway between the Gironde and Paris. The instructor's analysis: "Shows clear determination, good knowledge and diligence. Suitable as a platoon leader."

Life became easier for Hertel's staff officers when he went on leave in early August, but not before causing them untold misery in the days prior to his departure. On the first day of his commander's absence, Grimm calmly went about his duties, had a nap after lunch and caught up on some personal correspondence in the afternoon. As the ranking officer at mealtimes, he also put an end to Hertel's two-hour lunches: "As soon as the meal is finished, a cigarette is unhurriedly smoked and then everyone leaves whenever they want." Grimm even had time to visit a toy shop in town and buy a small doll and waterproof picture book to send home: Lotte was now seven months pregnant with their first child. Three days after Hertel's departure, he wrote to his wife: "Since I no longer have to breathe in [his] venom, I am a new man. Last night I went for a short ride with a few comrades and this evening we want to visit a lonely lighthouse keeper." Balmy nights were passed in the garden. Soon, circumstance and Hertel's pre-planning – whether unintentional or malicious – wreaked revenge upon Grimm. His small section was overwhelmed with work, yet Hertel had sent one man to school and Inspektor Zeller on a training course, and Feldwebel Bromeis fell ill. Three men were left to do the work of six. Grimm was thrown a lifeline when word came through that a new officer was on his way to take over the job. Grimm hoped that he would arrive before Hertel returned. No such luck. Hertel came back from leave on 4 September. The new Hauptmann, an "older gentleman" reported in a few hours later. This new arrival, 48-year-old Hauptmann Hugo Neddermann, did not bring the relief Grimm expected. Hertel told him that Neddermann would take his place but he was to remain where he was for the moment. Minor compensation for this unpleasant extension of his tenure was notification of his promotion to Oberleutnant, backdated to 1 August.

Two weeks out from his wife's due date, Grimm tried to arrange furlough so that he could be with her, but Hertel rasped: "Wait, Grimm, wait until you receive news of the birth." When Grimm persisted, Hertel sarcastically asked how he knew when the baby would arrive. Grimm replied confidently: "I expect to get a telegram on the 17th." As a father of two, Hertel knew better: "Well, you'll see, there's always unexpected delays with the first child." And that was that. Despite his burning desire to be with his heavily pregnant wife, Grimm was forced to stay at his desk, continue to work and put up with his grating boss. "I'll be glad when I'm on leave just so I no longer have to hear 'Grimm!' screamed out every minute of the day." Ultimately, Grimm's premonition proved to be the most accurate. Just before noon on the 18th, the welcome tiding reached him: "Gisela has arrived, both are doing well." The new dad was over the moon and close to tears of both joy and frustration. The adjutant, Leutnant Fritz, was on furlough, so he was filling in that role as well as his own, and Hertel declared that he could not go home until Fritz returned. Grimm was mortified. That was almost three weeks! Protestation and pleading had no effect. Grimm broke the bad news to his wife. Fortunately, friends were looking after her, so she was not alone. The days dragged by and Grimm was treated to another one of Hertel's explosive outbursts over a minor matter:

> Our "old man" was invited to see his boss last night. You should have seen the commotion beforehand! We have a barber and because it was Saturday and he wanted to buy a flexible shaft for his electric clippers, I let him go to a neighbouring town. I gave him two official orders to take care of at the same time. Tonight, of all evenings, the old man wants to have his hair cut but the barber is not here. What a shitstorm! I was in the orderly room and was called in at least ten times... When I went back to maison rouge, my comrades and I laughed our heads off!

Grimm boarded a train for home on 8 November. His daughter was already three weeks old when he first laid eyes on her. They were the lucky ones. As the war in Russia reaped a grim toll, countless babies throughout Germany were fatherless even before they were born. The Bodensee Division had so far been spared this tragedy, but not for long. When the Red Army launched a massive counteroffensive in December 1941, it was clear that forces would need to be extracted from France and sent to the eastern theatre, and indeed, small batches of men were taken in mid-December. For example, on 18 December, the battalion handed over eighteen men to bolster the 246th Pioneer Battalion. Another ten were withdrawn on 26 December, this time for the 211th Pioneer Battalion. It was stipulated that they be born after 1914 – thus younger soldiers – and be suitable for the war in the east. Finally, on 29 December, another nine men were sent to the 8th Pioneer Battalion. As a whole, the division handed over 39 NCOs and 717 men.

Grimm returned to the battalion on 1 December and was told by friends on the staff that he might be shifted to another post, away from Hertel. The rumours proved true. Mid-December, he assumed command of the light pioneer column, which transported the battalion's ammunition reserve, demolitions equipment, mines, barbed wire, compressors, power and hand tools, and all the other accoutrements required by the pioneers to carry out their myriad tasks. With three platoons under his control, Grimm was now a bona fide company commander, but more significantly for him, he had escaped Hertel's clutches. With his new-found autonomy came a new billet: he moved his possessions from maison rouge to a "fantastic home" that he temporarily shared with his predecessor, "but there is room for all."

The division celebrated Christmas in its old billets. The pioneers laid on a feast. "The Christmas party started at 1800 hours [on 24 December]," Grimm recalls, "and my men were in high spirits. Geese had been organised for months, so there was an eighth of a goose for everyone and it was eaten and we drank. Then I gave a speech worthy of the occasion, and remembered our fallen soldiers and especially our great leader… [Hertel] was our guest for a while and brought me a Santa Claus gift from the staff. A child's pop gun with a cute dedication… I barely recognised the old man because of his pure cheerfulness." Hertel also presented Grimm with unexpected recognition: "For excellent work on the staff and as a section leader, I was awarded the War Merit Cross Second Class with Swords. I'll wear it in my buttonhole until I earn the Iron Cross, which is every soldier's desire." After a further exchange of presents, a choir began singing carols and was quickly joined by everyone present. Alcohol, robust dialogue and frivolity made the evening pass quickly. Grimm left at one in the morning. He cleared his fuzzy head on Christmas morning by riding his horse through villages, forests, over the dunes and onto the beach. Reinvigorated, he set about mastering a stack of paperwork before renewing celebrations in the evening with his NCOs. More goose, more meat, roast potatoes, cauliflower and even real coffee. "I said to my men that we could never eat this well at home."

There was no chance to revel in the festivities: a transfer to a new sector had already been arranged. In the final days of 1941, the 305th Division handed over its Gironde coastal defences and shifted northward by train to take over a sector that stretched from the U-boat bases at Lorient to the port city of St. Nazaire. Broad golden beaches were interspersed with inlets and rugged headlands. Amongst the last to arrive were the pioneers. They celebrated New Year's Eve in their old billets. With the exception of Oberst Hertel and Leutnant Beigel, every officer in the battalion gathered at Grimm's house to ring in 1942. Top-shelf champagne, cognac and cigars accompanied a dinner of freshly-shot hare. No-one present had an inkling that the coming year would change their lives forever.

The new coastal defences were officially taken over on 5 January. Within two weeks the division began to bolster the defences by subdividing its tract of coastline into three sectors and allocating resources to each. The construction skills of the pioneers were frequently called upon. Absent from this labour, however, was Traub's 2nd Company: it had been assigned to the Kompanieführerschule (company leaders school) in Tours as instructors. Grimm was far less busy these days, his main task being the procurement and transportation of building material to the sites of defence construction. His new lodgings in a cliff-top cottage pleased him greatly. While he sat inside completing paperwork, the wind whistled around the house and storms rolled in off the Atlantic. His housemate, Dr. Erich Scheffel, was like an old friend, and their personalities instantly meshed in the most agreeable way. At the end of most invariably wet and cold days, they sat next to a crackling oven, sipping cognac and puffing away on cigars while Scheffel – a first-class raconteur – recounted stories of his life and veterinary practice back home. "He's a really marvellous character," Grimm told his wife, "and we hope we can continue to live together." Military necessity dashed their wish: at the end of January, Dr. Scheffel was transferred to another unit within the division.

The Bodensee Division's cushy occupation duty on the French coast was ended by a decree from 7 February 1942 that ordered the reorganisation of the division into an assault unit by 15 April. Most of the unit's original commanders, predominantly older officers, were transferred out of the unit in February and March 1942. This process started at the very top. Division commander Generalmajor Kurt Pflugradt was replaced by Generalmajor Kurt Oppenländer, recently promoted to General and almost recovered from a serious arm wound suffered on the Russian front. The unnerving rigidity of all fingers on his right hand caused many to think it was a prosthesis. All three infantry regiment commanders were changed, as was the artillery regiment commander. Many command positions in subordinate units also changed hands so that younger leaders were now in charge. Officers recovering from wounds suffered in the east were transferred to the division to take over battalions and companies. The transformation to an attack division brought an increase in men and weaponry. The conversion also affected the pioneer battalion. Transfusion of young blood for old had forced Hauptmann Traub from commander of 2nd Company to battalion staff (as special duties officer). Taking his place was Oberleutnant Grimm, as he explained to his wife on 5 March:

> In Riedlingen [in early 1941] I led a company, that of Oberleutnant Bloch, for extended periods of time. Today I was advised that in the foreseeable future I will most likely have to take over this company as commander. My current unit has worked its way into my heart but this agreeable change will be an improvement.

Whilst retaining control of the light pioneer column, he assumed temporary command of 2nd Company on 10 March, though it was an inauspicious beginning: "I rode to my company but arrived in the rain and was soaking wet. As long as clothes can be dried in a warm room, it's not too bad. How different it'll be at the front…"

The biggest upheaval in the 305th Pioneer Battalion was at the top. Oberstleutnant

Hans Hertel was replaced by "a young Hauptmann decorated with many war medals". Hauptmann Friedrich Beismann was a dashing leader fresh from the front. His dog-tag revealed that he was the inaugural member of 3./46th Pioneer Battalion and indeed, had led that company for well over three years: defended the Westwall during the Polish campaign, built bridges and cleared mines through the Netherlands and France in 1940, supported panzers driving south through Serbia and Greece and finally participated in the invasion of the Soviet Union. It was in this campaign that Beismann shone as a leader, not just in the art of military engineering but also as a combat officer.

Hertel's departure on 15 March roused mixed feelings amongst his officers. Good riddance to his controlling attitude and fussy habits, but his epicurean ways, particularly the pursuit of the finest wine and brandy, would be missed. Hertel was devastated by his expulsion from the battalion and suffered a breakdown. Rather than heading to his new assignment, he checked himself into the military hospital in Rennes. An examination by the division's sympathetic head doctor found him to be suffering from "neurosis of the cranial nerves, signs of exhaustion, a depressive mood, an inability to concentrate and is currently unfit for regular duties." Oberarzt Dr. Dopfer had requested a four-week stay at a health spa for Hertel; the divisional doctor concurred.

Granite-jawed Beismann was a no-nonsense firebrand who opted to maintain the status quo for now. The days of long lunches and guard duty on sun-drenched coastlines must soon end and he would then strip away the fat of a leisurely year of occupation duty. Grimm had barely settled in as company leader when Beismann came crashing into his life. Initial pleasure at the advancement in his career was swiftly tempered by a headsplitting workload and consequent personal disappointment when hopes of surprising his wife and child with unexpected furlough were dashed. No officer could be spared. The spectre of deployment on the Russian Front loomed.

For the moment, deep peace reigned in Gouarec. Base pay could be exchanged for French francs and local food was good and plentiful. Wine and cognac. Eggs were dirt cheap. Cakes and pastries were available in a small bakery in the afternoon and both the baker and German personnel respected the fact that everyone should receive some. Even the hotel housing the battalion staff offered modest food to locals and soldiers alike. "We simply could not comprehend that we had it so good here," wrote Grimm.

Beismann began to get the measure of his officers. Some he liked instantly, particularly the younger, more aggressive officers: Leutnant Beigel was a favourite. Others, like Traub and Klein, he respected as Great War veterans. He had reservations about Grimm. They dined together on 19 March and Beismann rattled off question after question about various matters. Both men were avid hunters: Grimm had even brought his sports rifle with him to France, so he invited his commander to go shooting. Grimm had his eye in, quickly picking off three wild pigeons, but was miffed when Beismann borrowed the rifle and returned it only after expending the few carefully husbanded rounds. Satisfied with what he had seen and heard, Beismann told Grimm that complete control of 2nd Company would soon be his. A few days later Grimm's fortunes took another unexpectedly favourable turn when he was assigned a new billet:

After a long ride, my [horse] brought me to a gorgeous château. What's more, the mansion is splendidly positioned in a vast, delightful old park with small lakes. Inside, however, the château, as usual, is old, old and ancient. As a hunter, I could work here. Game everywhere. Boar, deer, etc. It is a so-called hunting reserve, where people such as me would never be allowed to hunt... I'm waiting for the order to take over the company. I will then have to leave the château. I would gladly stay here for a few more weeks, making it easier to forget the beautiful beach [in Saint-Georges-de-Didonne].

– – –

A jingling phone broke the silence. Leutnant Fritz fumbled for his wristwatch and saw the time through foggy eyes: just after 0330 hours. Division was on the line. The battalion was told to quickly get ready to march because the nearby port of St. Nazaire was under attack. Fritz phoned through the order to all companies. Officers were roused from their slumber, NCOs instructed to get the men up. The mood was electric. The Allies had landed! It was 28 March 1942 and the British had launched an audacious amphibious attack against the massive dry dock at St. Nazaire. The pioneers kitted themselves for combat, readied their weapons, and were told to stand by. Hours passed before another call came in at 0815 hours: stand down, the raid had been defeated.

The billets of 2nd Company were still abuzz when Grimm officially took over the company in the evening, but he returned to his previous lodgings in the château, to sleep in luxury one last time. He considered his new accommodation in Gouarec to be "in last position" compared to his previous French billets. Nevertheless, the inevitable could not be delayed, so he moved in the following day, 29 March:

The company officers and rank-and-file from staff had private quarters in the village. Second Company was accommodated in a flour- and fruit warehouse on the edge of the village. The mill was about eighty metres away, somewhat lower down, on the channel that powered a waterwheel. Lying in between was a road and a small fish ladder with a courtyard in front of the mill. About twenty metres opposite the flour warehouse – on the same level – was the two-storey house of the mill owner, occupied by two elderly aunts and the nephew who ran the mill. My predecessor, Hauptmann Traub, had his quarters there.

My takeover of the company occurred during services in which the battalion order was read out by the Spieß. The detachment of Hauptmann Traub was celebrated in the hotel by the commander and officers, while I rode through the area with my Kompanietruppführer (company staff leader) to gain an overall picture. I heard that I was being looked for but was able to get out of it and go to the hotel later. The wine there was so powerful that no-one really noticed my absence. The handover of the company still had to be celebrated anyway.

It was already dark when my orderly and I carried my things into my new quarters in the miller's house. Everything was blacked out, so I could not find my way to the door without being escorted by the old ladies.

Two nights later, Grimm was ripped from his slumber by screams and banging on his door. Despite his grasp of the French language, he could make no sense of their frantic yelling, but he thought he heard the word "help". While pulling on his pants he called back: "Speak slowly so that I can understand you." The two old women burst into his room. "Monsieur, monsieur, le moulin brule!" The mill is burning! Through the open door he saw a flickering light on the lower floor, and upon emerging from the house, saw that the mill was well and truly ablaze and being fanned by the wind. Grimm ran next door to the company billets in the warehouse. "Everyone up! Platoon leaders to me!" Loudly, so that everyone could hear, he issued orders, and a practice alarm – usually signifying that enemy paratroopers had dropped – was sounded. Fuelled by ancient timberwork and volatile flour dust, the flames roared like a blast furnace and consumed the mill's interior. Explosions hurled flaming flour sacks out of the windows and high up over the roof, where they burst in the air and let loose a cascade of fire, like a fireworks display. The local fire brigade moved forward with hand pumps and the pioneers assisted with spades, compressors and buckets of water from the Plavet, yet it was obvious to all that this inferno was beyond control. The mill lay on the edge of the village but the stiff westerly breeze was pushing the flames towards adjacent houses. All of Gouarec was in peril. Villagers closest to the mill began saving their most prized possessions from the threatened houses. Drastic intervention was needed. "Beigel, prepare demolition teams," Grimm called out. "Clear everyone from those buildings," ordered Grimm, pointing to a small row of houses hunched helplessly in front of the conflagration. There was no need: those residents were already out, rescuing their furniture. Acting quickly, the pioneers lodged explosives against load-bearing walls, inserted detonators, wired the demolition charges and cleared the area. Crack! The sturdy stone houses collapsed amidst churning clouds of dust. The row of houses was sacrificed to create a fire-break. Grimm's heavy-handed solution worked: the fire was contained. When day broke he allowed his company to stand down, but kept his staff section and one platoon at the site of the fire. Hauptmann Beismann arrived mid-morning when the fire had flared again. Other officers arrived to watch the spectacle. One of them, fresh from the Eastern Front, was unimpressed by Grimm's decisive action, and remarked snidely: "Good, Grimm, your first independent decision!"

Despite the demolition of several dwellings, word spread swiftly through Gouarec that the small, round-faced officer had saved the village, instantly making him a popular figure. "Every resident greeted me on the street today," he wrote to Lotte, "even though I had not been particularly nice to them [in the past] because I had to chase them into their homes when [enemy] planes threatened." His hosts offered him prime position in the kitchen, in front of the hearth, and served freshly baked Breton pancakes. Even the miller, always on his guard with his German boarders, thanked the "Capitan" and his unit for their help. Grimm's deed also made the local newspaper.

– – –

Released from its coastal defence duties, 305th Infantry Division was designated army reserve and ordered to hasten the completion of its replenishment and training. Beismann put a bomb under his officers: unexpected visits and incisive queries were swiftly followed by orders for major alterations in the structure and training of the companies. There was no longer any doubt that they were destined for deployment in the east. Training was tougher. Live rounds were issued instead of practice ammunition. The men became familiarised with gunfire by sitting in foxholes while bullets flew overhead. Obstacles were blown up, hand grenades thrown and stamina increased by a gruelling exercise regimen. "I will lead my company into combat this year," Grimm explained to Lotte. "I have a lot of work, but at least now I know what I'm working for. After all, leading a company is a fine and great responsibility. I think back to my time as a recruit in 1934 and now find myself as the leader of a combat-strength company. Whatever I do, whatever I work on, it's for my company. It is even accepted that I will lead this unit into battle and because of that, no amount of work is too much." Grimm's pudgy body was tightened up by the training. The contrast to a year earlier was marked. Beismann forced his officers to be as fit, if not fitter, than their soldiers, and what was inflicted upon the leaders was done unto the men:

I work with my company incessantly, day and night. Sternness is called for and no
mercy is shown... Some see hard training sooner, others later. Out with all the
softness in our bones. This will be proven in combat! My commander knows no
softness, but he has been on all fronts and had success after success.

Beismann knew first-hand the correlation between tough training and front-line
performance. He pushed his officers harder. The closer the date approached to rail
transport for operations in the east, the more hectic became the training. An order was
enacted to get the troops accustomed to little sleep. At frequent intervals General
Oppenländer inspected the infantry and close-combat training, the laying of anti-tank
and anti-personnel mines, as well as mine clearance, activity on the water and
construction of makeshift bridges. On 28 March Oppenländer visited 2nd Company
training at Bon Repos and chatted with Grimm:

My company was inspected by the General today. I constructed a beautiful bridge
really quickly and all my superiors were delighted with the company's performance.
Well, don't ask how I crammed beforehand. For the present I know just one thing
and that's performance. Only this leads to success and congeniality and personal
consideration, even towards myself, is completely forgotten.

An advanced order to transport the division eastward arrived on 22 April, and the
official order on 28 April 1942. This news was inferred by the rank-and-file in different
ways. Gefreiter Willi Füssinger, working in the supply train of 1st Company, received a
clear sign when told to gather march rations for twenty days. The imminence of
forthcoming deployment in the east prompted all men to send home unneeded
baggage. Grimm dispatched a large trunk which included his dress uniform, but
Beismann insisted that all officers take their ceremonial swords with them to Russia.
Affairs were finalised a couple of days before departure. Oberleutnant Grimm ensured
that his life insurance policy would not lapse by arranging automatic payments with his
bank. He also sent home a small tin box containing a movie reel (showing the battalion
in training), photos, a spare wristwatch, his medals and some personal papers. He also
prepared his wife for hard times to come. "You'll have to wait a long time for mail," he
wrote, "so you must be patient." Although he was on active duty and away from home,
the unfathomable depths of Russia were alien and sinister when compared to easy-
going France. After dodging the subject for several paragraphs, he finally expressed his
love to his wife, asked her to keep talking to their seven-month-old daughter about him
and wished them good health "to make the many hard times easier." This same
sentiment was repeated countless times as the finality of deployment to the Eastern

Front weighed upon the division's men. Once affairs were in order, Grimm and his officers celebrated their forthcoming commitment with a night of drinking, resulting in some sore heads the next morning. Before heading to the primitive conditions of the Eastern Front, Grimm treated his officers to a final meal of chicken, roast potatoes, a green salad and wine. "We'll think back about this day more than once," Grimm predicted.

The men of 2nd Company marched through the dawn streets of Gouarec, backpacks and weapons over their shoulders, jackboots clacking on the cobblestones and echoing hollowly off the stone houses. Villagers watched silently as they filed past. There was no arrogance in this parade: it was no victory march – these men were off to fight in the East. The pioneers boarded a local passenger train bound for Loudéac. Lack of loading facilities at Gouarec's small station had compelled Grimm to send the company's vehicles and supply train to Loudéac the previous evening. The train's whistle blew at 0528 hours and as it gathered speed, the men watched the small village disappear from view behind some trees. The scenic trip through the bucolic Breton countryside was over before the soldier's made themselves comfortable. They disembarked at Loudéac and reunited with the supply train. All 210 men, vehicles, horses and other pioneer equipment were smoothly loaded onto 53 wagons and the train departed at 1546 hours in the direction of Le Mans. Grimm's 2nd Company was the last of the battalion's units to depart. As their train chugged through rural France, Hauptmann Häntzka's 3rd Company was already approaching Kharkov and Klein's 1st Company was a few days behind. The battalion's companies were destined to go into battle individually.

They rolled through the Marne valley emblazoned with spring blooms, along the upper reaches of the Mosel and into Germany at Efringen. Then began a whistlestop tour of their Fatherland: into the wonderful Nahetal, through the flowering gardens of the Rhinegau, up the Main and into the industrial areas of Saxony and Upper Silesia. They stopped at Landshut in East Prussia on 18 May to exercise the horses for an hour before continuing the journey via Przemysl. The train squealed to a halt outside every station and sat for hours. Finally, inevitably, they entered the vastness of the East. The countryside and villages had not yet awoken to spring. Melt-water lay on distant fields. The populace looked downtrodden, miserable. Traces of war multiplied the further east they moved: dead cities, burnt-out of buildings, tank carcasses. Grimm numbly watched the depressing scenes. Anxiety gripped his stomach. What awaited them?

CHAPTER 2

Bodensee

Train after train rumbled into stations west of Kharkov on 13 May. General Oppenländer had orders to assemble his division outside the city. Men and horses scrambled out of the narrow railway cars in which they had been confined for nine days. "The Russians have broken through!" The call crackled through the ranks like wildfire. Officers were briefed immediately and then they jumped on their men. Stiff from days of idleness, the troops strapped on their gear, shouldered weapons and left their excess kit with the supply trains. The dire situation demanded their immediate deployment.

Hauptmann Häntzka's 3rd Company arrived after noon. Although ostensibly the mobile component of 305th Pioneer Battalion, they offloaded only one car and 104 horses in Merefa; their remaining cars and trucks would arrive the following day. Häntzka was ordered to hightail his pioneers across the city and join the other divisional elements – infantry of Regiment 578 and some artillery batteries – that had arrived hours earlier. His platoon leaders, Leutnant Hupfeld, Leutnant Staiger and Oberfeldwebel Hahnenstein, harried their troops, and before long they were formed up. The pioneers marched in loose formation through the outskirts and into the streets of central Kharkov. This metropolis never failed to impress first-time visitors, "a city of Bolshevik character, modern skyscrapers standing next to wretched huts," remembered one officer. There was no time for sightseeing. With gunfire grumbling in the distance, they trudged towards billeting areas on the city's north-eastern outskirts. Oppenländer's fresh men were to be held back until the threat was analysed and the main Soviet effort identified, but the front greeted them sooner than expected in the form of columns flooding back. The retreating troops were mostly orderly, though some wore panicked expressions and a few even talked about Russian tanks steamrolling everything in sight. The newcomers had the distinct feeling they had been thrown in at the deep end. Then, a new order soon rang out: "Everyone turn around – March!" The mixed combat group, including Häntzka's pioneers, tramped back through Kharkov to Lyubotin. A hasty order then sent them south-westward to defend the Novaya Vodolaga area.

– – –

What had happened? The Bodensee Division expected to arrive to a stabilised situation on the Eastern Front. After recoiling in the face of the Soviet counteroffensive in December 1941 and enduring months of gruelling winter combat, the German leadership and troops managed to halt their foe in March 1942 and then stabilise the front. The stubborn retention of some sectors and the loss of others was revealed in the tortured course of the front-line, where deep Soviet penetrations dovetailed with German protrusions held by exhausted units. In the southern sector the most significant Soviet intrusion was the Izyum bulge west of the Donets. Hitler issued a directive in early April 1942 for the further conduct of the campaign. All available German and allied units would be concentrated for a decisive blow to knock the Soviet Union out of the war, but the dramatic weakening of the Ostheer (German army in the east) inhibited an offensive along the entire front. Forces would therefore be assembled for an attack on a limited sector with the objective of seizing the oil region of the Caucasus and obliterating the Red Army's southern wing.

The Bodensee Division was to bolster 6th Army, an army set to play a major role in the upcoming summer offensive scheduled to begin on 15 June 1942. Before that date, however, a suitable line of departure needed to be gained through a series of preliminary operations. The Kerch Peninsula in the Crimea required elimination, then the Izyum bulge straightened out and finally the powerful fortress of Sevastopol had to be taken. Elements of 6th Army, 17th Army and 1st Panzer Army were assigned for the Izyum operation (codename "Fridericus") – to commence after the Kerch Peninsula had been conquered – whose objective was to cut off and destroy two Soviet armies west of the Donets with a pincer attack. Slated for this operation was 305th Infantry Division. While the division was rolling eastward on trains, the first German blow had already fallen on 8 May: Generaloberst von Manstein launched Operation Trappenjagd (Bustard Hunt) in the Crimea, and with one brilliantly conducted strike, his outnumbered army completely destroyed the Caucasus Front (three armies) in twelve days, cleared the Kerch Peninsula and provided the rejuvenating elixir thirsted for by the Wehrmacht.

The Russian leadership was not willing to idly sit by and watch the Germans prepare for another offensive. The German operation to snip off the Izyum bulge was set for 17 May but the Soviets forestalled it with an attack of their own. On 12 May, preceded by an hour-long artillery strike and air bombardment, powerful infantry forces and vast numbers of tanks moved off in a dual pincer movement from the Volchansk and Izyum salients. German resistance was fierce but defences were gradually worn down by airpower and concentrated barrages. The deepest penetration on day one was ten kilometres. Early intrusions and the threat to Kharkov attracted German reserves to the Soviet's northern shock group. Meanwhile, the Soviet southern wing scored some spectacular successes in the first three days, pushing deep into the German rear and menacing Krasnograd. Both sides slugged it out in miserable conditions: heavy downpours turned the battlefield into a mud-pit. Conditions favoured the Germans because the bottomless earth prevented Soviet artillery from keeping pace with the armour packs, allowing key German bastions to hold out and preventing the base of the Soviet bulge from expanding northwards. The situation was so threatening by the third day that Field Marshall von Bock, Commander-in-Chief of Army Group South, maintained that it was too risky to implement "Fridericus" in its original form. Three armies had been grouped for the attack and one of them, General der Panzertruppen Paulus' 6th Army, was taking a battering and crying out for reinforcements. With coldblooded professionalism, the German leadership discerned the potential for an overwhelming victory in the chaotic situation, but it relied on husbanding the forces assembled for "Fridericus" and thus denying them to Paulus.

The Soviet spearhead approaching Krasnograd was hindered only by very weak German units and ceaseless Luftwaffe sorties. Powerful forces were also attacking northward in the direction of Kharkov, making the hole north-east of Krasnograd even wider. After much persuasion, three infantry divisions and one panzer division were released to Paulus for the defence of Kharkov. This was the situation facing 305th Infantry Division as it arrived in dribs and drabs. Its inaugural commitment to battle was therefore piecemeal over a wide sector.

– – –

Rainstorms turned the sandy roads to porridge. Häntzka's pioneers trudged south with the division's advance guard, commanded by Oberst Willy Winzer, towards the enemy. Rain drops tapped on their helmets and Zeltbahn ponchos. Wagon columns of broken-spirited men, rear echelon troops mainly, creaked past them in retreat. After forty exhausting kilometres on atrocious roads, they set up defensive positions on the southern and south-eastern fringes of Novaya Vodolaga. The sight ahead of them was depressing: dreary skies and undulating brown plains utterly devoid of grass and trees. Häntzka's pioneers were lumped together with an anti-tank gun company to form Kampfgruppe Althenn and, together with an infantry battalion, moved up and dug into the black soil. They stared into the distance. Nothing was to be seen of the German front-line, nothing except columns falling back towards Kharkov. The withdrawing procession withered away to small groups of stragglers, then individuals, until eventually, not a soul came past. It was eerily empty out in front. Something sinister lurked beyond the horizon. They all felt it.

To the right of Winzer's position was a yawning gap that stretched all the way to Krasnograd, a void of thirty kilometres into which Soviet tanks and cavalry threatened to push. On direct orders from General Paulus, parts of Regiment 576 were hustled aboard a fleet of hastily assembled lorries and driven to Krasnograd to help hold the line. Prevention of a further Soviet advance to the west was imperative. Generalmajor Oppenländer arrived in the embattled city in his staff car after nightfall. During the coming night (16-17 May), all Soviet attempts to take Krasnograd were thwarted.

Back at Novaya Vodolaga, Häntzka's men were unsettled by the quiet. Winzer's regiment had been positioned in the rear to protect the open right flank of 113th Infantry Division and, consequently, the battered 8th Army Corps. The 113th Division had repelled heavy attacks throughout the day, yet local penetrations at sensitive points compelled it to pull back its main resistance line during the night. The front was now only ten kilometres south of Winzer's group. To thicken the new line, Winzer was directed to move south the next morning (17 May) and set up a defence line anchored on the hamlets of Prikhnovka, Sapunov and Karavanskoye. Häntzka's pioneers would go with them. The first troops to arrive would create strongpoints along the line so that defensive readiness was attained from 0930 hours; units arriving later would fill the gaps. The mission was simple: "Cover the western flank of corps against attacks from the south and south-west; in particular, prevent a thrust by enemy forces on Novaya Vodolaga."

First Contact (17 May 1942)

Hauptmann Häntzka and his men slept deeply but were roused long before dawn. Reports of tanks filtered through from the front. As a precaution, a battery of Luftwaffe 88s drove out of Novaya Vodolaga first at 0300 hours. Departing fifteen minutes later was a mixed combat group of infantry, artillery and Häntzka's pioneers. The bulk of Winzer's regiment moved off at 0530 hours. The first light of day was accompanied by ghostly silence. Coming toward them were individual German vehicles retreating ahead of an invisible threat. A single line of thought ran through their heads: "When will we run into the Russian? How will he push forward? And why is he not here yet?" Häntzka's men reached the railway embankment near Sapunov toward 0700 hours without making contact. In the sticky heat, they shovelled out foxholes in the lee of the railway embankment and waited for the situation to develop.

The infantry battalion on the left wing was the first to meet the enemy. While marching into Karavanskoye toward 0900 hours, it encountered a pack of Soviet tanks, yet it took the village and knocked out a heavy Soviet tank in the process. The division's first kill. Binoculars scanned the south-western horizon. Anti-tank guns covered the main approaches. Nothing. Hours passed. Late in the afternoon they appeared; the spearhead of General Pushkin's 23rd Tank Corps. This was it, the division's first true test. The tanks trundled across the barren fields, halting occasionally to fire, then lurching forward again in clouds of exhaust fumes. Explosions bloomed around them but they continued to charge and pushed into Karavanskoye. Furious fighting erupted. The infantrymen were outgunned and fell back. By 1800 hours only the northern half of the village was in German hands.

Winzer's regimental group was in a tight spot: its right flank was wide open, the left only tenuously connected with the exhausted men of 113th Infantry Division. A frontage of about 14 kilometres needed to be held against mounting Russian tank and infantry assaults, and to accomplish this, he had just two reinforced battalions defending two strongpoints (Sapunov and Karavanskoye). The latter was already in trouble.

Just before twilight, Hauptmann Beismann and his battalion staff, together with Hauptmann Klein's 1st Company, disembarked at Merefa station, 20 km south-west of Kharkov. After almost half a year, Beismann was back on the Eastern Front for his second tour of duty. He knew better than anyone else in the battalion what awaited them as he had seen it all during his first stint in 1941: major river crossings; cracking bunkers in the Stalin Line; victorious advances across wide open plains; desperate defence; embittered positional fighting in the Perekop Isthmus, the gateway to the Crimea; the siege of Sevastopol in freezing sleet and plunging temperatures. One particular episode in late July earned him accolades: after crossing the Dniester near Yampol, his pioneer company became part of a motorised advance detachment which also included a bicycle company, anti-tank guns, artillery pieces and some flak. The detachment operated out in front of 46th Infantry Division, and when its leader was wounded several days later, Beismann assumed overall command. When the detachment came under fierce attack from two Russian rifle regiments, Beismann organised his troops for all-round defence on Hill 270 near Lubashnaya. For several days and nights, under the heaviest, most accurate artillery and sniper fire, he rallied his men time and time again to repel dense waves of Soviet soldiers screaming "Hurra!" Ammunition almost ran out several times and casualties mounted, yet the advance detachment held firm under Beismann's prudent leadership until relieved on the morning of the fourth day of the siege. His efforts were recognised with the Commendation Certificate of the Commander-in-Chief of the Army, a document awarded for extraordinary bravery on the battlefield, as well as being named in the army's Roll of Honour. Being awarded both was a rare distinction.

A messenger from Hauptmann Häntzka interrupted Beismann's reverie. He filled in his battalion commander on 3rd Company's actions and situation. Not long afterwards, a runner from Division brought orders. Beismann ordered Klein to get his men ready:

infantry weapons only, leave all pioneering equipment behind. They were granted a night's rest before deployment the following day.

– – –

Winzer's green troops in Karavanskoye were in a nightmare. Tracers zipped through their lines. Gravelly Russian voices bellowed from the blackness and tanks crunched through the village. As soon as one attack was repelled, enemy bogeymen charged furiously from another direction. Worst of all, they were cut off. Winzer extracted units from quieter sectors and formed a new group to break through to Karavanskoye. Häntzka's pioneers were part of this relief force. Apart from small-arms, they lugged anti-tank mines and grenades in preparation to face the armoured behemoths. Small pioneer teams bolstered the infantry. As they marched south, their target was clear: the village was backlit by orange smoke. Gunfire resounded through the night. Utilising darkness and the element of surprise, they struck the Red Army troops manning the cordon north of the village from the rear and scattered them. Contact with Karavanskoye was re-established and most of infantrymen pulled out. Back in Novaya Vodolaga, rumours were rife that the regiment had been torn apart and many in the rear echelons believed this hearsay. Gripped by "tank fear", some were so panic-stricken that they wavered and started to fall back.

Hauptmann Häntzka had firm control of his men, so the hysteria did not afflict them. Although Hauptmann Klein was senior in time-in-rank and age, Häntzka possessed more combat experience in the current war and that was why Beismann selected his company to be the first to arrive on the Eastern Front. Häntzka's entire adulthood had been devoted to uniformed service, first as a no-nonsense policemen in Saxony and the Rhineland, then transferring to the army in the mid-1930s: regulations allowed policemen to receive an equivalent military rank, so he began his military career as a Feldwebel. By the time the war started, he was married, lived in Duisburg and was a senior NCO in the 26th Pioneer Battalion. In April 1940 he welcomed his first child into the world, a daughter Ursula, but just a few weeks later he was urgently called back to duty due to the invasion of France and the Low Countries. The campaign commenced sedately enough for his unit, mainly building and improving bridges over the many waterways, but before long the pioneers were drawn into combat and Häntzka proved his mettle as a courageous and conscientious platoon leader. His performance was rewarded with the Iron Cross Second Class and a recommendation for an officer's commission. On 15 October 1940 he was promoted to Leutnant and joined the 305th Pioneer Battalion the next year. Further career advancements followed swiftly: Oberleutnant's pips were added to his shoulder boards by the end of 1941 and he was the natural choice to lead 3rd Company after its commander, Oberleutnant Oster, was transferred. Beismann submitted a glowing assessment and strong recommendation for promotion, so shortly before relocation to the Eastern Front, he was elevated to Hauptmann. Tall, powerfully built and with a ramrod-straight posture that came from a life devoted to uniformed service, Häntzka commanded respect from both subordinates and officers from other branches of the service, thus vindicating Beismann's decision to send in his company first.

Beismann's Return to the Eastern Front (18 May 1942)

As day dawned the "tank fear" that pervaded the rear elements of Regiment 578 was calmed. The Russians came in at 0335 hours. Häntzka and his men, held in reserve, keenly observed the proceedings, knowing full well they could be pulled into battle at any moment. The first attack, prepared by heavy artillery but without tank support, was thrown back. Owl-eyed observers had spotted Russian tanks and infantry assembling in nearby woods during the night; Stukas swept in at 0415 hours and plastered the forest. The pioneers were awed by the spectacle. An hour later the Russians came at them again, this time bolstered by four or five tanks, and again they were repelled, leaving behind a pair of smouldering wrecks. More probed the line throughout the morning. Six tanks were eventually knocked out. At this time, T-34s were greatly feared because the 3.7cm Pak 36s were nigh on useless, except if the gunner blew off a track or scored a lucky hit on a vision port or turret ring, and the few 5cm Pak 38s, while more potent, were few in number and spread thinly along the line. The sight of T-34s being blasted by the 88s, therefore, helped the inexperienced troops overcome their initial dread when faced by these seemingly invulnerable behemoths. And if all else failed, it was up to the pioneers to stop the tanks, but to do this they needed to get up close and personal.

By late afternoon Winzer's casualties were about fifteen dead, probably twenty to thirty wounded, but the Russian assaults had not ended. Another one, supported by six tanks, surged towards the western edge of Karavanskoye at 1900 hours. Forty minutes later the Germans emerged victorious. Prisoners revealed that many of their tanks were British-made Valentines and thickly-armoured Matildas. To the Germans, it seemed they had halted the armoured advance at Karavanskoye, but their repulsion of the Soviet attacks had coincided with new orders received by their foe. General Pushkin halted his offensive based on a new directive to switch the axis of his advance eastward. Hänztka's pioneers were shifted to a new sector during the night.

– – –

Trains bearing units of the Bodensee Division continued to arrive. The men were sent to threatened areas, with most of the fresh blood being infused into the teetering sector of 113th Infantry Division. After marching most of the day, Hauptmann Beismann's staff and Klein's company reached Borki late in the afternoon. This small village, nestled in a barren dale between a shallow hill line and a copse, sat in the path of a Soviet salient that protruded ten kilometres into the German line. Beismann ordered Klein to extend the left wing by establishing a series of defensive strongpoints along a 1500-metre front. Enemy armour posed the main threat, so tank destruction squads were equipped with plate-mines and hand-grenade bundles. A few anti-tank guns from 113th Panzerjäger Battalion were also positioned along the line. The sector was ready for defence by nightfall. An occasional artillery round howled in and outposts scared off several enemy patrols, but in general, the night passed quietly for the pioneers. Hauptmann Beismann received confirmation of his assignment in an order from 113th Infantry Division: his staff and Klein's company were now part of Sperrverband (Blocking Unit) von Wedel, an ad hoc unit consisting of the staff of 113th Panzerjäger Battalion, remnants of 113th Reconnaissance Battalion, a battalion of 88s, two artillery batteries and five 5cm anti-tank guns. Their mission was to prevent the Russians from pushing any deeper into the left flank of 113th Infantry Division by setting up strongpoints on high ground and in scattered villages. Each defensive point consisted of a tank-busting gun protected by mine barriers and infantry.

Red Giant and Hill 211.7 (19 May 1942)

Oberst Winzer was under pressure on both wings. The Russians penetrated further into the open flank on his regiment's right while on the left a Russian rifle company advanced along the railway embankment and reached Burov Station, severing the railway link to Novaya Vodolaga and the connection to their neighbours. Winzer reacted swiftly and launched a counterattack with a few dozen men, including some of his own staff. This surprise move threw the intruders back over the embankment and a connection with Regiment 260 was re-established. The Russians left behind twenty dead. Nevertheless, the regiment was forced to bend back its left wing. After a small German attack was called off late in the afternoon when it failed to make any appreciable headway, the Russians followed up at 1900 hours with powerful forces – a dozen or so tanks, including heavy types – and attacked Karavanskoye from the west and south. Several tanks broke into the village. Fires raged out of control at several spots. The frantic efforts of all officers succeeded in banishing the "tank fear" that threatened to grip the defenders. One of Häntzka's platoons was sent into the dread village where they went to work laying mines and hunting tanks with grenades and Tellermines. The village was held with their help and by dawn it was back in the regiment's control.

– – –

Beismann's pioneers arrived in the nick of time. Three days earlier, on 16 May, the situation at 113th Infantry Division was desperate. Russian tanks and infantry had pushed through a hole between this division and their eastern neighbour (62nd Infantry Division), creating a salient five kilometres wide and ten kilometres deep. They then approached Borki, seven kilometres behind the division's left wing. A glance at a map showed how deep the Russians were – and very little stood in their way. Any available German unit was thrown into their path. One of those was Panzer Battalion Heydebreck from 23rd Panzer Division, pulled out of the line north-east of Kharkov despite the heavy fighting there and rushed south to assist 8th Army Corps. The German leadership had correctly identified the threat posed by Russian tank forces spearing north towards Kharkov. The battalion and its twenty-five panzers reached the sector at noon on 17 May. Corps commander General Heitz told Oberstleutnant von Heydebreck to "act like a battleship and destroy every enemy tank that surfaces!" Von Heydebreck established contact with 62nd and 113th Infantry Divisions; the Russians had not yet broken through but gaps between German strongpoints were growing steadily larger and neither infantry division possessed any reserves. The panzer battalion went into action on the afternoon of 17 May and shattered enemy forces encircling some infantry in Ryabukhino. On 18 May Russian tanks attacked north both sides of Dzhgun Station and threatened the right wing of 62nd Infantry Division. Heydebreck's panzers destroyed six tanks in Borki itself and then attacked south in an attempt to close the hole between the two infantry divisions. Eighth Army Corps warned that "it must be reckoned that the Russians will continue to attack tomorrow, primarily with tanks." The biggest concern remained the gap between 62nd and 113th Infantry Divisions:

> [Both divisions] will close the hole on their inner wings. The cooperation of [Panzer Battalion Heydebreck] is provided for this purpose... As the Russians have repeatedly attacked the boundary between both divisions during the last few days, I expect that after the hole is closed both divisions will do everything to satisfactorily buttress it with artillery and tank-busting weapons.

This was the situation when Hauptmann Beismann and his men entered the picture. The balmy night passed quietly in Borki. Another company of pioneers arrived towards 0300 hours, their commander reporting to Beismann: "Herr Hauptmann, the

Armeepionierführer[1] has sent my company to reinforce you." The arrival of 3./672nd Pioneer Battalion proved timely because two hours later, at 0500 hours, Beismann was ordered to prepare an assault group formed from Klein's 1st, the new pioneer company and a recon squadron from 113th Recon Battalion. The plan was simple: after a Stuka attack on Kovalenkoff, move off at 0600 hours and, with the support of Heydebreck's panzer battalion, attack south via Kovalenkoff towards Hill 199.5. Beismann took command of the assault group.

The droning of Stukas caused the pioneers to gaze skyward as most had never seen them in action before. They rolled into their dives, sirens wailing, and screamed down a kilometre or so ahead. Booommm. Booooomm. The ground trembled and thick clouds of dust boiled up. Excitement rippled through the ranks. Plane after plane blasted the small settlement. "OK men, move out!" called the officers. The assault groups got to their feet and trotted towards smouldering Kovalenkoff. This was it, their first attack. Nervous tension gripped stomachs. Beismann had seen it all before. He strode upright, chin forward, pointing this way and that, even as an occasional bullet chirped past him. The men pushed forward rapidly. Weak resistance in Kovalenkoff was swiftly overpowered and the site declared secure at 0800 hours. Two men had been killed. The assault group relaunched its attack an hour later towards a minor hillock marked on maps as Hill 199.5. This second phase began smoothly: the troops advanced through a shallow valley that concealed them from view. Once they emerged from this hollow, however, they ran into ferocious resistance from State Farm "Red Giant" (Sovkhoz Krasnyi Gigant) two kilometres south of Kovalenkoff. Small-arms fire was deadly accurate and several pioneers slumped to the ground, mortally wounded. Soviet mortars, including 120 mm heavies, retarded the advance with skilfully placed volleys. Beismann knew it was fatal to stall in the open.

"Klein, get your men moving!" he hollered.

Beismann had no heavy weapons for suppressive fire and artillery support was not on call. More men were felled by the heavy fire. The farm would have to be taken with an old-fashioned infantry assault. Beismann launched his attack just as Panzer Battalion Heydebreck struck out to the east and attracted the attention of the Soviet defenders. Charging forward, his men approached to within penetration distance. Cover-fire, hand-grenades, a short dash across open ground and the buildings were reached. The pioneers made quick work of the defenders and the state farm was taken at 1030 hours.

---

1   Armeepionierführer = Army engineer leader, a senior officer on an army staff who gave advice to his
    commander-in-chief about pioneer operations. In this case, it was Oberst Herbert Selle at 6th Army.

The bodies of twenty-five Soviet soldiers littered the battlefield.

The pioneers thought resistance was fierce but it could have been much worse. Just as in Oberst Winzer's sector, the armoured units opposite Beismann – General Kuzmin's 21st Tank Corps – were being disengaged for redeployment on a new axis. Due to inadequacies in the Soviet chain-of-command, the withdrawal order reached Kuzmin about ten hours late and he only began pulling back his tanks during Beismann's attack. The units sent to relieve him were forced to enter battle intermittently against a resurgent German force counterattacking in precisely the sectors where tank corps units were being replaced.

Just before noon, Beismann set his assault group in motion from both sides of the farm – the recon squadron on the left and the two pioneer companies on the right – to seize their daily objective, Hill 211.7. Before that, however, they needed to take Hill 199.5. The scene ahead was bleak: nothing but bare black earth, still without grass cover after the harsh winter. As they moved out of the assembly area, heavy artillery fire blanketed the right wing of the assault group. There was little the pioneers could do but keep pushing on. Hill 199.5 was gobbled up by the advance but Soviet resistance kept strengthening. A hornet's nest was stirred up when Panzer Battalion Heydebreck pushed eastward into "Mushroom Copse", a small patch of woodland adjacent to Hill 211.7. Large-calibre projectiles screeched through the assault groups as anti-aircraft and anti-tank guns fired at individual soldiers. Forward momentum waned and it wasn't until 1600 hours that the assault group reached the Ryabukhino–Taranovka road. Under heavy fire from every type of enemy weapon, the detachment dug in with its wings bent back to cover both flanks. Russian snipers on Hill 211.7 took a steady toll. An assault troop operation by Hauptmann Klein's company, launched with the intention of taking the hill, collapsed in concentrated fire. In order to spare further loss of blood, Beismann was content to relinquish the hill to the Soviets for the evening.

Sentries discerned small groups of shadowy figures scampering towards them in the amber twilight. Those were Russian helmets! Several burst of machine-gun fire dispersed the intruders. Soviet patrols continued to raid Beismann's line but were easily repulsed. Despite the destruction of some large-calibre weaponry along the northern edge of the copse, Panzer Battalion Heydebreck did not succeed in bringing any perceptible relief to Beismann and his men. Mushroom Copse remained just as firmly occupied, with infantry skilfully nesting themselves along the tree-line and heavy weapons hidden deeper in the forest. Snipers also perched in the trees. Effective suppression was almost impossible. During the night of 19-20 May two small enemy patrols were repelled. Snipers and mortars fired ceaselessly through the night. Beismann handed control of the detachment to Major von Wedel, commander of 113th Panzerjäger Battalion. Two 5cm anti-tank guns were rolled forward and dug into the front-line.

This first engagement proved costly for Klein's 1st Company: one officer and five men dead, one NCO and eleven men wounded. Hauptmann Klein scribbled the dead men's names in his economically worded diary: "Infantry deployment. Leutnant [Kurt] Birk, [Pionier Albert] Bächle, [Pionier Willibald] Burger, [Pionier Karl] Schiebel, [Gefreiter Albert] Laur and [Gefreiter Albert] Dorer killed." Klein was unsettled: his company's first taste of combat, six men dead and twice as many wounded. And this was the first day of a long campaign. Worst of all, one of the dead was platoon leader Leutnant Birk, talented accordionist and future heir of his family factory, cut down by a piece of metal in the head. The only NCO wounded on this day, Unteroffizier Lorenz Traub, was a squad leader in Leutnant Birk's 3rd Platoon. Birk's death and his own wounding was still fresh in his mind sixty-five years later:

I clearly remember being wounded on the same day that Leutnant Birk was killed. I was just about to take cover in a foxhole when a shell landed, which killed Leutnant Birk and wounded me. Shrapnel hit me in the back and upper thigh. My wounds were quite severe. I was in hospital for well over a year, both in Zhitomir and back in Germany.

Traub recalls that two of the men killed on this day, Pionier Albert Bächle and Gefreiter Albert Dorer, belonged to his squad.[2]

The war diary of 62nd Infantry Division observed that "the offensive strength of the enemy appears to have been broken along the entire [divisional] front." Commanders on the ground also detected the change in Russian intentions but did not know what had brought it on. Senior commanders, however, knew full well why the Russians were spooked: Operation Fridericus had been boldly launched as originally scheduled on 17 May. Contrary to the opinion presented by Field Marshall von Bock, the Chief of the General Staff, Generaloberst Halder, was adamant that there should be no fundamental deviation from the original plan, even though the entire northern jaw of the pincer (Paulus' 6th Army) was absent. The operation started with only the southern pincer formed by Army Group von Kleist.[3]

---

2 In 1944, Traub was declared fit and sent back to the front, this time to the Weichsel bend in Poland. There, he suffered a bullet wound to the head, which was not too serious, but it required treatment in hospital. At war's end, he trekked west and was taken prisoner by the Americans but was soon released.

3 Army Group von Kleist consisted of a total of nine German infantry division, three panzer and motorised divisions and four Romanian infantry divisions, all of them still worn out from the winter battles and not yet fully replenished.

It unfolded as planned; surprise was total. Even after just a few days, the threat represented by Army Group von Kleist alleviated Soviet pressure on 6th Army, especially on the sectors held by 62nd, 113th and 305th Infantry Divisions. Timoshenko's Southwestern Front headquarters was aware from captured documents that the German command was preparing an offensive towards Savintsy and Izyum and that it was to have begun between 15 and 20 May. This intelligence, combined with Army Group von Kleist's attack, forced Timoshenko to the realisation that the Germans intended to thwart his offensive by pushing into his rear. Timoshenko implemented countermeasures, one of which was to concentrate a strong tank grouping deep in the bridgehead. He hoped two tank corps could defeat Army Group von Kleist and re-establish the southern face of the salient, but in doing so, he deprived his offensive of its main shock forces. By midday on 19 May, he realised that not only had his offensive failed, but the entire Southern Shock Group faced possible destruction. He therefore sought approval to assemble all possible forces to destroy von Kleist's group. Stalin agreed. Timoshenko thereupon ordered his own 6th Army and Army Group Bobkin to go over to the defensive along existing lines and begin regrouping for new missions.

The German leadership seemed to know of Timoshenko's new plans, possibly through radio intercepts. Hitler spoke to von Bock by telephone on the evening of 19 May and "quickly agreed it would now be a good idea to accomplish the entire objective of "Fridericus" by having Army Group von Kleist go the rest of the way and link up with the static 6th Army at Balaklaya." Instead of saving his army, Timoshenko's new plan was hastening its demise.

These high-level decisions had a palpable impact to the men in the trenches. The pioneers and infantrymen felt the impetus dissolve from the Russian attack and the scales began to tip in their favour.

The First Sturmtag (20 May 1942)

Hauptmann Beismann's staff and 1st Company continued their struggle in the shallow hills between Ryabukhino and Taranovka. At dawn, the right neighbour (Infantry Regiment 261) launched an attack on the wooded terrain south of Ryabukhino with Stuka support. Assault Group von Wedel, in which Beismann commanded both pioneer companies, was ordered to follow up this attack. On the right wing, 3./672nd Pioneer Battalion reached an embankment that connected two pieces of woodland, and despite volleys from heavy calibre artillery, began to roll up the enemy. This daring advance took them to within two hundred metres of Hill 211.7. Beismann introduced Klein's company and ordered them to storm the hill while the recon battalion

(Auf.Abt.113) screened the left flank. Both pioneer companies rushed up the slope, blasting and shooting away at any opposition. The Soviet defenders reacted fiercely and hand-to-hand fighting broke out. The pioneers prevailed. The bodies of fifty Russians were counted on this hill. "Prisoners were not taken," stated Beismann.

Once on the hill, the assault group was hammered by vicious defensive fire from the edge of the woods, despite supporting fire laid down by assault guns. Into this murderous hail drove a half-track towing an 88. Shielded by their vehicle, the crew set up the gun in an open position on Hill 211.7 and fired directly at the tree-line 400 metres away. This suppression effectively silenced the enemy guns along the wood-line but mortars ranged in on the gun and smothered it with dense volleys. Half of the crew became casualties, yet the gun remained in firing position, some pioneers even volunteering to haul ammunition and load shells. The Flak Leutnant aimed and fired the gun himself. After the remaining thirty shells were fired, the gun was hitched up and towed away. "It was an outstanding feat," Beismann reported. Two pioneers were wounded while helping out. To avoid further casualties, Beismann pulled back both companies about one hundred metres to the rear slope of Hill 211.7 and gave the order to dig in. An attack from the hill towards the tree-line would only be possible with the heaviest casualties, if at all, because even while sheltering in foxholes, the pioneers – particularly Klein's men – suffered high losses from mortars and snipers perched in trees. The fire eventually subsided and some Soviet units were observed falling back from the southern edge of the copse. Beismann sensed an opportunity. He initiated an operation against the tree-line with a platoon of pioneers and a pair of assault guns. The assault detachment approached to within a hundred metres of the trees without being fired on. Maybe the enemy had withdrawn? Just at that moment, heavy fire from snipers, two machine-guns and mortars suddenly erupted from all sides. The assault guns reciprocated, pumping high-explosive rounds into the woods while the pioneers sought cover. Machine-guns on Hill 211.7 laid down suppressive fire. Armour-piercing shells shrieked past the assault guns. The anti-tank guns also targeted individual pioneers. The assault guns trundled forward, stopped, fired, then darted forward another few metres. They blotted out one muzzle flash after another. Their crews were deafened by the constant ringing of projectiles against their armour. Beismann knew it was futile to continue. The wounded were picked up and he called off the operation. Although he was well-acquainted with the fighting qualities of the ordinary Russian soldier, he was impressed by the extraordinary tenacity of the enemy troops in the copse. In the process of covering the disengagement of the storm group, the assault guns had driven to within twenty metres of the trees, stomping out resistance at the closest range, but Russian snipers continued to fire at them, and both assault guns returned with their scissors-periscopes and gunners' sights shattered.

Two Russian tanks crashed out of the copse at 1500 hours and trundled towards Klein's sector. Tank destruction teams were waiting in deep foxholes with demolition charges. "The men exuded calmness," recalled Beismann. The tanks veered before reaching the German line and proceeded to drive around in front of Klein's positions, menacing and intimidating the Germans, sometimes drawing so near that a few men held their breath for fear of choking on the exhaust fumes. Then, unexpectedly, they turned back. An attack was not forthcoming. Bombardments by heavy mortars and large-calibre artillery resumed but did not lead to any casualties amongst the pioneers because the crest of 211.7 concealed their rear slope positions. On the other hand, observers posted on top were nearly wiped out twice, once by concentrated mortar fire and then by machine-gun bursts from one of the tanks. "An officer of 1st Company met a hero's deaths in this way," reported Beismann. Casualties were dreadful. Two NCOs and three men from Beismann's staff were wounded, but the heaviest losses occurred in the two pioneer companies. The attached 3./672nd Pioneer Battalion lost twenty-five men (8 dead and 17 wounded) while Hauptmann Klein grieved the deaths of an officer (Leutnant Lutz) and two men, as well as the wounding of five NCOs and fourteen men. Again from Hauptmann Klein's terse diary: "Infantry deployment. Leutnant [Siegfried] Lutz, [Obergefreiter Erwin] Vischer, [Gefreiter Otto] Weisser killed at Borki. Nine men killed, 32 men wounded." In two days, 1st Company had lost half of its officers and a third of its combat soldiers. The death of Leutnant Siegfried Lutz was a heavy blow, not just for the company but for the entire battalion. Well over six feet in height, he was one of the tallest men in the company and one of the most popular. As an architecture student, he was a natural choice to be one of two officers in the battalion trained to build bridges using the zinc-coated metal components of Kriegsbrückengerät 'B'[4] which were assembled like a construction set. When one of these bridges was to be built – a certainty during the upcoming summer campaign – everyone would be looking to the two officers who had the knowledge. One was Lutz: the other was Leutnant Birk, killed the previous day. Oberleutnant Grimm was staggered when he learned of their deaths – both were good friends – and he kept the news from his wife for months, so as to spare her from the pitilessness of the war in the east. Only in early August did he let her know: "I don't know whether I've already told you that Leutnant Birk – who lived with me in l'Oasis in Gouarec and where we got to know Gerti and the other ladies – was killed during the

---

4   Kriegsbrückengerät 'B' = military bridge equipment 'B'.

first days [in Russia]. Leutnant Lutz, who was first in my company in France and later transferred, also fell a day later. I often hosted the U-boat crew with Kurt Birk."

This attack was acknowledged as a "Sturmtag"[5] for 1st Company ("20.5.42: Attack on the "Snake Forest" near Borki, Kampfgruppe Beismann"). Soldiers needed three such days to be awarded the Assault Badge, but they were harder to accrue than may first appear. Determining precisely why a certain day qualified as a Sturmtag is unclear – 20 May was counted, yet 19 May was not – but the intensity of the fighting appears to be a factor. Criteria for bestowal of the badge, however, are quite clear: a soldier must have participated in three attacks on three separate days, in the forward line, with weapon in hand. Regarding the attack on 20 May, an officer from another company would later write that "it is known to me that all members of 1st Company participated in this attack, and [it] was acknowledged as an assault day because everyone participated in the penetration." Eyewitnesses also saw Hauptmann Klein at the very front. Feldwebel Erich Maier, a clerk on battalion staff, found himself near Klein during the attack and later testified that Klein was personally involved in close combat.

– – –

Further east, the daring German plan to cut off all Soviet forces west of the Donets was coming to fruition. Army Group von Kleist had narrowed the neck of the salient to just eighteen kilometres. The six bridges over the Donets River – upon which the Soviets relied for their survival – fell into German hands one by one. The Luftwaffe prowled the shrinking bottleneck, blasting anything that moved; even if Soviet forces caught inside the salient reached this exit, they would almost certainly have been slaughtered. By the end of the following day, it was only thirteen kilometres wide. The sole escape route for Timoshenko's forces was closing rapidly.

A Breather (21 May 1942)

Winzer's regiment received a straightforward order: attack south-eastward and reach Staroverovka. After quickly moving into assembly positions, Combat Group Winzer[6] launched its attack at 1330 hours, reached Staroverovka around 2000 hours and formed a small bridgehead over the Berestovenka River. Russian resistance was minimal and casualties very light. It was obvious that Russian resolve had disappeared, and based on aerial reconnaissance, so had many of their units. Spotter planes observed

---

5   Sturmtag = assault day.
6   Kampfgruppe Winzer comprised Infantry Regiment 578 without 9th and 10th Companies but reinforced by a flak battery, one Nebelwerfer battery, 305th Panzerjäger Battalion, Häntzka's 3rd Company, a platoon of assault guns and two artillery battalions from Artillery Regiment 305.

Russian columns rolling east. The crisis was over for 305th Infantry Division.

Hauptmann Beismann, however, detected little change in enemy behaviour. No operations took place during the night of 20-21 May but his positions were blanketed by well-aimed machine-gun and rifle fire, as well as occasional mortar volleys. The fire died down from 0400 hours. It was soon discovered that visible movements were no longer fired on, except if the Russians felt threatened, so throughout the morning the pioneers found it possible to recover their dead. Mid-afternoon a connection was established with the left neighbour (62nd Infantry Division). When elements of that division moved up to the same level as Beismann's assault group, Russian resistance flared up to its previous intensity, in turn causing heavy calibre artillery to bombard Beismann's left wing. Hauptmann Klein's last officer, Leutnant Buchner, was hit by shell fragments but he remained at the head of his platoon. In this situation, Beismann obtained an order for the relief of the two pioneer companies. Despite sustained enemy fire, both companies disengaged without loss and pulled back to Zelenyi for a rest, while simultaneously protecting the artillery positions from tank attacks. The following morning Klein's company was set in march towards Novaya Vodolaga.

Grimm Arrives (22 May 1942)

Regiment 576 attacked at 0300 hours. In the sultry post-dawn hours, the regiment swung to the south-east and set about occupying the Berestovaya between the villages of Vlasovka and Medvedovka and forming a bridgehead on the opposite bank. Resistance was weak. Winzer's Regiment 578 continued its attack at 1300 hours. The regiments linked up south-east of Staroverovka in the afternoon. The gaping hole was finally closed. Both regiments then pushed south-eastwards in a joint attack and gained bridgeheads over the Berestovaya shortly before nightfall. Häntzka's 3rd Company rebuilt the bridges over the creek during the night so that they could bear the weight of horse-drawn artillery. For the first time the Bodensee Division came together as a whole under the leadership of Generalmajor Oppenländer.

– – –

While battalion staff and Klein's 1st Company were recovering from their baptism of fire and Häntzka's men were strengthening bridges, Leutnant Homburger's light pioneer column and Oberleutnant Grimm's 2nd Company were only just arriving on the scene. The only hiccup during transportation happened when the train carrying Grimm's company skipped off the rails in partisan territory. "We reached our final destination of Merefa at 0400 hours in the morning," Grimm wrote in his diary:

We had to wait until we could be driven along to a loading ramp. We detrained from

1000–1230 hours, then received rations. It is oppressively hot. At 1430 hours we marched to [Novaya] Vodolaga. Oberzahlmeister Gabelmann gave us the first reports from the front about 1st and 3rd Companies. Leutnant Lutz was killed yesterday.

The men of 2nd Company were shocked by the casualties of their brothers. Grimm and his officers found out on 22 May, their men the following day. Grimm put a brave face on it in a letter home: "The other companies have already been 'blackened by the smoke of battle.'" The battalion's officers, particularly those who joined in 1941, were a tight-knit group and knew each other intimately. Leutnant Lutz had once been part of 2nd Company. Leutnant Beigel wrote to his girlfriend: "Leutnant Lutz, my good ol' [Gefreiter] Laur and Leutnant Birk gave their lives for this victory. I can only console poor Adeline…" Beigel's mention of Adeline – the girlfriend of either Lutz or Birk – shows just how well these men knew each other. Officers and partners had fraternised with other couples when based in Germany, and after transfer to France, officers on furlough would visit their fellow officer's girlfriends/wives to deliver gifts and letters. German formations were still recruited regionally at this stage of the war and entire units usually came from a small geographic area, so customs and dialects also tied the officers and men even closer together, and in some cases, provided instant bonds. Such was the case when Oberzahlmeister Gabelmann, long-time battalion paymaster, was transferred to a new posting and his replacement had no opportunity to acquaint himself with all commanders prior to deployment. Grimm only met the new paymaster a few hours after arriving at Merefa and made an instant connection: "I was introduced to a new colleague, Oberzahlmeister Keppler (son of a trade school teacher), and this in southern Russia. That was a bit of fun." Grimm and Keppler were both born and raised in Kirchheim, a small town south-east of Stuttgart, hence Grimm's astonishment at befriending a fellow Kirchheimer in far-off Russia.

However, there was no time for socialising. Battle was in full swing and German units were shifted at short notice to sectors of the greatest need. The initial relief at being able to properly stretch after ten days on a train quickly turned to dismay for Grimm and his men. After loading their horse-drawn carts and hoisting heavy packs, they began marching on stiff legs towards Novaya Vodolaga just eighteen kilometres away. A march of this distance would normally have been a doddle but Grimm's men were about to experience Russian conditions first-hand. The air was hot and thick with moisture, the sun blazed down and any sort of breeze was absent, but the worst thing was the "road". It was just a broad stretch of sand. Every step was an exertion because boots sank to the ankle. Faces were red and sweaty. Tracked vehicles churned past the column, kicking up dense clouds of dust that lingered over the road and adhered to perspiration-soaked skin and uniforms. Both man and beast struggled through the ordeal. Night fell. Escorted by chirping crickets, Grimm's men staggered into Novaya Vodolaga an hour short of midnight. They had averaged just over two kilometres per hour. Grimm declared that "eighteen kilometres was an effort equal to fifty." He later wrote that "after getting off the train, a race was on. A dubious foretaste right from the beginning. Heat and knee-deep sand with earth and dust. There are no roads, simply inconceivable." The men collapsed on the bare ground and fell asleep in an open field. They were granted six hours of rest. The officers and NCOs attended to essential matters before collapsing into unconsciousness.

As in all armies, non-commissioned officers were the backbone of a unit, forming the link between officers and enlisted personnel. They were responsible for executing orders, providing advice – particularly important for junior officers who lacked practical experience – and ensured that the soldiers were doing their job. Grimm was blessed with a fine set of senior NCOs. His Kompanietruppführer (company staff leader) was Oberfeldwebel Lorenz Locher, born the same year as Grimm but with considerably more experience. Locher was the sixth of twelve children, including six sons all serving in the Wehrmacht, and was an excellent marksman, keen sportsman and expert in all pioneering specialities, including assault boats. Leader of the 3rd platoon was Feldwebel Willi Platzer, a tough Austrian from Linz who, as an unmarried careerist, had seen more action than most in the battalion. A calling as a baker in Linz's old quarter ended in 1934 when his boss hanged himself and Willi, aged 18, volunteered for the Austrian Armed Forces. After the 1938 Anschluß with Germany, Platzer retrained, became a pioneer, and experienced the campaigns in Poland and France. It was due to his leadership qualities that he was given command of a platoon.

– – –

General Oppenländer issued an order in the evening for an attack over the Berestovaya River valley the next day. The division was expected to reach the Orel Creek between Nizhne-Orel and Dimitrovka. The regiments would initially gain an intermediate hill line from which they could move off to attack the first objective – another range of hills – as soon as ample artillery had been crossed over the watercourse. After capturing the hills, the Orel Creek was to be reached in one go and bridgeheads quickly formed. Winzer's Regiment 578 was on the left, the main effort was with Krüder's Regiment 576 on the right. To the south was open space. This lengthening flank would need to be covered as the advance progressed, but both regiment commanders were warned that friendly forces – panzers and Romanian troops – may appear from that direction.

– – –

The battle was being decided further east. Having concentrated his armoured and motorised forces, von Kleist relaunched the offensive on the morning of 22 May. While 16th Panzer Division and 60th Motorised Division struck out north-westward, 14th Panzer Division pushed north and linked up with the Balaklaya cornerstone, a small town stubbornly held by the Austrian 44th Infantry Division throughout the winter and during Timoshenko's offensive. Timoshenko's protruding armies were cut off but the question was whether this thin German dike would withstand the Russian flood. Early the next morning, elements of 23rd Panzer Division met up with the armoured spearhead of 16th Panzer Division, thus broadening the original narrow cordon to a sixteen kilometre-wide barrier across the mouth of the bulge.

Marching... (23 May 1942)

A new phase began. As part of 8th Army Corps, the Bodensee Division would pursue enemy forces retreating to the south-east, but the western and southern faces of the Soviet pocket were already collapsing inward even without German pressure. The Soviet units ahead of the division withdrew to the Orel Creek in order to shorten the line and free up forces to form a new reserve. The chase of a beaten foe was launched at 0815 hours. With just weak Russian rearguards ahead, the division was only troubled every now and then by individual tanks, artillery and mortars. The first objective was reached by Regiment 578 at 1300 hours and by Regiment 576 forty-five minutes later. A pause was called while artillery was brought forward. Rain began to fall in the afternoon, exacerbating an already challenging task, but the attack was relaunched at 1700 hours. Oberstleutnant Krüder's regiment advanced with one battalion echeloned to the right rear to protect the southern flank. Rainfall and the gathering darkness caused great difficulties bringing heavy weapons forward. The regiment failed to reach the Orel Creek by dusk but it pushed on regardless through the miserably wet night. Oberst Winzer's units made better progress. The torrential downpour had transformed the dirt fields into a swampy morass and there was not a road or bridge to be seen, yet Winzer's men slogged on, reaching the Orel valley almost without a fight.

The pioneers marched more than they fought on this day. After strengthening the bridges over the Berestovenka, the 305th Pioneer Battalion (without 1st and 2nd Companies) was kept at Oppenländer's beck and call from the beginning of the attack. Hauptmann Klein was ordered to get his company to Berestovaya village by evening, as was Grimm, though he had a much greater distance to cover. Despite exhaustion, he still found time to make an entry in his diary:

At 0500 hours we marched on to Staroverovka (25 km) on similar road conditions [as the previous day]. The divisional command post is there. After a one-hour break we had to continue on to our battalion and 1st Company. It was another twelve kilometres to Berestovaya. We dug ourselves in and pitched tents. We've been marching for thirteen hours.

Results Matter (24 May 1942)

With Regiment 578 at the Orel and Regiment 576 imminently so, Division issued a new order: advance over the creek in a south-east direction and gain the high ground between Mironovka and Grushino. Deepening the bridgehead was the first step so that artillery could be positioned to cover the objective, but to do that, bridges required strengthening. Enter the pioneers. The overnight downpour had turned roads into swamps and creeks into rivers. Day dawned dull and miserable; the rain had set in. Klein's and Häntzka's men laboured in the glutinous muck and the neck-deep water of the swollen creek to reinforce the rickety bridges. Soon, half-tracks and horse-teams were carefully dragging artillery pieces over the creaky spans. With the big guns in position and ranged in, the pursuit recommenced. Winzer's regiment spearheaded the attack. Oberstleutnant Krüder was ordered to cross the Orel Creek north of Nizhne-Orel and follow up the attack of Regiment 578 by staggering his right-hand units to protect a lengthening southern flank. Before this could be implemented, however, Krüder needed to get his regiment across the swirling brown torrent, and the task of helping him fell to Oberleutnant Grimm. Shortly after dawn, his company marched to the front, towards Shlyakhova, the current location of the command posts for Division and Regiment 576. Grimm was told find a crossing for Regiment 576. His small motorised team pushed forward in driving rain to explore crossing sites. What he found was mushy riverbanks and no bridges. The weight of responsibility gnawed at Grimm. His team had advanced ahead of Krüder's unit so that it had crossing points ready and waiting, but the reconnaissance took many hours. His liaison detachment then became bogged in the churned streets of Nizhne-Orel. Any delay was intolerable, so Division diverted the bulk of Krüder's regiment northward to use Winzer's bridges. Grimm's first mission was a failure. The atrocious conditions handicapped him right from the offset but Beismann was unsympathetic: in war, results matter. In Beismann's eyes, Häntzka and Klein had already proved themselves under combat conditions by successfully completing all tasks, while Grimm did not yet have any runs on the board. This inauspicious beginning marred their professional relationship. Sidelined by events, 2nd Company took no part in the pursuit.

Meanwhile, the advance was making rapid progress. At 1445 hours, after non-stop movement, Winzer's regiment reached the daily objective, the high ground just east of Timchenko, and dug in. Protecting Winzer's left flank was Oberst Voigt's fresh Regiment 577, a vital necessity because the right wing of the neighbouring 113th Infantry Division was stymied by ferocious Russian resistance in the Yefremovka area. With all objectives achieved, a new order arrived from corps: Regiments 576 and 578 will continue moving east at 1700 hours and reach the railway line. The tired, filthy troops set off. The constant rain softened everything; motor vehicles remained stuck in the mud, horses could barely pull the ammunition wagons, and the field-kitchens and supply trains lagged far behind. Obsidian darkness fell but both regiments pushed on. Leading elements of Regiment 578 approached the railway embankment near Likhachevo railway station at about 2130 hours but it was not until midnight that the main units tramped in. Following behind was Regiment 577, still tasked with covering the open left flank. Krüder's regiment was unable to reach the objective.

– – –

Timoshenko's armies were compressed into a pocket measuring 48 x 27 kilometres. Intensive fighting flared up on 23 May as Soviet forces tried desperately to bust open the German cordon. Timoshenko issued an order for Group "South" (created from units of 6th Army, 57th Army and Army Group Kostenko trapped in the pocket) to employ its main forces to deliver a blow against the German cordon, penetrate the ring of encirclement and systematically bring its forces across the northern Donets River. The renewed German attack on the morning of 24 May prevented Group "South" from regrouping as planned, so the intended operation was stillborn. The commander of Group "South", General Kostenko, decided to break out by delivering an attack in the general direction of Chepel.

Pursuit (25 May 1942)

Despite soft ground and greasy roads, all regiments captured vast tracts of land in their pursuit of a collapsing foe. Regiment 578 surged far ahead, crossed the Bereka and formed three bridgeheads. Winzer found it easy to make the decision because "the Russians were battered and finished." German pressure from all directions was herding the Russians into the Bereka River valley.

The role of the pioneers in this advance was vital but unglamorous. Each company was attached to a different regiment to strengthen roads and bridges. Oberleutnant Grimm was again allocated to Krüder's Regiment 576 and bad luck plagued him once more: Oberstleutnant Krüder had shifted his command post from Point 160.3 to whereabouts unknown. Misplacing a regimental command post would not get Grimm into Beismann's good books, so together with his HQ section leader Oberfeldwebel Locher, he searched for Krüder throughout the morning. No luck. The fluid situation meant Oberstleutnant Krüder was on the move. Grimm was eventually forced to call battalion and tell them he could not find Krüder or his HQ. Beismann ordered him to rejoin the battalion, so at 1500 hours 2nd Company marched to Aleksandrovka in "pig's weather" and set up camp. Beismann clearly lacked confidence in Grimm because he immediately attached a two-man radio team from his own staff to 2nd Company. After a dressing down, Grimm was sent forward to the Orelyka Creek to survey crossing possibilities.

– – –

Disorganised Soviet attacks came in against the German cordon all day and continued after darkness fell. By the light of swaying flares, massed Russian infantry attacks surged toward the German positions, but whichever way they turned, they were struck by blistering fire. Linking arms and screaming "Urraaa," they charged forward again and again, only to be scythed down by intense machine-gun fire. Tanks bounced through the chaotic scene but they, too, were stopped cold. Sometimes an attack crashed over the first German line before being hurled back. Such was the fury of Russian desperation that German soldiers were found with their skulls caved in or beaten or stabbed to death. The German breakwater held, and those Russians who survived the ghastly spectacle slunk back into the Bereka River valley.

Grimm's First Fatality (26 May 1942)

In contrast to the grisly horrors on the eastern side of the pocket, the night passed quietly for the men of the Bodensee Division in the west. The rain cleared, the sun came out and the roads quickly began to dry and harden. The infantry regiments moved out of their assembly areas at 0400 hours while the pioneers worked behind the lines. Grimm's men were roused at 0300 hours. A quick breakfast, weapons and kit left behind, then immediate departure with tools and construction material. Their task was to repair and strengthen two bridges on the road into Aleksandrovka, bread and butter work for the pioneers. Four hours later they were done. Hot coffee awaited them, arranged by their commander, then they marched into a copse near Sudanka at midday. Oberfeldwebel Locher and two men were detached to Regiment 576 as a liaison party to avoid a repetition of the previous day's "wandering command post" fiasco.

The noose around Timoshenko's men was pulled tighter. Regiment 578 pushed south-east along the Bereka and reached Babinovka without fighting. Klein's company was set to work improving the bridge in the village. That the battle was a foregone conclusion was now obvious even to soldiers in the supply trains, like Gefreiter Willi Füssinger: "The enemy is gone, but we are marching forward victoriously and the ring is quickly contracting around the enemy. A dreadful catastrophe is looming for the Russians."

Further south, Regiment 576 reached the Bereka between Krasivoye and Otradava and set itself up for defence. Stukas hammered the encircled Russians from early morning. A desperate attempt to break out to the east came within seven kilometres of success. Such efforts were not restricted to the eastern face of the pocket. Grimm's pioneers witnessed a smaller attempt, as Gefreiter Heinz Rinck noted in his diary: "Surprise Russian tank attack at 0900. Army flak drove into position." Grimm excitedly recorded what he saw in a letter home:

> I experienced a close-up view of how a Red tank attack with sixteen tanks was smashed. The whole thing was an unparalleled spectacle. Directly in front of us was a swampy area, so the chaps could not reach us. Our defence shot up two of them, just like me bringing down birds... I saw all of this from approximately 500-600 metres away, as if I was sitting in a cinema. Of course, I forgot the fact that Stukas would drop their eggs right on the tanks, so I had my nose in the dirt and did not see the impacts.

After the vicarious excitement of the tank attack, the company was assigned to mine clearance. Grimm brought his company forward to the regimental command post where he was ordered to locate and destroy all mines near a vital intersection and the approach route to the regimental command post. His men began the delicate task at 1330 hours. Ironically, most of the mines were German, dropped by the Luftwaffe days earlier. The devilish devices lay scattered on the ground but care was taken nonetheless as the boggy earth may have swallowed a few. Half an hour later, seventy mines had been removed, stacked and then blown up by Leutnant Hepp. The search continued for another six hours with only seven more being found, but not before one exploded and wounded Gefreiter Karl Pfleiderer and Pionier Friedrich Sträßle. A truck quickly evacuated both men but Sträßle, suffering multiple injuries to his head, chest and abdomen, died en route to the aid station, and so became the company's first fatality. Gefreiter Karl Pfleiderer was hit in the buttocks by shrapnel but would recover from his wounds and rejoin the company months later. Sobered by the accident, the rest of the company returned to its bivouac at dusk.

– – –

By the afternoon Russian forces were compressed into a sausage-shaped pocket measuring three by sixteen kilometres and began surrendering in droves, though in contrast to Russian armies trapped in pockets in 1941, this one included a major force of uncommitted armour, two complete tank corps in fact. They represented the final Russian hope. A shock group surged forward just after dark, capturing Lozovenka and advancing six kilometres, but it was eventually halted in scenes of carnage reminiscent of the previous night. The German 17th Army, which had taken over the front on the Donets facing east, observed "with astonishment" that during the entire ten-day encirclement, Soviet forces outside the pocket made virtually no attempt to relieve their trapped compatriots.

Road Building (27 May 1942)

The day dawned bright and sunny, with light cloud covering the sky. Road conditions were good. The smoke-shrouded Bereka valley seethed with explosions as German artillery pounded the pocket. Luftwaffe air strikes came in all day, particularly against a group still trying to break out. At 1115 hours General Oppenländer was ordered to cross his division over the Bereka and capture or destroy enemy forces that were wheeling south-east into the area between Lozovenka and Bereka. To cope with expected traffic, elements of the pioneer battalion were set to work repairing the bridge over the Bereka in Mikhailovka. It was completed at 1500 hours and the division resumed its advance: two regiments (577 and 578) and the artillery moved towards Krutoyarka with the aim of pushing deeper into the pocket and applying pressure to the rear of powerful Soviet forces poised for a break-out.

Oberleutnant Grimm, still attached to Regiment 576, was directed to reconnoitre crossings over the Bereka River in the regimental sector but none were available, so once again Krüder's regiment was forced to use another unit's bridge. After crossing at Mikhailovka, the regiment headed for Lozovskiy. Grimm's company also marched to Mikhailovka, arriving there at 1800 hours to improve the approach and exit routes of the heavily used bridge. An amazing sight greeted them: encamped in a riverside meadow were 16,000 Soviet POWs. Grimm tried to explain it to Lotte: "That many people – tens of thousands of them – have never been on the Cannstatter Wasen[7] and I was astonished: pretty good soldier material... Then I heard on the radio that this was a small experience!" The prison cage provided an abundant source of free labour, so Grimm borrowed 130 prisoners and used them to carry forward logs for the corduroy

---

7   The Cannstatter Wasen is a 35 hectare festival area on the banks of the Neckar River in the part of Stuttgart known as Bad Cannstatt.

roads being constructed either side of the bridge. The pioneers laid forty-five metres of log roads by 2100 hours, ate a meal brought to them by the company field-kitchen, and marched back to their encampment in Sudanka. They slumped into their tents, exhausted, at 0130 hours.

All eastward movement was halted by 8th Army Corps at 1900 hours. The division's role in the Kharkov battle was over. In preparation for Operation "Wilhelm" in early June, the division began to immediately organise the transfer of its units into assembly areas north of Kharkov. Four march groups were created and, as usual, the 305th Pioneer Battalion was split up. The entire division needed to redeploy within a week.

Victory (28 May 1942)

The spring battle of Izyum-Kharkov ended in bright morning sunshine with an extraordinary German victory. Two Soviet armies and parts of two others, with twenty-two rifle divisions, seven cavalry divisions and fifteen armoured and motorised brigades, were encircled and, for the most part, destroyed. After the count was finished, Army Group von Kleist and 6th Army found that they had captured 240,000 prisoners, over 1,200 tanks and 2,600 artillery pieces. Loss reports assembled by the Red Army General Staff after the operation recorded Soviet human losses as 13,556 dead, 46,314 wounded and sick evacuated to hospital and 207,057 who went into captivity, a total of 266,927. Also lost were 652 tanks and 4,924 guns and mortars. These war-time calculations have since been revised upward: the now officially accepted total figure is 277,190 men. Amongst that number were many senior commanders and staff officers, including well over a dozen generals and scores of colonels. Timoshenko's Southwestern Front had been significantly weakened and the Donets recaptured as a base for Hitler's upcoming summer offensive. After the severe winter crisis, the initiative was squarely back in German hands.

There was no time to rest on laurels. Units not needed to mop up the pocket were immediately shifted into assembly positions for a series of offensives. The Bodensee Division was one of the first to move. Their route passed through battlefields from previous days. Fresh flowers were laid upon the graves of fallen comrades.

In Good Spirits (29-31 May 1942)

After being on their feet for fourteen hours the previous day, Grimm's company revelled in the luxury of a sleep-in and some free time to attend to personal hygiene, make repairs and clean weapons. For many, it was the first opportunity to let their loved ones know that they were still in good health. Grimm dashed off a quick postcard to

Lotte: "I'm alive and well, in good spirits even. I cannot wish for more. I'm using my time to shave and wash and am now darning socks... The weather is varied. Otherwise there is a lot of news." His company broke camp at midday and departed an hour later, destination Berestovaya. Grimm jumped in a sidecar and drove on ahead to reconnoitre the route. The roads were firm and dry so, foreseeing no problems for his men, he waited for them in Berestovaya. This unexpected opportunity allowed him to compose a lengthier letter that would be enclosed in an envelope, away from the judgemental eyes of the censors:

My first operation is behind me. My company was not heavily involved in the fighting, so it came off unscathed. In contrast, the two other companies were really blackened by gunpowder smoke. The radio will tell you where we are... At the moment I'm sitting in a small house which seems to be masonry. The rest of the houses are made from wattle-and-daub with mud and cow dung. The roof is straw and when we spend the nights in our foxholes or tents, the second-best house is dismantled in order to get straw. I'll never sleep in one of these houses.

Meanwhile, the bad weather returned, dumping huge amounts of rain and pureeing the roads. His company's progress slowed to a crawl. He immediately left his dry quarters and rejoined his men. They only reached Lozovaya where they spent the night in farmhouses. On the morning of the 30th they decamped at 0400 hours and marched another twenty-eight kilometres. That evening, for the first time, they heard the radio broadcast about the battle of annihilation near Kharkov. The knowledge that they had participated in such a momentous feat-of-arms injected the men of the Bodensee Division with a warm mix of superiority and invincibility.

The next day, in front of Stanichnyi, 2nd Company bumped into their lorries and Grimm's command car, an ex-British Morris, which had been separated from the company during the trip to Russia. Their footslogging days were over! The company climbed aboard the vehicles and drove to the Merefa railhead. They arrived at 1100 hours and set up camp in a forest. The horsedrawn supply train did not arrive until 2030 hours. As compensation for making the combat soldiers wait so long, the company cooks laid on excellent rations.

- - -

The pioneer battalion – and with it the entire Bodensee Division – emerged from its baptism of fire slightly blackened but forged into a harder weapon. Their introduction to the Eastern Front could not have been more difficult. The transition from comfortable trains to a wavering front-line occurred within twelve hours, sometimes less, with little chance to acclimate to the conditions or the impending crisis. Their first impression came from the sight of German soldiers retreating and these scenes imprinted themselves firmly in their minds. The unit's performance during such a critical situation clearly demonstrated the Wehrmacht's skill in transforming toneless occupation divisions into competent combat formations. The installation of experienced officers in command positions was the prime reason for the division's achievements. The second factor was the order in which the unit's were dispatched from France: those with battle-hardened leaders arrived first. Oberst Winzer was the most seasoned infantry commander, hence his arrival in the first transports. The same applied to the pioneer battalion – the company commanders arrived in order of experience: Hauptmann Häntzka first, Hauptmann Klein second and Oberleutnant Grimm last. Häntzka operated independently, with little supervision from Beismann, and acquitted himself admirably. Klein's company endured the toughest fighting but there is little doubt that Beismann's presence and direction guided the men through the bloody encounter. Grimm's company was kept out of the fighting, whether by design or happenstance is not known, although circumstantial evidence suggests that Beismann did not yet have confidence in Grimm's leadership abilities. The coming months would put those doubts to the test.

Even as the battle of Kharkov was drawing to a close, the German high command decided to capitalise on the victory by eliminating two Russian groups still poised near Izyum and Volchansk. The stronger of the two – 28th Army near Volchansk – was first on the chopping block. Tasked with this operation codenamed "Wilhelm" was Paulus' 6th Army. The objective was to destroy the Russian 28th Army through a double envelopment and shunt forward the starting positions for the planned summer offensive over the Donets, up to the Burluk and into the area north of Volchansk. In doing so, a better launch pad would also be gained for the subsequent attack against the Russian force near Izyum. The Bodensee Division had passed the acid test in the battle of Kharkov and was now earmarked for Operation "Wilhelm".

1-9 June 1942

June began with a morale-boosting march through Kharkov to reach designated assembly areas. The procession was akin to a victory parade, although the nasty cobbled roads took off some of the gloss. Hauptmann Häntzka's company had joined Marschgruppe I the previous night, so they were the first pioneers to stride across the city on 1 June. Following about two hours behind with Marschgruppe II were Hauptmann Klein's men and Leutnant Homburger's light pioneer column. Both march groups camped for the night near Russkoye Lozovoye. In the meantime, the third and fourth march groups closed up to the city's southern fringe. Oberleutnant Grimm's company trudged through deep sand in blazing heat for seven hours and set up camp in the gardens of Pokotilovka, seven kilometres outside the city. Once night fell, Russian planes buzzed overhead and German searchlight batteries swept the skies. Officers attended to paperwork by shielded candlelight.

Grimm's men began marching from the Udy bridge at 0800 hours on 2 June and trekked through Kharkov an hour later. The impressive sight of Red Square, even in its current state of dilapidation, never failed to elicit comment from the soldiers, nor did the juxtaposition of modern high-rises with adjacent wooden shacks. This jarring contrast only served to further strengthen the German mindset that Communism was morally bankrupt and far from "communal." It was plain for all to see how the Bolshevik elite worked and lived compared to the peasants and workers. As they marched through

Kharkov in warm sunshine, ominous storm clouds darkened the eastern horizon. On fresh orders, Grimm and his company separated from Marschgruppe III shortly after reaching the city. In driving rain from a cloudburst, the company reached Lemnoy at 2330 hours, about twenty-five kilometres north of the city, after completing a forty-kilometre march.

Tight scheduling compelled the division to begin the process of relieving other units up front. All movements would be restricted to the night-time hours, now a strenuous undertaking because thundershowers had liquefied the "roads" and the thermometer had plunged. It was vital that the Russians remain unaware that a new division was being inserted into the line, so the relieving regiments were ordered to man sectors with the same strength as the former garrison and deploy heavy weapons in identical positions. Combat activity (barrages, harassment fire and so forth) must continue in the previous manner. Secrecy was paramount. And under no circumstances should members of the division fall into enemy hands: patrols and raids were strictly forbidden, even if they were "previous habits" of the old garrison.

The four days of 3-6 June were a time of hectic organisation for General Oppenländer and his staff. The division was required to occupy positions fifteen kilometres west of Volchansk, south of the Russian-occupied village of Murom. Late on 2 June a battalion each from Regiments 577 and 578 moved forward to relieve units already in the line. They marched at night in rain and on the worst roads – Klein's and Häntzka's companies had been fighting a losing battle all day to improve them – and both infantry battalions failed to reach their destinations. Grimm's men were also affected by the deluge. Just after midnight, their cosy nest in Lemnoy was hit by pouring rain and every tent, officers included, was awash with an inch or so of mud and cow dung. After a miserable night, they departed at 0700 hours. The bottomless roads were strewn with bogged lorries. Slogging through the clinging mud was arduous, so Grimm was glad that their destination, Bolshiye Prokhodiy, was just five kilometres away, but those five thousand metres took an exhausting five hours to cover. Unlike some officers, Grimm had foregone his elegant officer's riding boots in favour of enlisted man's "Knobelbecher" (dice cups) and he never regretted his choice:

I've been able to walk in them really well so far and already have a few hundred kilometres behind me without getting a blister. That is really great. You wouldn't believe how important this is. To be "march-sick" is a bad thing.

He had been momentarily tempted to wear his officer's boots and even tried them on, but swiftly pulled them off. The Russian front was no place for finely tailored leather.

Upon arrival, his men slumped to the ground for a quick bite to eat, but there was no rest for Grimm: he had been ordered to reconnoitre crossings in Chernyakovka. In the afternoon his company was split into platoons to carry out vital repair work on a pair of bridges. "My company's endless foot-race was interrupted and we carried out a task so typical for my branch of the service," Grimm explained to his wife. "It's always hard. I am – in my opinion I have to be – really hard upon my men. Until now I have achieved every objective, that means I've fulfilled all of my tasks." Despite their commander's rigour, the men were able to set up camp while the sun was still in the sky, a luxury few of them appreciated before their introduction to the Eastern Front. "Today, we were able to pitch our tents before nightfall," Leutnant Beigel told his girlfriend. "I can hardly describe what that means: washing, shaving, eating dinner comfortably, reading mail, and so on… The sun was shining today and everything is good."

On the evening of 3 June, all three pioneer companies and the light column were positioned close to river crossings that required upgrading. The battalion's assignment was simple: reinforce bridges to handle heavier traffic. In the meantime, under the cover of darkness, the relief of infantry units in the line continued. Guns and rocket-launchers from two artillery battalions were also moved into position. All movements were completed before sun-up and went unnoticed by their opponents.

Grimm's pioneers set out at dawn in drizzling rain for Chernyakovka in order to improve and strengthen its bridge's capacity from four to eight tonnes without stopping traffic. To get there, however, they needed to enter an alien world where even nature worked against them, as Grimm explains:

I had the task of getting my company to a command post. After I crossed a creek, ahead of me was vast, slightly undulating countryside without trees or anything else in the distance upon which I could get a bearing. All that could be seen was earth. I knew which point I had to aim for on the map, so I set the march compass and reached my point exactly after going cross-country for sixteen kilometres. No road, not even the most primitive dirt track or anything else, was available as an orientation point. The fields, especially the untilled ones over which no vehicles have driven, are the best roads here. When it rains, then a lot of German swearing is heard. Many times I just left my vehicles where they got stuck. The black soil is like clay, it builds up under the mudguards and axles and the machine can no longer get through. A downpour can prove disastrous. All that can be done is to remove all the mudguards and dig out the muck with a bayonet. After one hundred metres, the exact same thing happens. Then the sun comes out, hot like at the Bay of Biscay (though often hotter), you can drive again after a few hours and by the next day everything is dusty and you look like a Negro. Then it can happen that a thousand men are directed to a well from whose depths the water – reasonably dirty, of course – can be drawn. The many horses are also thirsty. I therefore have a wash every three days or so. It is better at the moment.

Mastering the elements would come in time. For now, Grimm and his men got to work strengthening a bridge while traffic still rolled over it. Stripped to the waist, they hauled mud-encrusted lumber, while others stood naked in the dirty water, positioning the beams. Everyone dived for cover when planes screamed in and strafed the bridging site. Artillery bombardments crashed down irregularly, hindering the work, but not too seriously and no casualties were suffered. The weather remained unsettled, roads and paths were again bottomless, and all units had their hands full carrying out supply runs. Infantry operations were temporarily postponed but the pioneers kept working as every bridge was needed to bear the weight of war machinery. Grimm's men completed their task and returned to their camp-site before sunset.

Everything was still sodden on 5 June but the sun occasionally broke through the dark clouds. Klein's 1st Company began the next stage of bolstering the load-bearing capacity of the Kolokiye bridge from sixteen to twenty-two tonnes. Grimm's men scrambled aboard lorries at 0300 hours and were driven to their work site. Once again Russian fighters strafed them as they worked on the bridge. Remarkably, nobody was harmed. Grimm was called to a conference at battalion HQ to discuss the upcoming offensive, primarily the pioneers' role in getting the troops over the Murom Creek. Beismann ordered his three company commanders to send out reconnaissance patrols that night. So, while his men bedded down in their tents, Grimm moved off into the rainy gloom with Oberfeldwebel Lorenz Locher to scout crossing sites. Hours later they returned, sopping wet and plastered in mud, but with the information they needed.

– – –

On 5 June the division issued detailed instructions for Operation Wilhelm. It was a classic envelopment operation: the southern assault group consisted of "Gruppe von Mackensen" with 3rd Panzer Corps and 51st Army Corps, while the northern group, which would attack near Volchansk, was formed by 8th Army Corps. The three divisions between these two assault prongs would pin down the Russians. The Bodensee Division was on the southern wing of 8th Army Corps and tasked with attacking just south of Volchansk in the direction of Belyi Kolodesi. Oppenländer encapsulated the division's task in one sentence: "The division will push through the enemy positions to the Donets, force a crossing and carry the attack forward to the hills south-east of Belyi Kolodesi." To do this, it would break through the Russian position with Oberst Voigt's Regiment 577 on the right and Oberst Winzer's Regiment 578 on the left, and after attaining a favourable intermediary line, push forward to the Donets in one go and force bridgeheads near Staritsa-Prilipka (Regiment 577) and Leski (Regiment 578).

On 6 June 8th Army Corps announced that "Wilhelm" would begin at 0255 hours the following day. Units were forbidden from moving into their assembly areas before 2100 hours, which left little time, but the element of surprise needed to be maintained. General Oppenländer shifted his command post forward to Borisovka, not far from Hauptmann Beismann's HQ. Reinforcements appeared: some fearsome 88s and a battalion of assault guns. Oppenländer stipulated that the assault guns be first to cross the Churkin bridge, the only one in the division's sector that could bear their weight.

The pioneers moved out at 0230 hours. To prepare for the attack across the Murom, Grimm's company marched to a copse near Novaya Dernevya. This hamlet consisted of a few houses nestled beside a picturesque lake embedded in a deep, wooded valley – little wonder Oberst Winzer had installed his headquarters there. However, his sanctuary was shattered when the pioneers arrived in the sweltering heat with chainsaws and axes to prepare lumber for bridge construction. The woods hummed with activity.

Hauptmann Klein's company constructed roads and improved the so-called "Häntzka bridge". Klein was never very effusive in his diary but he jotted a minuscule note under the 6 June heading: "2 Lt." These three characters recorded the arrival of two new officers, Leutnant Erwin Hingst and one other.[1] Each of his three platoons now had an officer in charge.

Threatening clouds moved overhead, opened up and quickly softened the roads. This prompted 8th Army Corps to postpone "Wilhelm" for twenty-four hours. The Luftwaffe and panzers needed better weather, but the driving rain did not keep the pioneers under shelter. Grimm dispatched teams to reconnoitre routes to the Murom in the sector of Regiment 577. The night and following day passed quietly. Leaders looked skyward, not for enemy planes or shells, but at meteorological conditions. The tropical heat and thick clouds did not bode well but the rain held off all morning. At midday on 7 June, however, a steady drizzle started and turned torrential an hour later. This variable weather was playing havoc with the timing of the attack. It had to be postponed yet again, but once more the pioneers were out in the foul weather. After

---

1   Possibly Leutnant Johannes Lindner.

nightfall Grimm sent out two patrols under the command of Oberfeldwebel Locher and Feldwebel Grassl to continue examining approach routes and crossing points over the Murom. Pitch darkness, pelting rain and the menace of enemy patrols lurking in the gloom aggravated an already onerous task. Locher's and Grassl's teams carefully checked the selected routes and found four mines, which they cleared.

Grey skies had set in so deeply on 8 June, with little prospect of improvement, that Beismann passed on word to his company commanders that the attack was postponed for another three days. No matter, that gave the pioneers more time to complete their myriad tasks. Lumber already collected and dressed was hauled forward to depots just behind the front. Grimm's senior platoon leaders, Leutnants Beigel and Hepp, each took out a patrol in the afternoon. Five mines were detected and lifted. The weather cleared in the evening and a strong wind picked up, but the roads were still quagmires. During the night, Beigel's 1st Platoon reinforced a four-tonne crossing on the approach road to the planned bridging site over the Murom Creek.

The troops waited and waited for the attack. The sun emerged on 9 June and began to quickly dry out the doughy ground which gleamed like silver. The topsoil was transformed to dirt and dust by evening, but the ground continued to be soft in low-lying areas. With a firmer footing, Operation Wilhelm was rescheduled for 10 June and the division issued final orders. The objective remained to hurdle the Murom, reach the Donets as quickly as possible and form bridgeheads.

Forward Russians posts on the opposite bank of the Murom, judged to be relatively weak, were just two-hundred metres from German positions. A German war correspondent attached to the division for the assault, Oberleutnant Clemens Graf von Podewils, described the scene in front of him:

> From a ridge of hills, which runs up to the front, enemy mortar positions and foxholes dug into the reverse slope on the east bank of the Murom Creek are recognised with binoculars. There is no firing. The enemy does not move, but the assembly of our troops cannot have escaped his notice. Our columns move to the front over open ground. A belt of oak forest at least offers some cover from artillery.

German assault troops silently filed into assembly positions under cover of darkness. After completing their preparations in the copse near Novaya Dernevya, Grimm's men brought forward all remaining construction material, as well as inflatable rafts, into the vicinity of the bridging site. Beigel's 1st Platoon and Hepp's 2nd moved forward into the assembly area while Feldwebel Platzer's 3rd Platoon remained behind in the billeting area. Grimm and his HQ section trudged to the front just before midnight.

Wilhelm: Day One (10 June 1942)

The infantry huddled in small groups and waited for the signal to attack. In front of them, in the dark, the pioneers were already hard at work. Hours before the artillery roared out, Beismann and his companies worked within earshot of the Russians, preparing bridges over the Murom, particularly the main one at Churkin. Grimm inspected the bridge in his sector and determined that the spongy road to and from the reinforced bridge could not withstand the traffic of two battalions, so he reinforced it with wooden planking. The artillery opened up from all barrels at 0250 hours with fifteen minutes of drum-fire on the wooded area east of the creek. German aero engines and cracking bombs vibrated the sky. Five minutes into the barrage the infantry moved off, utilising the remaining ten minutes to close with the enemy. On the right was Oberst Voigt's Regiment 577, on the left was Winzer's Regiment 578 and following behind was Oberstleutnant Krüder's Regiment 576. Two flamethrower sections (each of one NCO and four men) from Klein's 1st Company were subordinated to both leading regiments; Voigt and Winzer were instructed to deploy them in urgent situations only.

Precautions taken to keep this attack secret seemed to have worked because surprise was complete. The Murom was overcome in ten minutes at first light. The infantry pushed through the woods, stormed up the slope and gained the first objective. Many Russians surrendered, others lay dead in their shelters, only a few fled over the barren heights. As the infantry reformed for the second phase, the pioneers worked feverishly to complete the Churkin crossing because heavy weapons were required to support the infantry advance. Häntzka's 3rd Company supplied most of the workforce, while Grimm's 2nd Company erected several smaller bridges across the creek. The company diary recorded that Leutnant Beigel's 1st platoon finished twelve metres of bridge, while Leutnant Hepp's 2nd platoon completed eleven. Beigel's platoon finished their bridge at about 0430 hours, Hepp's at 0600 hours. Häntzka's men also finalised the Churkin bridge around this time. Two batteries of assault guns had priority but their weight was too much for the bridge, so they were sent south to the Sereda bridge in the neighbour's sector. After this small detour, they were met at the eastern end of the Churkin bridge by a pioneer team assigned to escort them and clear mines. The first motorised vehicles across the Churkin bridge were Leutnant Homburger's lorries with lightweight bridging sections needed to span an anti-tank ditch in Staritsa.

With all bridges completed, pioneer details moved off to join the infantry regiments. These patrols, each led by an officer, were attached to every front-line battalion with the main task of deploying mine clearance teams. Leutnant Beigel, accompanied by Unteroffizier Rauschenbach's squad, went to Regiment 577, while Leutnant Hepp and

Unteroffizier Pauli's squad headed to Regiment 578. After handing over the Churkin bridge to a company of road construction troops, Häntzka's 3rd Company and half of Klein's men closely followed Regiment 577, while battalion commander Beismann – as ordered – remained at Oberst Voigt's command post during the course of the attack.

After reforming and attaching heavy weapons, the assault troops swung back into action. It was a question of speed: the Donets was barely fifteen kilometres away and their mission was to get across the river and gain footholds. On the left, well-camouflaged snipers in the tree-line picked off some of Winzer's marching soldiers, but the main body of Regiment 578 kept advancing while the insidious marksmen were dealt with by smaller detachments. The Russians were soon broken and they came out of exquisitely laid-out defensive positions, hands above their hands, without offering resistance. On the right, Regiment 577 made even swifter progress.

Leutnant Beigel's patrol had the task of securing the Donets bridge near Staritsa. The infantry were far ahead and Beigel needed to catch up. He left Unteroffizier Rauschenbach and his squad to follow up as quickly as possible while he jumped into a BMW motorcycle-sidecar and zoomed off into the vacuum behind the leading troops, a dangerous void not yet cleared of enemy personnel. Beigel's cavalier attitude was rudely shaken when he approached what he thought was a group of German soldiers. Realising the error, his driver, Gefreiter Wilhelm Danner, skidded to a halt. A quick glance behind… more Russians. They were hemmed in. The pair leapt off the bike and bolted towards some nearby shrubs, bullets zipping around their heads. One round gouged a furrow in Danner's right arm but they made it to cover and went into hiding. The motorcycle-sidecar was plundered for anything useful. Unwitting rescuers arrived on the scene and scattered the ransackers. Beigel and Danner were lucky: instances of lone soldiers or small groups being ambushed and bestially murdered were common lore amongst the Germans. They remounted their machine – miraculously still in working order – and soon caught up to the forward troops. Beigel was briefed by a battalion commander. Unteroffizier Rauschenbach's squad appeared an hour later. At 0745 hours, together with the infantry, Leutnant Beigel reached the bridge and went straight into action. With the threat of instant death hanging over them, the pioneers dismantled the fuses or simply removed the explosive charges and heaved them into the water, and in the nick of time, too, because a few exploded, sending up steep fountains right next to the bridge. Strangely, Beigel mentioned nothing of his daredevilry in a letter home, his usually enthusiastic tone replaced by an enigmatic terseness:

I could not write the last few days. Now I'm in a forest with my men sitting around
a camp-fire and drying our clothes. I've experienced all sorts of things in the last few

days. Pretty much everything has worked out all right. I'll tell you about it later.

His act was of great significance because his patrol was acknowledged as the very first "Sturmtag" (assault day) for 2nd Company: "10 June 1942, Officer recon patrol of Leutnant Beigel. Seizure of the bridge near Staritsa."

While Voigt's Regiment 577 captured the Staritsa bridge intact and seized another one near the village of Grasskoye, its sister regiment, Regiment 578, had none. Winzer's leading units reached the river at 0900 hours but all bridges in his sector had been blown. This eventuality had been planned for: river crossing equipment was not far behind. Hauptmann Beismann ordered all inflatable boats to Regiment 578.

The soldiers were disappointed by the river. Images of cool green water danced in their minds as they tramped through the heat, yet what stood before them was muddy waterway. It was probably for the best because there was scant time for a refreshing dip; a bridgehead needed to be formed on the other side of the river on a wide, almost imperceptibly rising plain. Two of Winzer's battalions crossed over in inflatable rafts, while a third followed up over a bridge in the sector of Regiment 577. Once sufficient forces were over, the regiment continued its advance at 1530 hours with the objective of gaining the road to Volchansk east of Sinelnikovo. It did so by 1800 hours.

The morning was hectic for Grimm and his men but their duties lessened as the day progressed. Towards 1000 hours the Murom bridges were handed over to a construction company and Grimm's company marched to Staritsa, arriving there towards 1500 hours. "The day was very hot and exhausting but everything worked out," wrote Grimm. The pioneers of the other companies worked until much later. Hauptmann Klein noted that a mine detection team from his company lifted fifteen mines from the Staritsa area, but there were obviously more, as a war correspondent vividly described:

The narrow bridge, whose destruction had been prevented this morning, was the only crossing point over the Donets. The traffic banking up there was made worse because it had to be confined within the white tape used by the pioneers to mark lanes through the minefields. A cloudburst came down and transformed the road into a morass. Finally the sentry succeeded in opening a passage for us and we got over the river. There, two explosions in quick succession shook the air, two columns of black smoke rose up in front of us. In the crush, a baggage wagon had swerved from the road, over the white tape and on to a mine. One horse was ripped in half, the other one fell to the ground, bleeding. Shreds of intestines lay splashed in the road grime. Fire leapt from the burning wagon to an ammunition vehicle. Small-

arms rounds and hand grenades exploded, as if combat was in progress. The rain drummed down but it was unable to quell the flames which rose ever more luxuriantly from the jerricans. Now the coloured smoke signals also caught fire and a multi-coloured radiant corona formed over the fiery glow, violet, orange, toxic-green. From the west, out of receding clouds, the sun shot dazzling rays, but in the east, a rainbow arched in front of the withdrawing thunderstorm.

The division had attained its objective of a bridgehead south of Volchansk. In a new order issued during the night, the division briefly analysed expected enemy resistance: "The enemy, with seemingly weak forces, including a few tanks, stands in front of the division. Several tanks are behind the fronts of the forward regiments. Scattered enemy units are in the forests south of Staritsa." About two-hundred prisoners had been taken. In return, divisional casualties were twenty killed and about 110 wounded. One of the latter was Leutnant Hermann Hupfeld, a platoon leader in Häntzka's 3rd Company. The son of a famous Protestant theologian, he had celebrated his 25th birthday a few weeks earlier amidst the turmoil of the Kharkov battle. While clearing a minefield near Churkin, an anti-personnel mine exploded as he was removing its fuse, ripping four fingers from his right hand and embedding dozens of metal and wood fragments in his face and eyes. After being stabilised in a field hospital in Churkin, he was transported to Kriegslazarett 2/541 in Kharkov, arriving there in a severe state of shock, and died at 1800 hours that same evening.

Overall progress was gratifying. Occasional rain did little to halt 3rd Panzer Corps further south after it crossed the Burluk River courtesy of two captured bridges and began the advance upstream. Eighth Army Corps did even better north of Volchansk, taking three bridges on the Donets and passing north-east of Volchansk by late afternoon. Forming the inner edge of this northern pincer was Oppenländer's division.

Wilhelm: Day Two (11 June 1942)

The division relaunched its attack at 0230 hours. Rain had fallen during the night and stopped and started throughout the day, so progress was heavy-going and required great physical efforts by the infantrymen. Fortunately for them, the Russians were withdrawing. The pioneers had little face-to-face contact with the enemy on this day because all companies were busy maintaining the Staritsa bridge and approaches. The fact that they were in the rear did not remove them from harm's way. Grimm's men had just started improving the bridge's exit road when eight Ratas roared in low along the river, machine-gunning visible movement and releasing bombs over the bridge. On return runs they also targeted the bustling village of Staritsa. The pioneers huddled under shelter and watched the stubby-winged fighters buzz by. By some miracle, Grimm's company was spared any loss.

Up front, the spearheads set a blistering pace. Leutnant Hepp and Unteroffizier Pauli's squad were assigned to Oberst Winzer's Regiment 578 for pioneering matters. With the help of an attached assault gun battery, the regiment surged along the railway line toward Belyi Kolodesi. Meanwhile, a few kilometres behind, individual Russian tanks slipped in behind the German spearheads, terrorised the supply trains and took a few prisoners. The tank threat was taken seriously at the Staritsa bridge because if it fell, divisional forces east of the river would be greatly endangered, so elements of Klein's and Grimm's companies were taken from their construction tasks and set to work protecting the bridge. Feldwebel Platzer's 3rd Platoon laid two mine barriers, while elements of 1st Company laid another in an arc east of the bridge. This precaution proved justified when news arrived that the neighbouring 79th Infantry Division had fallen back in the face of Soviet tanks.

Regiment 578 took Belyi Kolodesi at 1600 hours and established defensive positions on the high ground east of the village. Rain began to fall again. Russian resistance stiffened and sporadic counterattacks were launched, all of which were driven off. Winzer's men now waited for the panzers to push up from the south and close the ring.

After attaining its objective (the line Yurchenkovo – Belyi Kolodesi), the division relinquished some of its attached artillery assets. Daily casualties recorded in the divisional diary were 10 dead and 50 to 60 wounded, but these figures are incomplete. The full toll would only be discovered after the battle. Häntzka's 3rd Company suffered a painful reverse near Petrovka when several positions were overrun by Red Army tanks. Oberpionier Ernst Mittelmeier was killed instantly when shot in the head. Most ran for their lives. An unfortunate few were bundled into captivity, hands over their heads, and were found eight days letter, shot dead. One of them was Gefreiter Josef Böhm, a bespectacled 21-year-old farmhand, who was executed soon after capture. Häntzka sent a small detachment to the site and buried Böhm on the spot. It was a similar story with Oberpionier Richard Bauer and Pionier Clemens Mayer. A couple of others remain missing to this day. Enquiries revealed that one of them, Pionier Emil Betting, was wounded when he disappeared. Gefreiter Hermann Ruthardt could not be found after the attack and six days later Häntzka informed his family that he had gone into Soviet captivity.

– – –

The infantry of 8th Army Corps had achieved their objective by reaching Belyi Kolodesi but the motorised units of 3rd Panzer Corps due to meet them were mired forty kilometres further south after a downpour engulfed them in mud. The chance of a quick victory was slipping away. Intelligence reports showed that the Soviet 28th Army was abandoning its front west of the Donets and marching east. The panzer corps was ordered to forget everything else and close the encirclement the following day.

Wilhelm: Day Three (12 June 1942)

Grimm's company camped for the night in a forest three-hundred metres east of the river. Even though everything was wet, they managed to kindle camp-fires and attempted to dry out their sodden uniforms. They headed back to the bridge in the morning and recommenced their work. The weather was warm and sunny but secondary roads were still spongy.

There was still no sign of the German armoured spearhead said to be approaching from the south, but worrying indications that the Russians were starting to reorganise themselves appeared. Resistance stiffened appreciably and German casualties mounted. General Oppenländer wanted to relaunch his attack in the afternoon but Russian forces boldly pushed into his assembly areas. These audacious thrusts were shattered by defensive fire. The German assault set off at 1630 hours with the aim of gaining the ridge of high ground just east of Baksheyevka railway station. The infantry soon moved outside their protective umbrella and were entirely on their own. The Russians resisted sturdily at first and German casualties were grim; the fact that the enemy's were far heavier was of little consolation. Regiment 578 pushed up to the heights four kilometres east of Belyi Kolodesi in falling darkness and took the high ground.

Klein's 1st Company shifted to Belyi Kolodesi in the morning and experienced much of the front-line unpleasantness. Grimm's company was spared, but mid-afternoon he was ordered to pack up and join Klein. Time was insufficient for a foot march, so Grimm's company was shifted forward in platoons on two lorries running a shuttle service. Leutnant Beigel felt like he and his men were coming to the rescue: "We were driven to the front. We have to help the infantry." The supply train, as usual, marched under its own steam, rejoining the rest of the company at 0200 hours. Hauptmann Beismann moved his command post into the village, as did the divisional staff and several other units. Leutnant Beigel did not think much of Belyi Kolodesi: "It's very unpleasant here in this village. Nothing but collapsed houses, no beautiful meadows, mud everywhere." Beismann informed his company commanders that they would be deployed early the next morning as assault troops.

As 3rd Panzer Corps slogged northward, Soviet columns headed south-east past Belyi Kolodesi out of the pocket. Number of prisoners was the yardstick by which victory would be measured, yet Russian troops were steadily trickling out, like an hourglass. If the Russians escaped, the entire operation would be pointless. Time was of the essence.

Wilhelm: Day Four (13 June 1942)

The division's new objective was to veer south-eastward, gain local bridgeheads over the Khotomlya Creek and make contact with elements of Group Mackensen approaching from the south. Elements of Klein's and Häntzka's companies were subordinated to Regiment 577 to help seize a crossing near Novaya Aleksandrovka. On the left wing, Winzer's regiment would swing out further to the south-east, roughly following the course of the railway, while Regiment 576 covered the division's right flank.

The division moved off at 0300 hours along the railway line. Resistance was weak and the number of prisoners and deserters increased. The sun was out, the air warm, and road conditions improving, as were German spirits. Winzer's regiment, again protruding far out in front, chased the retreating Russians along the railway embankment and by mid-morning had taken Prikolotnoye railway station, twenty kilometres east of Belyi Kolodesi. Voigt's Regiment 577 arrived at the Khotomlya Creek at roughly the same time. Reaching towards the regiment from the south was 3rd Panzer Corps. After fighting their way through several lines of Russian tanks, all of them hull-down in a desperate attempt to blunt the southern pincer, the lead panzers made contact with Regiment 577 and theoretically closed the pocket, though in reality it was porous. Increasing numbers of Russian planes bombed and strafed German units in an effort to delay the advance and enable their ground troops to escape. Some got through but most were caught in the thickening net thrown across their exit. News of the link-up spread swiftly. Discounting the counterattack during the Kharkov battle, this was the division's first offensive and after just three days success seemed imminent. The excitement was palpable and contagious. Beigel wrote home:

> Our brave infantrymen have closed the ring by meeting up with the panzer troops coming from the south-east [and throwing] the enemy back with real gusto. The Russians are fleeing like maniacs. From time to time there is stubborn resistance but it is violently broken. Even the "Stalin organs" cannot help them. We always notice that the Russians are never far away! Just now there were eight or nine Ratas that only fired with their onboard weapons. They don't have many bombs. I don't think the war will last very long...

Grimm viewed events differently. Any romantic notion of a gallant war and dying a hero's death dissipated after several weeks at the front. He kept outward displays of emotion in check and tried to sustain this façade in letters home, albeit unsuccessfully, because in spite of attempts to assuage his wife's fears, his downbeat tone and silence on combat issues conveyed far more than intended:

I actually have to force myself to put pen to paper. I won't write anything about the fighting because why should you have unnecessary worries. A normal man simply cannot imagine what Russia is like. The natural difficulties opposing the army are incredible, yet these are all overcome. We have been attacking for several days and with regards to deployment, I've had luck with my company. Hopefully it will remain so in future. The weather changes every day and we struggle against the mud. You'll soon hear about our successful action on the radio; this time I was a little to the north. I have not taken any photos. I don't even want to capture these memories on film. It's better if they are forgotten.

The lot of 2nd Company was not too bad for the moment. They had time to wash and put up their tents before the rain fell. The other pioneer companies were up front. Hauptmann Klein's men bivouacked for the night on Hill 206.0 near Novaya Aleksandrovka; several planes swooped in, strafing and bombing madly, forcing the company to scatter into narrow foxholes, but it suffered no casualties.

Operation "Wilhelm" was not yet over but already 8th Army Corps issued orders for 305th Infantry Division to assemble north of the Volchya River for a future operation. After an intense week of preparations and non-stop advance, the troops hoped for at least one day off, and most received their wish.

Wilhelm: Day Five (14 June 1942)

The pioneers pulled back from the front-line early in the morning. Hauptmann Beismann's orders were as follows: "The pioneer battalion will assemble in Volchansk-East by the evening of 15 June. The 1st Company (in bivouac half way between Belyi Kolodesi and Volchansk) is to be set in march so that it does not interrupt the march of Regiment 577. The 3rd Company will depart before Regiment 577."

Just as Klein's and Häntzka's companies were rejoining the battalion staff in Belyi Kolodesi, Grimm's company – designated as the point unit of Regiment 576 – was marching off. They arrived in Volchansk at midday, having covered the thirteen kilometres in three hours. All men were now well-practised in the art of establishing a camp-site. As always, Leutnant Beigel shared quarters with his orderly and a Feldwebel. First they dug a square pit. "A lot of sweat is needed to excavate this hole," remembers Beigel. The tent was placed over the top and the hole lined with straw. Everyone was then free to attend to personal matters. A Feldwebel lanced a blood blister. Orderlies fetched water and food for their officers. This domestic harmony was spoiled when a cloudburst dumped on them for a few minutes. Grimm was in the middle of writing a letter; he picked up his correspondence and sought cover. Apart from discomforts such as this, Grimm and his company had led a charmed life so far. Minimal casualties, no combat and very little interference from Russian planes and artillery. Not so for the other two companies. Klein's 1st Company, now in residence in Grimm's old billets in Belyi Kolodesi, were forced to endure another Russian aerial attack.

Wilhelm: Day Six (15 June 1942)

The ring was closed, the objective of Operation "Wilhelm" was, to some extent, achieved, and the German line pushed forward to the Burluk. The trapped Red Army units were quickly mopped up. Almost 25,000 prisoners were taken, yet a sizeable part of the Soviet 28th Army had escaped. The division was proud to have measured up to all requests in its second battle. The sight of Russians fleeing or meekly surrendering filled the men with confidence. Leutnant Beigel's sentiment – "I don't think the war will last very long" – was held by many. Beigel reiterated his belief in a letter home a few days later: "I believe that the war will soon be over. Today a deserter told us that they'd had nothing to eat for five days. Other reports are also favourable for us." Despite such effusive letters and glowing reports on radio and in newspapers, his girlfriend thought otherwise and told him so. Her exact words are lost but Beigel's reply is not: "You believe that the war with Russia will last another two years? That's totally out of the question. I've often written to you how weak they are. We all think it'll be over this year." He expounded on this train of thought in a subsequent letter: "Things aren't so bad now with the Russians. Even an enormous army like theirs must run out of strength after the heavy blows of the last few months."

What caused such optimism and triumphal certainty? They had been on the Eastern Front for just over a month, yet they strongly believed victory was certain. Veterans who had experienced retreats during the previous winter did not think this way. Was it because they were untarnished by defeat? Experiencing victories was undoubtedly the main factor. Two battles, two successful outcomes. Leutnant Beigel's confidence was also due to the fact that 2nd Company had suffered extremely low casualties: one dead and two wounded. Unlike Beigel, Grimm was not ebullient, he was simply relieved to have withstood his first trial: "If I'm honest, I must admit that I'm glad I have endured my Russian baptism of fire and now know battle. Having something like that hanging over you is horrible, at least it was for me."

Beigel was glad to see that his men had withstood their first tests: "I must declare that my men are 'fantastically up to scratch', both in battle and also with regards to organisation. It's very reassuring for me." This same feeling pervaded the divisional staff. The inexperienced Bodensee Division had outperformed the veteran formations on either side.

- - -

The weather was bad and roads sloppy but the Volchansk region seemed like paradise: the small houses and cottages were clean, food was plentiful and the populace friendly. Man and beast could relax, weapons and ammunition were replenished and some replacements arrived as casualties amongst the infantry units had been particularly high. The timetable for the summer offensive was tight and permitted little time for rest. The division was ordered to immediately take over the Nezhegoly bridgehead fourteen kilometres north of Volchansk. Leading the way was Oberstleutnant Krüder's Regiment 576. Oberst Voigt's Regiment 577 would move into the bridgehead the following night (16-17 June). To hide this relief from the enemy, the protocols implemented at Murom Creek were observed. Hauptmann Beismann was ordered to carry out reconnaissance and then move his battalion forward to Voznesenovka. In the afternoon three patrols – each led by the company commanders – were sent to reconnoitre bridges in Petrovka [Nezhegol]. Gunfire crackled and artillery boomed. The three officers carried out their task in the middle of an artillery duel. Fighting flickered around Petrovka until the evening hours. Thundershowers added to the misery. The bridges stood but needed repairs and roads were in poor condition. Muddied and drenched to the skin, the patrols returned at midnight.

The First Iron Crosses (16-23 June 1942)

Shedding their sodden uniforms, Grimm's patrol turned in for the night, yet a dry spot to get some sleep was denied to them, as Gefreiter Rinck tersely noted in his diary: "Rain. Tents full of water." Why put up with such appalling conditions when the houses and huts of Volchansk were available? With many units billeted in and around the town, space was at a premium, yet a company commander like Oberleutnant Grimm could easily have had a roof over his head. Nevertheless, even he chose to remain outside:

I went into a Russian house, or rather a hut. The walls are made of braided sticks smeared with a mixture of mud, cow dung and straw, overhead is a thatched roof. The thatched roofs are of great benefit. In winter they are fed to the horses and now they are used in the tents as bedding straw. The houses are whitewashed inside, so sometimes they seem pretty neat. The floors are always packed earth with some sand scattered on top. If the enemy allows it, I sleep in the car. I'll never sleep in one of these houses.

Leutnant Beigel also turned up his nose: "A tent is much nicer than the most beautiful Russian hut." Both officers never gave their reasons for shunning Russian accommodation. Stories about bed-bugs, fleas and lice prompted many Germans to sleep outside but the biggest factor was a huge dose of chauvinism. National Socialism drummed into the German national conscience that Slavs were sub-human, beneath them, and these huts and houses were tangible proof. Grimm set up his command post next to a hut and was able to observe the habits of the villagers, recording his observations like a naturalist:

Behind the house is a pile of cow dung and the civilian population have dug a deep hole behind every house that is as large as a cesspit... The population has no crockery, photos or anything else. Perhaps a few pots and a bucket, but their largest asset is a farmhouse. Small children scuttle around the villages en masse! I don't want to go into the house. The women sit out the front and pick lice off each other.

Even a rain-soaked night could not convince Grimm and Beigel to move indoors.

While the pioneers tried to sleep, the miserable wet night of 15-16 June was filled with activity as Bodensee units relieved foreign units. Hauptmann Klein jotted the following in his diary for 16 June: "Volchansk. Rest (rain). Roads impassable." His final sentence succinctly encompassed the problem the pioneers needed to solve over the coming days. Grimm's men broke camp and set off in the afternoon for Petrovka; their fourteen kilometre trek promised to be taxing. They trudged off into the gloom, accompanied by the drumming of rain on their heads and the suction of greasy mud on their boots. Leutnant Beigel was in charge of the column. Ten and a half hours were required to cover the distance: the utterly exhausted and filthy pioneers did not stumble into Petrovka until 0230 hours.

The rain eased on 17 June and held off for the rest of the day. Groundwater quickly drained and roads began to dry out. The motorised elements of Grimm's company – his headquarters section and the MuMa[2] – arrived in Petrovka early in the morning, but not without some trouble:

The weather has been really bad the last few days and we are truly stuck in the mud. One day it was so bad that we could only move on foot or with horses. During the march I left all of my company's motorised elements behind. A cloudburst was enough to turn the roads into torrential streams, although there cannot be much talk of roads. There is no firm subsurface anywhere, not a stone for miles around. Everything is just black dirt. I got my jeep here with snow-chains on the tyres... It seems the weather is improving. The hot sun will quickly dry the deepest mud, sometimes so that the dried surface can be driven on while the soil below is soft and moist.

Once, while trying to outsmart and outrun Soviet artillery observers, Grimm raced along one of these chocolaty roads as fast as his staff car could move. The long-distance guns landed shells in front, then either side and behind. Gefreiter Nuoffer, his driver, did not need to be egged on by his commander: the accelerator was flat to the floor. A three-kilometre stretch of road was under observation and to the Soviet observers the German vehicles were like ducks in a shooting gallery. Grimm's jeep shuddered violently as it crashed over deep ruts and boggy patches. He was fortunate to possess an ex-British Morris car reconfigured for the Wehrmacht and it was more robust than German makes:

A normal car would have disintegrated into its component parts if it had been driven in such a way through the pot-holes and ditches at full steam, like we were. At one point we stopped because water and dirt had been forced into the engine. The hot machine quickly dried the dirt and we got going after a few seconds. Some men in foxholes watched us and laughed. Artillery fire is not too bad when you calculate that it won't quite be a direct hit. The incoming projectile can be heard and you can lie down. On the other hand, mortars are terrible.

Tales of bottomless mud made it into many letters home. Gefreiter Füssinger from 1st Company wrote: "We must throw away everything unnecessary so that the horses can make it through." When it reached the point that even panzers could not crawl through the mud, Füssinger took matters into his own hands and traded four cigarettes for a light Russian wagon. Leutnant Beigel tried to explain the mud to his girlfriend:

---

2   MuMa = Munitions-und-Maschinentrupp (ammunition and machine troop), part of the 23-man company section. This troop contained three medium trucks and two trailer-mounted air compressors.

At the moment we're reconstructing bottomless roads. If only it wasn't raining any more! Ten minutes of rain makes all work useless. Therefore the roads should be made ready as quickly as possible. It's often enough to drive you to despair. We had gorgeous weather yesterday, everything was reasonably firm, and today it rained the entire morning. What that means cannot be described. You'd have to experience it for yourself.

Road maintenance occupied the pioneer battalion for the next week. It repaired the approach and exit roads to the eastern and western bridges of Voznesenovka and tried to keep the lateral connections both sides of the Nezhegoly in a navigable condition. Culverts inside the bridgehead were also strengthened to permit vehicular use.

Oberleutnant Grimm was doing paperwork and filing combat reports on the morning of 18 June when summoned to battalion staff by Hauptmann Beismann. While there, he received a pleasant surprise:

In the miserable mud I had to go back to battalion staff again today and take care of different matters. It was dicey and the Russians lobbed some "luggage" from a great distance. There were also some other vehicles on the road. While at battalion, the Iron Cross Second Class was pinned to my chest... On the way back my Iron Cross was furnished with dignified splashes [of mud].

At lunchtime in Petrovka, Grimm awarded the first twelve Iron Cross Second Class to members of his company. All three of his platoon leaders – Leutnant Beigel (1st Platoon), Leutnant Hepp (2nd Platoon) and Feldwebel Platzer (3rd Platoon) – were amongst the recipients. In combat situations, as in life, cream always rises to the surface, and so the company's ablest NCOs were decorated: Unteroffizier Erwin Bub (2nd Platoon), Unteroffizier Adam Pauli (2nd Platoon) and Unteroffizier Fritz Rauschenbach (1st Platoon). The other two companies had already received a round of decorations for their performance in the Kharkov battle but a few more were allocated for the latest victory. Hauptmann Klein bestowed Iron Crosses to five of his men as did Hauptmann Häntzka. The latter had a sixth to award, but this was done in private: the certificate and medal were posted to the family of Leutnant Hermann Hupfeld, mortally wounded on 10 June.

Openly hungering for medals was definitely poor form and therefore avoided by leaders who wanted respect from their men, but few would deny that receiving such visible symbols were a form of validation, that they were doing their job properly. Leutnant Beigel wrote:

I was awarded the Iron Cross yesterday. Most of my recon patrol also received one.

My men genuinely earned them. Even if the Iron Cross is awarded too often, I don't think so when it comes to infantrymen and pioneers... The main thing is that the individual is conscious that he has truly earned it.

After describing the circumstances in which he received the medal, of which he was immensely proud, Grimm made a coy request to his wife:

I ask that you not talk about my medal too much. The main thing is that I have already proven myself in a short amount of time... When I think back about my time in France, when I was the only one of the officers without the Iron Cross, then I have to smile today. I know I am just sensitive in many respects. I did not want to experience this war without combat.

The near-hopeless struggle against the elements continued. As soon as progress was made, the heavens opened up, liquefied the earth and put the pioneers back to square one. The rain was constant, although it did let up for a while on 20 June, as Beigel cynically noted: "This evening, as a special exception, it didn't rain. We're glad about that." One bright spot was the fact that they were based in one location for almost a week. The opportunity to forage for food was exploited, as Leutnant Beigel reports:

We've made ourselves comfortable, that is, we've chased the chickens out of the house, even swept it, scrounged a table, set it up and then ate like royalty. You would say like savages if you knew how we were eating. A squad from my platoon butchered a 300-pound calf two days ago. Thirteen men, with me as the fourteenth, have completely devoured this calf in two days. We cooked the meat in buckets that had been washed out. Every man could eat as much meat as he wanted. We also roasted some of it. Three days ago every one of us had a goose for dinner. You can of course imagine that the cooking was totally primitive.

Wet weather returned on the morning of 21 June. Although the higher-ups thought the day was quiet, it did not seem that way to those on the receiving end of a Russian aerial attack. Leutnant Beigel regaled his girlfriend with such a tale:

There was a murderous air raid just before nightfall. I was out and about and able to watch the spectacle from a hill. A Russian fighter dropped a light bomb right next to the house in which the company command post had been set up. Three of our NCOs were several metres away and were flung to both sides like a house of cards. The house burned and our car received a bullet hole through the roof and windshield from a plane's cannon. Otherwise, nothing at all happened during this fuss. The Russian fighter was brought down a short while later by an Me-109. Such experiences shape our lives in a variety of ways. Ah, there's so much to tell...

Despite a bomb landing right next to his command post, Oberleutnant Grimm remained silent on the subject when he wrote home, no doubt to prevent Lotte from worrying about him.

– – –

The second preliminary operation prior to the main summer offensive began on 22 June with the cover-name "Fridericus II" and its purpose was very similar to that of "Wilhelm". Due to the unfavourable weather, the attack originally scheduled for 12 June had postponed several times and seriously affected the beginning of Operation "Blue", which Hitler had initially set down for the 15th. "Fridericus II" went off without a hitch. The Russian Izyum group was smashed in a five-day attack, the small town of Kupyansk on the Oskol was reached, as were the lower reaches of the Oskol, and another 22,800 prisoners went into the bag. As during "Wilhelm", the premature retreat of Red Army forces seemed to indicate that the Soviet leadership was trading ground for time. Casualties remained within tolerable limits, improved starting positions were gained for the main operation and Timoshenko's troops were weakened, but these successes were paid for with a delay in the replenishment of the divisions involved that harmed "Blue".

The division's men always knew another offensive was in the offing but the arrival of dry weather, assault guns and more artillery pieces proved its imminence. In order to occupy better starting positions for "Blue", the division was commanded to conduct a limited strike on 24 June to push the front-line a few kilometres forward from forested hills into open ground. During the night of 22-23 June, Regiment 577 shifted into the left side of the bridgehead. Regiment 576 was already there. The threat of mines ensured the deployment of the pioneers. Hauptmann Beismann assembled his company commanders for a conference on the morning of 23 June. "Gentlemen, prepare your units for an attack," he stated bluntly. Klein's and Grimm's companies were allocated to Regiment 576 and Häntzka's men to Regiment 577. Grimm was content: the commander of the infantry battalion he was to support was Major Braun, an officer he admired and with whom a good working relationship had been formed. Grimm formed five mine-detection teams, each with a group of prisoners to help clear mines. At 1100 hours his company moved from Voznesenovka into the assembly area of Braun's battalion in Petrovka. Assault guns rolled into the bridgehead at midnight. Divisional units not participating in the attack were ordered to keep the airwaves free.

Men Versus Tank (24 June 1942)

Soft pastels washed a pre-dawn sky unblemished by dark clouds. As planned, the assault troops, accompanied by assault guns, moved out at 0300 hours. Forward observers and radio teams from divisional artillery escorted the infantry. Gunfire soon began crackling through the Nezhegol River valley and rapidly escalated into an all-arms engagement. By 0540 hours Regiment 576 had approached Height 202.1 against patchy resistance. Oberleutnant Grimm and his command section stuck close to Major Braun. Only one of Grimm's platoons was up front: Feldwebel Platzer's 3rd Platoon was split up amongst the rifle companies to look for mines. The other two platoons, under Beigel's command, tagged along with Braun's rear elements. Hauptmann Klein's 1st Company supplied mine-detection teams to the other assault battalion of Regiment 576. No mines were found. The pioneers penetrated into the woods north-east of Petrovka and supported their infantry comrades in treacherous forest fighting. Despite tanks, artillery fire, numerous skirmishes amongst the trees and strafing by Soviet fighters, Platzer's pioneers and Grimm's HQ section experienced no losses.

Against weakening resistance, Regiment 577 reached the northern edge of the forest, kept pushing on and reached the daily objective by 1000 hours. Sporadic attacks by enemy tanks were repelled the rest of the day. Another tank, caught behind the lines by the German advance, caused havoc, and once again the pioneers were in the thick of it. Leutnant Beigel's report clearly shows how difficult it was for a tank to be destroyed by footsoldiers before the introduction of the one-shot Panzerfaust:

The rest of 2nd Company, consisting of 1st and 2nd Platoons as well as the vehicles, were moved forward under my command into the attack sector of II./Inf.Rgt.576. Marching ahead of the company were three heavy anti-tank guns (5cm) and two light infantry guns with ammunition vehicles… About 1800 metres south of Point 188.8 I heard the sound of firing flaring up from an eastern direction which could only be coming from tanks. I equipped 1st Platoon for tank destruction. About an hour later, at 0955 hours, the platoon was near Point 188.8 when the call came through from the vanguard: "Pioneers to the front!" The head of the march group was about 200 metres east of Point 188.8. I led the platoon past horse teams that were pulling back. A tank had apparently come face to face with our march group.

Situation at the advance guard: about fifty metres in front of the advance guard, a tank – approximately thirty tonnes in weight – was in a four to five metre deep depression that ran through the forest from left to right, across our direction of march. The tank was attempting to drive along the bottom and on to an elevation in order to gain a field of fire over the column. Its forward progress was hindered by the trees. The first anti-tank gun was in position on the route of advance but it had no field of fire on the tank. The second anti-tank gun was about twenty metres back, on the other side of the road, but could no longer be turned around by its crew because the road was a morass... I detailed men from the reserve group to the second gun crew so that the guns could be moved into firing position. I worked my way along the left side of the road, up to 3rd Squad, in order to observe the tank and deploy the tank-killing team. Approaching the tank by staying to the left of the road was not possible because this area was under fire. I ordered the squad to stay put and take over security. Once it was seen that the anti-tank guns could not fire, I decided to destroy the tank in close combat.

Meanwhile, Gefreiter Schingnitz, armed with a 3 kg demolition charge, struck out on his own to the right of the road and sprinted up to the tank to destroy its tracks. After this first explosion was unsuccessful, Schingnitz then fetched a second demolition charge. This detonation only twisted a drive wheel. While I was moving ahead on the right, Schingnitz ran up to me and reported on his detonations. We moved up on the tank from the right. I placed an anti-tank mine – which had been fitted with a fuse igniter – on to the armoured deck close to the turret and ignited it. Detonation failed, whereupon Unteroffizier Bub, who had also come forward, placed his 3 kg demolition charge on top of the anti-tank mine and ignited it. This charge, combined with the anti-tank mine, cracked the armour plating and ripped the turret out of its housing. A tongue of flame erupted from the tank. Until this point in time, the tank had been firing continuously with its cannon and machine-guns. It began to burn. Shortly before this explosion, Feldwebel Grassl and Gefreiter Weid had destroyed the tracks with an anti-tank mine and a demolition charge, so the tank was unable to move. After about five minutes, a few of the crew tried to fire out of the tank with handguns and then escape. They were mown down by the covering detachment. A short time later a huge detonation erupted in the tank, completely destroying it... During the destruction of the tank, Gefreiter Steinacker was fatally wounded in the abdomen by a burst of machine-gun fire from the tank. In addition, Gefreiter Frank was lightly wounded on the right foot by shrapnel.

Beigel could not contain his excitement in a letter to his girlfriend:

[It] was a very eventful day. It was a large set-piece attack carried out with extraordinary dash by our infantry. At the beginning, the edge of a forest was stormed with shouts of "Hurrah!" Russians in the forest who were not knocked off scattered. I led my company forward with the vehicles up to the waves of infantry in order to help if something or other should happen. In front of me were two anti-tank guns and two artillery pieces. Suddenly, I heard: "Pioneers to the front!" The horse-drawn vehicles were sent back straight away and we worked our way up to the head of the column. There, a 34-tonne tank blocked the column's path. It was in a depression between the trees and was not able to fire at us directly with its cannon. Because of the bottomless road, the heavy anti-tank guns could not be moved into position. The tank, firing into the tree tops above us, tried to move out of the depression in order to "pester" us. I've always had a good sixth sense, so I was already prepared to destroy the tank with explosives. I had already heard the tank firing an hour earlier.

All hopes were placed on us pioneers. A quarter of an hour later the tank was completely destroyed. First we blew up its tracks, then together with an Unteroffizier I placed a mine on top of it and then it went up in flames. The crew managed to get out of the burning vehicle but were bumped off by my covering force. I can tell you that you've never heard such a "hurrah" come from everyone's throats as when the ammunition of the colossus went up in flames.

A Gefreiter in my platoon has proven himself to be particularly brave and daring. And one of my men was done in. That was our experience yesterday, of which I am immeasurably proud. I'm eager to know what the commander [Beismann] thinks. Is it enough for the Iron Cross First Class? Please tell my parents about this. They'll be happy about it...

The company's first assault operation warranted acknowledgement as an "assault day". For those fighting under Leutnant Beigel: "24.6.42: Destruction in close combat of a T-34 tank in the forest near Zavodnyi." And for the rest of the company under Oberleutnant Grimm: "24.6.42: Assault on forest north of Petrovka [Nezhegol] with subsequent forest fighting and tank attacks."

Busy Hiatus (25-27 June 1942)

Nuisance raids by Soviet bombers and fighter-bombers plagued the division throughout the morning of 25 June, while individual tanks chipped away at the forward position. Hauptmann Klein's company laid several anti-tank minefields to help shield the front-line from these armoured intruders. Grimm's company was also called upon to deal with tank forays. After marching back to Krasnaya Sarya just after dawn, Grimm was instructed to form five tank destruction teams and position them on the defensive sector of Regiment 577. After completing their assigned tasks, all companies of the 305th Pioneer Battalion assembled south of Zavodnyi.

In addition to bolstering the infantry line, the pioneers were assigned a mundane task: cutting wood. One of Grimm's platoons prepared lumber for the infantry to use in their dug-outs while 1st Company readied enough to build ten metres of bridge. The job was not without peril. Artillery and mortar barrages, as well as waspish enemy fighters, constantly tormented the woodchoppers.

– – –

With two preliminary operations successfully completed, preparations for Operation "Blue" were in full swing and the men of 305th Infantry Division were set on track to their destiny at Stalingrad. Initially, the city bearing Stalin's name was but an interim objective vital to the capture of the Caucasian oilfields. During an operation into the Caucasus, the enemy could not be left idle on the eastern flank or in the rear of such a thrust, so Hitler's first aim was to destroy the Russians in the large Don bend, occupy the Stalingrad region and thus gain favourable opportunities to protect the flank and rear. Some of the German force, indeed most of it, must remain in this region and along the Don, while mobile forces continued southward into the Caucasus. The destruction of the enemy in the large Don bend was to be achieved in three double envelopment attacks, one following the other, north to south. The division of this phase into three consecutive operations was based on two factors: first, Hitler wanted to start his offensive as early as possible and not wait until all assault units had been concentrated; second, the Luftwaffe was just not strong enough to support the entire attack front that stretched from Kursk to the Sea of Azov, a distance of almost seven hundred kilometres.

In the first operational phase, two assault groups would break through and then push to the Don: Army Group von Weichs[3] from Kursk to Voronezh, and one hundred kilometres south of them, 6th Army from the area north of Kharkov to the north-east. While the three panzer corps of Army Group von Weichs and 6th Army thundered east

---

3 Army Group von Weichs consisted of 2nd Army, 4th Panzer Army and the 2nd Hungarian Army.

to the Don, the inner wings of both assault groups, consisting of infantry divisions, would slice into the rear of the enemy caught in between and encircle them. Eighth Army Corps, and with it the Bodensee Division, belonged to the southern pincer.

After completion of the first phase, all three panzer corps would be concentrated under 4th Panzer Army for a push along the western bank of the Don to the south-east, while 1st Panzer Army simultaneously advanced east from the area south of Kharkov. These two armies would link up and destroy the enemy trapped in a pocket east of Kharkov. For the third operational phase, the chain of command would alter: Army Group South would split into Army Groups A and B, the former taking control of all armies between the Sea of Azov and the Donets, as well as in the Crimea. The remaining armies would come under the command of Army Group B. Both army groups would strike the enemy in the large Don bend, cross the Don and gain control of the Stalingrad region. Army Group B would protect the northern and eastern flanks of the operation by forming a defensive front from Kursk across to Voronezh and then along the Don down to the landbridge west of Stalingrad. To assist in holding this lengthy front, it would be supplied with the 8th Italian Army and the 3rd Romanian Army. Upon completion of this phase, the pre-conditions would be produced for the fourth stage: a southward push over the lower Don into the Caucasus by 1st Panzer Army with 17th Army on the western flank and the 4th Romanian Army on the eastern flank. The Caucasus Mountains had to be taken before the onset of winter, so Hitler's entire scheme was under the tightest time constraints right from the beginning.

General Oppenländer issued his orders for "Blue" mid-afternoon on 26 June. His units were instructed to form up during the night of 26-27 June for departure at 0230 hours. The final decision as to whether or not the attack would take place needed to be made. Codewords: Dinkelsbühl = attack will be carried out as planned; Aachen = attack stopped. At 1600 hours, barely forty minutes after the division sent out its order, "Aachen" was received from corps because continual thundershowers had greatly softened the roads and ground. The weather warmed up on 27 June but road conditions were still appalling.

No time was more frenetic for the pioneers than the days leading up to an offensive. Grimm's company had to fell timber for the construction of a temporary bridge and the manufacture of fascines. It was then attached to Regiment 576. Later, at a conference in Krüder's command post, Grimm was told to support an infantry battalion with mine removal teams and tank destruction troops. He selected Platzer's 3rd Platoon for the job. Soon after, word was received that the attack had been postponed, so the company moved to new positions:

At 1400 hours the company took up security in the corner of the forest west of Zavodskoi [Zavodnyi] and expected to get a breather. We did not get a reassuring feeling as we approached this piece of woodland in plain sight and received mortar bursts along the tree-line. An infantry squad we met there suffered casualties in this barrage. Even while the platoon leaders were being instructed to improve the positions, the pioneers were cutting down approximately 15cm thick trees with power saws, preparing the timber for use and covering the positions. The condition of the soil and the will of the pioneers to obtain shelter meant that this work progressed unexpectedly quickly. Without further interruption by the enemy, shrapnel-proof stalls for the horses and vehicles were completed before nightfall. Our Hiwis[4] built their own protection with diagonal shrapnel-proof walls, covered with plenty of sticks and earth and open on one side – interestingly enough, completely deviating from our construction method.

On the afternoon of 27 June, Field Marshall Fedor von Bock, Commander-in-Chief of Army Group South, consulted his army commanders about the possibility of beginning the offensive. Generaloberst von Weichs considered it feasible, Generaloberst Paulus thought not because of continuous heavy rain. Bock sent the codeword "Dinkelsbühl" to Army Group von Weichs and "Aachen" to Paulus' 6th Army. Operation "Blue" was finally getting underway, albeit with only its northern pincer.

Waiting for "Aachen" (28 June 1942)

Russian forces became more active as darkness closed in on the evening of 27 June. Rumbling engines throbbed through the blackness. German sentries concentrated hard, cocking their heads to try and locate the source of the gurgling diesel motors, but the rolling dales made it hard to pinpoint. All of a sudden, artillery crashed down and everyone ducked into shelter. Next it was rockets. The surprise barrages were downright unpleasant, especially because troop movements were taking place: reserve battalions were moving forward to occupy a second line, while assault battalions called up their reserve companies. Engine noise increased at 0100 hours from the area north and west of Krasnaya Polyana. Keen ears pointed out that the Russian tanks seemed to be moving westward. All regiments went into full defensive readiness. A fusillade of rockets pounded the main combat lines of Regiments 578 and 577 and about ten tanks, supported by a battalion of infantry, attacked Regiment 577 from a depression. Grimm's

---

4   Hiwis = Hilfsfreiwillige (volunteer helpers), Soviet POWs and deserters who worked for the Germans.

men were in position behind the line:

> Heavy artillery and mortar fire started to come down at 0230 hours and we heard
> the whoosh of rocket-launchers, known as Stalin organs. Heavy fighting must have
> developed in the division's left sector and we heard from some infantrymen that the
> Russians had attacked. The company stationed tank destruction squads along the
> tree-line and in the village. They were armed with hand grenades, demolition
> charges and anti-tank mines. The situation calmed down and 3rd Platoon
> remained on call for the regiment.

Beigel and his men were asleep when the cannonade began:

> Last night we moved further forward and bivouacked in a gorgeous oak forest. This
> morning an infernal artillery barrage ripped us from our slumber. The Russians
> attacked with infantry and several tanks. He did not achieve anything, however. At
> 0330 hours we were again able to crawl out of our foxholes.

The night-time assault was a minor inconvenience for Grimm's company. Their
field-kitchen turned up in the morning with warm food, "which we'd had to do without
all too often," noted Grimm, then they had time to take care of personal hygiene, do
some washing and clean weapons and equipment. Up front, the experience was
terrifying. Ominous black machines approaching in the gloom, tracks grinding,
engines roaring. Fortunately, cool heads prevailed and the attack was parried.

Hauptmann Klein's 1st Company was camped in the same forest as Grimm, about
two-hundred metres away, but suffered more under the bombardment, as Gefreiter
Füssinger attests:

> In a large forest in which our baggage train and I are accommodated, we got hit by
> an artillery bombardment and particularly by planes and tanks. You'd think that
> no-one would make it out because we had to spend all day in foxholes. Nevertheless,
> our weapons checked the enemy, all his tanks were knocked out and enemy
> artillery was forced to retreat. And all this with terrible rain and thus terrible filth,
> sleeping in holes that soon filled up with water. How nice it would be to have a roof
> over my head again and to sleep in a feather bed.

– – –

Army Group von Weichs launched "Blue" early in the morning after a half-hour
preparatory bombardment. Progress was rapid. The spearhead of 4th Panzer Army,

General Werner Kempf's 48th Panzer Corps, covered the sixteen kilometres to the Tim River by midday and captured a railway bridge intact. That afternoon it completed another sixteen kilometres and crossed the Kshen River, leaving only open ground between it and Voronezh. Russian resistance varied: obstinate in some places, chaff-like in others. One thing was certain: there had been no premature retreat. By nightfall, Kempf's panzer corps had covered yet another sixteen kilometres, the last few in driving rain.

The weather in 6th Army's sector was dry throughout the day despite thick cloud cover, but the roads in the forests and valleys were still sloppy. The codeword for 6th Army was therefore "Aachen," signifying another 24-hour postponement.

Still Waiting (29 June 1942)

Russian artillery, mortars and rocket-launchers repeatedly dropped barrages on the division's positions during the night of 28-29 June. Oberleutnant Grimm's company was on the receiving end:

Heavy artillery and mortar fire came down at 0240 hours and we were in no doubt that the Russians were keeping tabs on our corner of the forest. When the sentries heard Stalin organs being launched, they yelled out "take cover" and took shelter themselves. About an hour before daybreak, the time had come. The first salvo roared overhead. The second kept us waiting only a few minutes but it landed in our forest. Where could the Russian radio post in our rear be from which he was so accurately directing their fire? The third volley came in just over our heads: we heard the howling approach of the rockets, the bursting of the shells in the tree-tops and on the ground, while the ear-splitting crack and flash pressed our bodies to the ground. As soon as the horrific experience was over, we got ourselves out of this stupor, wanting to know whether we were unscathed. After this strike, which came from two rocket-launchers, it rather suddenly went quiet. In our exemplary positions, which were widely spaced, we suffered unexpectedly high casualties. Obergefreiter Waldmann was killed. Ernst Kempter took some shrapnel in the stomach while Gefreiter Fritz Wüstner was wounded in the hands and feet by shell splinters. Two Hiwis were killed and three wounded. Ten of our best draught horses were wounded and three subsequently died. A motorcycle-sidecar was heavily damaged. Most vehicles received some shrapnel but remained operational. Fortunately, we had the rest of the day to repair the damage.

Pionier Kempter would recover from his stomach wound and re-unite with the company in Stalingrad. Gefreiter Wüstner's shrapnel wounds were more serious: the index and middle fingers on his left hand were sliced off and the rest of his hand left so damaged that the surviving fingers were permanently stiff. His feet were not in much better shape. In fact, his injuries were so dire that he was still in hospital a year later. Grimm was sorry to lose Wüstner: as a deputy squad leader, he had repeatedly proved himself in combat situations, was courageous and would have received the Iron Cross a few weeks earlier but for the fact that comparatively few had been allocated for Grimm's company. In August the following year, Grimm saw to it that Wüstner finally received the long-overdue medal.

– – –

Rain fell on Army Group von Weichs until noon. 48th Panzer Corps churned through the mud and consolidated its hold on the land-bridge towards Voronezh, while 24th Panzer Corps formed bridgeheads across the Kshen River. Tasked with covering the southern flank of the armoured thrust was the Hungarian 2nd Army, but it could not get past the Tim River. As one historian wrote, "it was being held up less by the rain or by the enemy than by its command's ability to stage a coordinated attack."

In the afternoon, 6th Army consulted its corps commanders. All reported that roads in their sectors were passable. With that, Field Marshall von Bock issued the codeword "Dinkelsbühl". The decision was met with relief: the troops were in limbo and sporadic harassment fire from mortars and artillery was inflicting pointless casualties.

Carnage (30 June 1942)

Once again the Russians fired into German assembly areas with rocket-launchers and caused considerable casualties even before the attack began. Artillery and mortars joined in. The "jump off" at 0230 hours therefore amounted to a release of nervous energy. Rocket-launcher inflicted casualties upon Voigt's and Winzer's regiments, though they were silenced when an entire Stuka wing – upwards of 80 dive-bombers – concentrated on Krasnaya Polyana and nearby tree-lines at 0300 hours. Effectively supported by artillery and the Luftwaffe, the attack swiftly gained ground, and despite predictions that the fiercest Russian resistance could be expected around Krasnaya Polyana, the village was pushed through in one stroke by Regiment 578.

On the right wing, the spearhead of Regiment 576 entered the sparsely wooded area south-east of Alkhimovka. In support was 1st Company, closely trailed by Grimm's 2nd Company, both tasked with assisting the infantry in forest combat. Grimm's company was marching behind the leading assault wave, helping the heavy weapons of Regiment

576 move forward, when two lines of mines were detected in a forest cutting one kilometre south of Alkhimovka. It was Grimm himself who found the minefield:

Detection is not purely a matter of luck and it was no exception in this case. Important prerequisites had to be met: good training in the installation of minefields; a precise knowledge of Russian mines and fuses, which I had because I gave lectures about them in France; acute observational powers of a huntsman and the best eyesight. We never used electrical mine detectors in action because we only ever came across wooden box mines. The most suitable implement was the mine probe. I had a wooden stick, sharpened at one end – all of us did – kept handy by being stuck down inside my boot. In forested areas I preferred the mine probe to the submachine-gun, particularly recommending it for mine detectors who moved on alone up front.

As company commander, I had the mission – supported only by my HQ section – of reconnoitring a route so that the heavy weapons of the infantry regiments could be brought forward into the front-line to advance with the infantry. The only way through was a forest cutting. The approximately 15-20 metre wide cutting must have been avoided by the infantrymen during the attack. I presumed it was mined after determining that there was no other passage through the forest. When detecting mines, you move forward alone or only with a large gap to the next man. The cutting here was too narrow for a second man. Normally, minefields are watched over by those who laid them, that is to say, with infantry and heavy weapons. The possibility of the enemy doing this here was poor and was thus not the case. Why they didn't do it, I don't know.

Dawn broke quickly after I'd moved about fifty metres along the forest trail. I could make out two faint cart tracks from which fresh steppe grass was sprouting due to recent rain. I moved forward just off to the right of the track, which had neither been driven on nor walked on since the rain two days earlier. There was fighting in the woods a few hundred metres ahead, but no longer near the cutting, yet I had the feeling that I'd walked into a trap. While I was standing still, like a wild animal taking in the scent, scanning the cutting and the edge of the forest for signs of the enemy and his mines – which could normally be recognised because they were laid in dry ground with wilted grass over them as camouflage – I spotted a 2cm wide strip, with hairline cracks on both sides, running across the track a pace or so in front of me. I crouched down and scraped away the dirt and grass with my hands, revealing a large wooden anti-tank mine which would certainly have reacted to a man's weight. The enemy hoped that individual riflemen would not step on a mine so that the minefield would not be recognised. Then, tanks or heavy weapons could fall victim to the mines.

I looked around to see if I was on the edge of the minefield or in the middle of it. The camouflage on the mines did not permit the extent of the minefield to be determined. I called for my HQ section but lost a lot of time because contact had been lost. It was common knowledge that the Russians were quite capable of camouflaging their mines excellently but they only ever laid their minefields in strict accordance to a mine plan, in a chequerboard pattern, with precise distances in between that were large enough based on experience that the explosion of one mine did not send up the entire field. Up to that point we had only encountered anti-tank mines, which also reacted to riflemen. I no longer know exactly but I think they were called TMD-40, had almost twice as much explosives as our T-mines and were wooden boxes that were practically invisible to our electronic detectors. These had an incredibly simple fuse: when pressure was applied, a small piece of wood that had been inserted – similar to a matchstick – was sheared off.

I pulled out grass, as if I was weeding, and stripped away the camouflage on the closest mine. Oberfeldwebel Locher, an experienced mine expert, met up with me first and then cleared a lane in the wheel track furthest to the left. I found a second mine barrier. The rest of the HQ section arrived, gave us cover and began to mark out the field. In the forest to the left and right, on both sides on the minefield, were stacks of mines, probably a sign that the enemy had not finished laying them. The rattling of tracks from up ahead forced us all to stand up. Oberfeldwebel Locher was the first to see that it wasn't an enemy vehicle; by waving steel helmets and shouting loudly, the [half-tracked] vehicle halted in front of the narrow lane. [A] serious mine incident for the company was avoided at the last second.

Little did Grimm know that a "serious mine incident" had only been delayed. After marking out the minefield, Grimm handed it over to Hauptmann Klein, whose company eventually removed 35 anti-tank mines from the cutting. Grimm's company marched on, faithful to their task of helping Krüder's heavy weapons move forward.

They approached another forest defile:

The company took rifle fire when it entered the forest cutting, as well as several mortar impacts. A line of mines was found across the road… Rauschenbach's squad, with one NCO (Unteroffizier Rauschenbach) and four men, received the job of clearing these mines. They were wooden box mines, laid in a chequerboard pattern.

Having left these five men behind to clear the mines, Grimm and the rest of his company marched towards Belyi Kolodesi through a seemingly endless cornfield. A short while later, at about 1000 hours, a massive explosion rang out behind them and two immense pillars of dirt and smoke billowed up into the sky. Everyone watched in awe. Grimm later wrote, "we were already a few kilometres away when we saw the two giant mushrooms of smoke shoot into the air." The tragic truth was revealed later that night when a member of Rauschenbach's mine clearance team reached the company after walking all day. His report stunned Grimm. After 125 mines had been cleared and placed in two stacks ready for demolition, he gathered the team's weapons, clothes and equipment and moved them to a safe distance. At that instant, a monumental explosion hurled him to the ground. A split second later a second detonation, bigger than the first, shook the earth. Although dazed and deafened, he immediately comprehended what had happened: a stack of forty-five mines inexplicably detonated and triggered the second, larger stack. Three men were vaporised by the explosion: Unteroffizier Fritz Rauschenbach, who had celebrated his 25th birthday just the previous day, Gefreiter Michael Schiefer and Gefreiter Karl Seifert. The two survivors searched for any sign of their comrades but all that could be found were ragged scraps of cloth and flesh.

Then a second devastating blow struck Grimm's company: a report arrived with news about a direct artillery hit on a squad (the 9th) from Platzer's 3rd Platoon attached to Regiment 576. The sobering details were entered in the company diary:

[It was] split into three mine-detection and removal teams, as well as three anti-tank combat teams. The Fuss squad, located with the reserve company of III. Bataillon, suffered a direct artillery hit while in jumping-off positions on the edge of the forest one kilometre north-east of Zavodnyi. Four men (Gefreiter Beck, Gefreiter Schlereth, Gefreiter Friedrich Schneider and Pionier Reinauer) were killed and five men (Obergefreiter Fuss, Gefreiter Schöffmann, Gefreiter Raue, Gefreiter Sparrer and Gefreiter Prüll) were wounded, while two prisoners were killed and one severely wounded.

When Grimm and his men received this news, "hunger, thirst and stress were forgotten. Not another word was said. The Russians did not disturb our thoughts. No-one thought about sleeping." Less than three months earlier, after learning of the death of a childhood friend, Grimm consoled Lotte – and himself – with these words: "His life was crowned with a Heldentod (hero's death). This is the most beautiful death. It is very hard for his family because Fritz was everything to them… Every soldier must be able to look death in the face." The Wehrmacht was inculcated with this fatalistic belief. Its soldiers were lionised as courageous and self-sacrificing and their deaths placed within the context of the German people's future. A heroic death to gain "living space" in the east would further new life. Now that his men were being scythed down by the randomness of war, Grimm's attitude slowly changed. He realised there was no heroism in being blown to shreds.

One of the men wounded by the direct hit, Gefreiter Hans Sparrer, remembers that "three or four Russian Hiwis were carrying T-mines." The artillery shell triggered secondary explosions of mines and explosives being carried by the pioneers and their Hiwis, thus accounting for the fact that almost the entire squad was felled. As for his own wounds, Sparrer recalls: "I received shrapnel wounds in the back and right leg. I returned to the company on 6 or 7 August." Squad leader Obergefreiter Eugen Fuss took a piece of shrapnel in his upper left arm, while Gefreiter Heinrich Prüll was struck in the face and right thigh by shell fragments. All the wounded were taken to an aid station in Petrovka. Only Sparrer and Gefreiter Prüll would return to the company prior to Stalingrad. The direct hit had wiped out the entire 9th Squad – "Beismann's fiercest squad", as Sparrer called it – but the news got worse. Another man from the same squad, Gefreiter Willi Kaul, died in the same Petrovka dressing station from injuries suffered in an unrelated incident, while one of the men wounded in the direct hit, Gefreiter Georg Schöffmann, succumbed the following day. This one squad lost six dead and four wounded in a single day. The company's grisly toll was compounded by the three dead from the mine incident. Grimm rightly wrote that his company was "shellshocked" by these casualties.

The day was acknowledged as an "assault day" for the company: "30.6.42: Attack and forest fighting near Alkhimovka." Even though Grimm's unit was the last pioneer company to arrive on the Eastern Front, it was the first to accrue three assault days. Soldiers that had participated in all three attacks were eligible for the General Assault Badge, though the pace of operations caused delays in such non-essential paperwork. Grimm would not lodge the first claims for the badge until late August.

The death of Unteroffizier Rauschenbach presented Grimm with a conundrum. Rauschenbach's family was duly informed but explicit details were withheld in order to spare further pain. Sometimes, however, such tactful handling caused problems.

Grimm explains the ramifications of denying next-of-kin the whole truth:

> *The mine incident with Unteroffizier Rauschenbach was terrible. As far as I recall, he was the only son of a war widow, which I only discovered later during the subsequent exchange of letters. In the following correspondence I provided his mother with the home addresses of all witnesses – because of heavy casualties, naturally also with mine – but unfortunately in my diary I did not write down the names of the company members who had survived and reported the course of events to me... Frau Rauschenbach handed the matter over to her local Nazi Party district leader. The report of witnesses and the provision of all private addresses brought no further answers. I asked the district leader to communicate to Frau Rauschenbach what I had withheld from her out of consideration. I was then insulted in the worst possible way, as a corpse robber, among other things, because we were not able to return an engagement ring[5] as part of his personal effects... Naturally I did not write any unkind words to Frau Rauschenbach about her insults that had arisen from the pain she was suffering.*

Grimm was not the only one struggling with this matter. Leutnant Beigel received a letter from Rauschenbach's mother and the difficulty of disclosing the horrific details is apparent:

> *I've got a difficult letter to write today. The mother of my Unteroffizier Rauschenbach wants information about how her son was killed and whether he'd left any final words for her. Now I have to write to the poor lady that he was completely ripped apart by mines. We were only able to find small pieces of him. It is no secret in this case because all his belongings, which he had on him (watch, wallet, etc.) were all blown to smithereens. It's difficult for me to console this lady. I've already spent a long time writing this letter, and I have to keep revising it. I have to comply with many such requests because it is not written in the initial notification how the man died.*

Unteroffizier Rauschenbach was an original member of the company, and indeed, had been an integral part of it. Leutnant Beigel truly lamented the loss of Rauschenbach: "I lost my best NCO. In addition to him, two of my best men. That's fate." Beigel's

---

5   Rauschenbach was not married, but a tradition amongst some German soldiers was to carry or wear the rings of their deceased fathers.

attitude may seem callous but fate is what he and many others believed in. His girlfriend obviously worried about his safety, yet Beigel was convinced that his fate was different to Rauschenbach's:

You worry too much about me. I keep telling you that nothing will happen to me. You know that I'm careful if the situation allows it. I've used up all of my recklessness. Besides, the fate of a soldier is in God's hands, He who controls everything. When fate is predetermined, nothing can be done to resist it.

Most of the southern Germans in the Bodensee Division were pious Catholics and as soldiers in imminent threat of death, their belief in divine providence strengthened, allowing them to endure the terror of war on the Eastern Front. If their time was up, there was nothing that could be done. Grimm, one of the few Protestants, was of the opinion that the destiny of the individual was no match for the cruel randomness of war. After receiving news that an acquaintance had been killed on the Eastern Front, Grimm wrote to Lotte: "Yes, the fighting is hard and whoever is here knows how necessary it is and that individual destinies play no role at all." Another example of that was taking place nearby. Leutnant Berthold Staiger, a platoon leader in Häntzka's 3rd Company, was physically and emotionally traumatised by another mine incident:

During the advance into enemy territory, a bridge over which our troops would push forward had to be held. Together with my Feldwebel Janzik, I went with an advanced troop to look for mines on and around the bridge. This was an extremely difficult task to undertake because we guessed they would be dangerous "pull igniters", as we called them. Despite every precaution, there was suddenly a violent explosion from a mine not too far away from me. Feldwebel Janzik was torn into unrecognisable shreds. I have been unable to banish this image from my mind. It was simply dreadful. That both my eardrums had burst was only noticed later when I saw shells exploding but did not hear the detonations or the cannon fire of enemy tanks. I'd been completely deafened and was put into hospital until my hearing partially returned.

– – –

In a progressive attack against weak Russian resistance, both Regiment 577 and 578 reached the heights north of Trud at 1330 hours. Regiment 576 reached Belyi Kolodesi at 1530 hours after overcoming mines and obstinate resistance by infantry and scattered tanks in the forests east of Krasnaya Polyana. The diary of Grimm's 2nd Company picks up from the point where they witnessed the twin mushroom clouds: "The company combed through the village [of Belyi Kolodesi] under the protective fire of an [artillery] battery. In doing so, a lorry was captured, eight prisoners were also taken, and in addition, a number of Russians who did not lay down their weapons in their hide-outs were shot." Leutnant Beigel relished this action: "The Russians are running. I captured a brand new lorry with two squads. We bumped off four Russians and captured eight. That was a blast!"

The infantry had covered thirty-five kilometres during the attack, but barely had the men collapsed to the ground in exhaustion when an order arrived for the push to continue. The Russians must be kept on the run; every Landser knew that the harder they chased, the more blood would be spared in coming days, which is why they now gritted their teeth and pressed on. Deep gullies required wide detours. After overcoming a final flurry of resistance in front of the objective, the regiments pushed on but the Russians detected the baby pincer movement and fell back. No motorised infantry units were available, so the Russians could not be cut off. The divisional history summed up the general feeling: "It was again confirmed that the pursued always runs faster than the pursuer." As night fell, the designated line was held by the three infantry regiments. They were proud of a quite remarkable achievement: over fifty kilometres in attack across rugged terrain.

While 6th Army cut deeply into the Russian defences, Army Group von Weichs struggled in a rain-soaked quagmire. The position of von Weichs' panzer corps in the north and 6th Army in the south indicated to the Russian command that another encirclement was developing.

– – –

Until 29 June, Grimm's 2nd Company had only suffered three dead and six wounded, light casualties for almost six weeks on the Russian front, but one day later the tally jumped to twelve dead and ten wounded, with one of the wounded dying the following day. The company had left France with 210 men of all ranks, meaning that 10% casualties had already been suffered, though it must be remembered that these were inflicted on the combat elements. This figure is placed in context by the fact that Grimm's company had seen far less combat than most other companies in the division.

The soldiers of the Bodensee Division had grown accustomed to a new way of life on the Eastern Front. Extreme weather, lengthy marches, then long periods of waiting, followed by violent surges of combat. The Red Army seemed to be fleeing headlong in front of them, afraid to stand and fight, so a victorious mantle still shielded them, and would continue to do so for some time. A sense of purpose and ascendancy coursed through the veins of the German army in the East. They felt ready to take on any challenge.

CHAPTER 4

With the southern pincer in motion, "Blue" was now in full swing. Hoth directed his 4th Panzer Army straight at Voronezh. The roads were choked with bogged supply columns. The panzers also struggled in the mud, so much so that infantry headed up the advance of 48th Panzer Corps. In the afternoon the right flank division of the panzer corps turned south to begin the link-up with leading units of 6th Army near Stary Oskol. Russian forces would be ensnared in a pocket west of the Oskol River.

The Oskol is Reached (1 July 1942)

The forces of 8th Army Corps moved out bright and early to reach the Oskol River and cut off the fleeing Russians. General Oppenländer's men were exhausted from the previous day's effort but they were on their feet, ready to march remorselessly towards the river and another victory. Winzer's regiment made a precise "right face" and began a vigorous pursuit over broken grasslands dotted with woods and rustic villages nestled in valleys. Resistance was non-existent. His infantry soon reached the Oskol near Chernyanka. The first bridge exploded as the leading units reached the western bank. An emergency bridge right next to it, still standing due to a poorly-executed demolition, was immediately used to cross the river. Hauptmann Häntzka and his men were not far behind and arrived just in time, when their skills were most needed, because the second bridge flew into the air and was beyond salvation. Häntzka's pioneers ferried over small groups of men in inflatable rafts while some elements forded the river. Resistance was weak. Patrols pushed further east.

By late afternoon, 6th Army had crushed the right half of the Southwestern Front west of the Oskol and seized a bridgehead. Soviet units west of the river were retreating so fast that von Bock was doubtful enough could be pocketed at Stary Oskol to warrant swinging 6th Army north, so he talked to Hitler about letting the army go north-east instead, "to cut off what is still to be cut off" by trapping the Russians between the inner flanks of 6th Army and 4th Panzer Army somewhere further east. General Paulus believed the Russians would not let themselves be encircled west of the Don, so he wanted to head due east. In any case, Russian resistance was definitively broken. The Bodensee Division had put another forty kilometres behind it during the pursuit and formed a bridgehead near Chernyanka. A crossing of the Oskol was scheduled for dawn. Small and large inflatable rafts were brought up during the night and suitable sites reconnoitred.

Bridging the Oskol (2 July 1942)

Small batches of men and equipment continued to cross the river throughout the night. The main crossing, begun after the sun had risen, ran smoothly. Anti-tank guns and infantry guns were ferried over; horses and gun-limbers utilised a ford. Work commenced immediately on construction of a bridge. Hauptmann Beismann had been ordered the previous night to have the bridge ready by 1600 hours on 2 July, but both Klein's and Grimm's companies had been left behind by the hectic advance. Their initial task of strengthening minor crossings on the supply roads was halted when word came through just before sunset that a major bridge was needed across the Oskol:

> An order for 1st and 2nd Companies arrived unexpectedly on 1 July to construct a bridge with the Kriegsbrückengerät [military bridge equipment] 'B'... Since the formation of the pioneer battalion in Riedlingen at the end of December 1940, no possibility existed to train the men on this bridging equipment which, with the exception of the covering on the bridge deck, consisted of zinc-coated metal alloys and was assembled like a construction set, in contrast to a makeshift bridge made and knocked together from dressed timber. The first company under Hauptmann Klein had registered the painful loss of two platoon leaders (Leutnant Birk and Leutnant Lutz, both killed in May) who had been instructed on this equipment...

After a few hours of shut-eye, the two pioneer companies moved out in pitch darkness. Hundreds of pioneers climbed aboard the bridging column of army-level engineers for a wild night-ride through the Russian countryside. They reached the river in dawn's golden glow. Grimm moved at the head of the column, reconnoitring approach routes to the bridging site for the unwieldy truck-trailer combinations. Häntzka's 3rd Company and elements of 162nd Pioneer Battalion were already hard at work. Grimm describes the scene:

> The pioneers dismounted and looked around incredulously. They had reckoned on heavy enemy interference and a river similar to the Donau. In front of them they saw the flat riverbed of the Oskol, a bridging site cleared on this bank and other pioneers busy in rubber boats on the water recording a profile of the river bed, exactly as on training exercises back home. Best of all, no enemy interference. Downstream was an elevated, partially destroyed wooden bridge with lengthy elevated entry and exit

approaches that constricted the flow of the river. Weapons were hastily laid aside at the ready, platoon leaders received their instructions and the first pontoons were brought down into the water. Leutnant Beigel was ordered to construct ferries and as a precaution reported to me the number of men trained on the B-Gerät and how the bridge sections were floated into place under the given conditions. The performance of our pioneers surpassed all expectations. It showed the good state of training on the water, on the Donau in Riedlingen, on rivers in France and on the coast. Despite the ropes still not being all coiled up and the bridge railings not yet affixed, the inspection and handing over took place shortly after midday.

Grimm's assertion that the bridge was completed just after noon is slightly misleading. The divisional war diary states that the 74-metre long 8-tonne bridge (later strengthened to 24-tonne capacity) was finished by 1415 hours, but even this time gives the wrong impression because a bridge is useless until traffic starts flowing. Vehicles started crossing at 1800 hours. Grimm hastily employed every spare man to improve the approach and exit points to the bridge. Two assault guns did not care to wait in the traffic jam, so they attempted to ford the river downstream of the bridge. The first one made it across but the second became stuck mid-river and endured an unexpected bath until it was hitched up to the first gun and pulled out.

Correspondents and cameramen followed in the wake of the advance, snapping scenes of defeated Russians and victorious Germans. Gefreiter Friedrich (Fritz) Vorherr, a motorcycle driver on Grimm's staff, wrote to his parents: "I was filmed by the Wochenschau [weekly newsreel], a few times actually, and the pictures will appear in the newspapers in fourteen days to three weeks, so keep an eye out." Vorherr's family routinely went to the movies but did not see their Fritz in any of them or in any magazine.[1] Vorherr was not imagining things because Leutnant Beigel also wrote home:

Nice pictures of us are being shown in the newsreels. Several men from my platoon have already received prints from film captures, which are really great. Keep an eye out for your Wiggerl[2] when he was there during the bridging of the Donets.

---

1   This is the first of many excerpts from Vorherr that will appear in this book. Gefreiter Josef Zrenner, a clerk on Grimm's HQ staff, recalls that "Fritz was a great comrade of mine but he had a quick tongue and always called it as he saw it." This frankness makes Vorherr's letters historically invaluable as he never sugarcoats the truth: he openly writes about what he's thinking, feeling, experiencing. Reports and letters by officers carry a veneer of propriety, even when written to wives and loved ones. Any signs of petty politics are hidden in veiled references. Vorherr, on the other hand, lays himself bare in his letters.

2   A pet form of the name "Ludwig".

Fresh forces were continually fed across the river and the bridgehead rapidly expanded to the eastern outskirts of Chernyanka (Novo Ivanovka). A goods train entering the railway station was happily received. Other trains in the sidings were laden with rations and clothing. Large oil tankers were also secured. The amount of booty was huge: at the station was a supply depot with fruit, flour and other foodstuffs. All units restocked their field-kitchens and fodder stores. Ordinary soldiers also indulged in some "foraging." A few Russians were also captured. The triumphant German infantryman were not impressed by what they saw: "The prisoners, 250 in number, were all different races of Russia and made a stupid, apathetic and – in some cases – a non-humanlike impression." Aerial reconnaissance spotted Russian columns retreating to the south-east and north-east. The only resistance came from the Red Air Force whose planes sporadically bombed the crossing sites. Towards evening the division had enlarged its bridgehead ten kilometres to the east without noteworthy resistance.

Traffic Control (3-4 July 1942)

The significance of the German bridgehead was not lost on the Soviets. Their planes energetically attacked the bridging site and its approach roads all night and intensified their raids in broad daylight, sometimes just a single bomber, other times a dozen. The entire division was being funnelled across the bridge, so traffic control and oversight of the bridge was entrusted to the pioneers. Grimm and Hauptmann Klein were designated bridge commandants, taking turns in three-hour shifts. It was during one of his stints that Grimm experienced the first-hand terror of a raid:

Vehicles banked up at the bridging site. The approach and exit routes were in increasing need of improvement. Corduroy roads were broadened. During attacks, vehicles and limbered guns could quite often not leave the roadway. Courageous and brave, the drivers kept their horses calm: there was no shelter. Bombs fell out of the night sky and caused the first losses. The pioneers worked feverishly. There were only a few men on the bridge because the excellent B-Gerät needed almost no maintenance, despite frequent heavy usage. Early in the morning, the first daylight raid – with American bombers – came in. Three waves, each with three bombers, attacked the bridge just as there was a hold-up on the other bank. Standing there in the middle of the bridge, I heard Flak shooting and ordered: "Clear the bridge! Take cover!" I found no cover on the other bank because the foxholes, meant for a single man, were occupied by several as a result of the traffic jam. I didn't find any cover and the bombs of the first wave flew down towards me, the second and third waves releasing their bombs precisely above the bombs of the first wave. I lay down on the bare ground and opened my mouth,[3] but because of the three tiers of bombs, I could not fix my eyes on the bomb that was going to take my life, as I had already concluded. The soft, sandy soil allowed a few men within sixty metres of the bombs to survive. Sand was forced into our mouths, noses and ears and we were only able to find our way around slowly in the gloom and clouds thrown up by the explosions. Pioneers were already rushing towards us to offer help. The men in the fully occupied foxholes were buried alive, without exception. The extent of the casualties was not assessed for a long time. If the carpet of bombs had fallen about fifty metres towards the middle of the bridge, it would have completely destroyed it. Even while the rescue operation was in full swing on the other bank, the crush of vehicles again rolled across the bridge: the aerial attack had halted the crossing for barely ten minutes. Also, during the attack, three nearby lorries of 1st Company loaded with mines were blown into the air, but it did not hinder the crossing... My company suffered no casualties. The pioneers of 1st Company were on the other bank, on the exit from the bridge, and they suffered heavy losses... I saw many dead and severely wounded.

Gefreiter Rinck recorded the grim death toll in his diary: "Russian aerial attacks ten times, each with nine planes. Bridge not hit. Sixty-five dead." Grimm had wisely spread out his company billets to minimise casualties. Inflatable rafts and various bits of equipment were destroyed. The approach road was struck by two direct bomb hits but the craters were swiftly filled. Apart from keeping the crossing open and traffic flowing, Grimm was once forced to wield his powers as master of the bridge:

I was called across to the bridging site on the other bank. Through an oversight, some severely wounded men – including several with stomach wounds and missing limbs – were guided to the bridge where the stretcher-bearers learned that absolutely NO traffic was allowed in the opposite direction. I had to decide between a fairly certain court-martial and assisting the wounded. I had the crossing stopped but had

---

3   The most dangerous aspect of a blast shockwave is what it does to lungs. The natural instinct during a crisis is to take a deep breath and hold it, but this turns the lungs into a pressurised balloon. When the pulse of air hits the body, that balloon bursts, causing massive damage to the lung's lining. Victims drown in their own blood. The best way to avoid that fate is to keep the mouth and air passages open.

to make use of all my powers as bridge commandant by running on to the bridge with several men, submachine-guns in hand, and forcing the oncoming vehicles to halt. Still ringing in my ears were the calls of "a lunatic is coming!" In a few minutes, however, I was saved by the shout of "Bridge clear!" You have to be lucky.

The pace was far less hectic for Grimm's platoon leaders. They improved roads and levelled craters, but all in all, the routine was more relaxed. Beigel even found time to catch up on his correspondence:

We've been allowed to get six hours sleep today. We fought, then marched a long way and constructed a bridge yesterday. We had one and two hours sleep the last two nights. We're more pleasant to be around today because of it. If the Russian planes don't whack our beautiful bridge, then there won't be much work today.

The night passed quietly for those in the bridgehead on the eastern bank. The infantrymen spent the morning preparing to renew the advance. Regiment commanders were informed that "the enemy ahead of 8th Army Corps seems to only have weak forces at his disposal." The plan was for 305th and 376th Infantry Divisions to advance along both sides of the Userdets River in a south-east direction. General Oppenländer was warned that 40th Panzer Corps would cut right across his path. The attack set off at 1500 hours with Regiment 578 on the right and Regiment 577 on the left. Hauptmann Beismann was ordered to have every available pioneer follow Regiment 577. The two spearhead regiments each deployed a reinforced battalion up front, the rest followed about an hour behind. The main bodies maintained large distances between units in order to allow elements of the panzer corps to pass through. The neighbouring division lagged far behind. Towards midnight, Oppenländer halted the march after fifteen kilometres. Russian soldiers, some of them sleeping, were dragged out of houses, so surprised were they by the speed of the advance.

Overall progress on 3 July was satisfying. Success after success was reported. Eighth Army Corps and 40th Panzer Corps had forced the Oskol and were on the verge of encircling a small Russian group. A larger body was cut off further to the north-west. The aim – to trap the Russians west of the Oskol – was partially successful. Elements of nine rifle divisions were surrounded and about 40,000 prisoners eventually fell into German hands. Russian forces that eluded the pincers were scramming faster than the Germans could advance. At Army Group von Weichs, the panzer and motorised divisions had gained freedom of movement and rushed towards the Don west of Voronezh. Bock sent a teletype to Weichs and Paulus: "The enemy opposite 6th Army and 4th Panzer Army is defeated."

As resistance was negligible in front of Regiment 578 and strengthening in front of Regiment 577, Oberst Winzer was only permitted to advance once Oberst Voigt's regiment had moved up to the same level. Regiment 577 set off at 0230 hours on 4 July and was level with Regiment 578 five hours later. With a united front, the division moved off. Its tiny mobile spearheads lanced deep into enemy lines. Russians flooded back to the left and right. Red Army units offered no resistance. The objective – high ground between the Userdets and Kamyshenka Rivers – was reached just before noon.

Terror in the Cornfields (5 July 1942)

Despite Russian forces retreating towards the Don as if their tails were on fire, the division warned its units to expect strong local resistance. The goal was to force a crossing over the Tikhaya Sosna (the "silent" Sosna) near the twin village of Alekseyevka-Nikolayevka and form a bridgehead on the southern bank. Resistance was expected along the Sosna watercourse. An advance detachment was ordered to seize a crossing near Ilinka in a surprise attack. The division moved off at 0230 hours. Out in front was the spearhead battalion of Regiment 578, the main body following some distance behind. Hauptmann Beismann was told to attach his companies and river crossing equipment to the advance detachment and Regiment 578. Although Russians were streaming back, it soon became apparent that Regiment 578 had pushed into the midst of a large defensive position still under construction near Podseredneye. Over a hundred bunkers were identified. The position extended back to the Sosna; a broad freshly-excavated tank ditch completed the complex. Podseredneye itself was a fortress, every house a bunker. Still, the attack quickly gained ground. A bridge over the Sosna fell intact into the regiment's hands at 0930 hours and the first foothold gained. A few hours later the bridges in Alekseyevka and Alekseyevka-Nikolayevka were captured and another bridgehead created. The defensive perimeter was pushed out to some high ground just south of Alekseyevka. A large amount of booty lay in the bridgehead; a massive supply depot with incalculable quantities of grain, flour and other foodstuffs; oil and petrol tankers; and various ammunition dumps, though they were ablaze. A train fully laden with food was triumphantly brought in while an ammunition train was blown up. Five-hundred prisoners were turned in to the division's POW collection point. The Russians could not believe that the Germans were already over the Sosna: several lorries drove unsuspectingly into the city and were halted by guns in their faces.

Both neighbouring divisions likewise forced the Tikhaya Sosna. But there was to be no halt. An order came through from 6th Army: "Forward!" Oberst Winzer immediately dispatched recon detachments southward to secure a string of villages and hills.

Grimm's company was similarly tasked after its arrival. First and second platoons reached the Ilinka sector at 1300 hours (Feldwebel Platzer's Third Platoon had remained behind to improve a bridge). Beismann ordered Grimm to reconnoitre bridging positions near Ilinka. This patrol became embroiled in the type of small encounter that happened umpteen times but never warranted a mention in higher-level reports. Grimm assembled a small team: six men (one officer, an NCO and four enlisted men) in two motorcycle-sidecars and a solo motorcycle. In the first sidecar combination was Grimm with map, compass, binoculars, carbine (preferred over the MP-40) and the 08 pistol that he wore at all times. The driver, Gefreiter Vorherr, had a rifle slung across his back. The second sidecar combination consisted of Oberfeldwebel Locher on the pillion seat with a semi-automatic Mauser Gewehr 41 (the company only possessed three of these), pistol and binoculars, in the sidecar was Gefreiter Schenkel with a carbine, the driver, Gefreiter Umele, had a slung rifle. The final member on a NSU solo motorcycle was Gefreiter Ley, who carried a carbine. Because of petrol shortages, no other vehicles could be taken along. Ley was not originally selected but was in the area, head bare, engine switched off, relaxing after the march. Nobody knew why he was there but when he asked to join the patrol, Grimm acceded, but reminded him to maintain a large distance to the bike in front. Grimm signalled his tiny convoy to move out:

I reconnoitred the bridges near Ilinka and drove through the village to its western exit, where elements of the Radfahrschwadron and the supply train were located. Because I had no overview of the terrain from this spot, I drove forward approximately 1600 metres in the direction of the signalman's shack. Approximately 200 metres west of there I spotted two enemy groups with the naked eye. I stopped the vehicles, climbed out and observed with binoculars. At this moment, from about fifty metres to the south-west and south-east, I received a blast of close-range fire from two heavy machine-guns and rifles from right in front. I ran, searching for cover behind the shack, while my recon team followed me. Gefreiter Ley was fatally wounded in the hail of fire. While I gathered my team behind this shack for defence, the heavy machine-guns fired on it from both sides with explosive rounds, but my team could not retaliate from the dead angle. Effective fire could not be laid down and I made preparations for the expected hand-to-hand fighting. On account of the swelling, well-aimed machine-gun and rifle fire, and the fact that enemy soldiers were closing in on my rear, I decided to fall back from behind the shack with my recon team by abandoning the motorcycles and moving to a 25-metre-long garden fence, which we managed to do without casualties despite the vicious fire. I was able to eliminate the aimed fire of the enemy in the adjacent cornfield with my recon team. I found three well-established field positions in this cornfield that had been abandoned. Various Russians were scattered about the cornfield, however. A section of attacking Russians was destroyed by Oberfeldwebel Locher with a quick burst of fire. Other Russians were eliminated in hand-to-hand fighting.

Gefreiter Vorherr, Grimm's driver, was separated from the rest of the group when the shooting started. It was a day he would not forget:

My best friend, Heinrich Ley, was killed. We had to abandon our motorcycles and scoot off through a cornfield. We didn't care about them. It was a difficult day, I was lucky to make it out in one piece. We were on reconnaissance with two motorcycle-sidecars, me out in front with the commander, the Zugtruppführer [Locher] behind me with the second driver and messenger. In the rear was Ley on a solo motorcycle. We stopped at a railway station about 800 metres outside the village. The commander got out, took my rifle, observed the buildings with binoculars. Barely had he got back in when we took fire from two machine-guns. We jumped off the bikes, me to one side, the others into the cornfield. I had just taken cover when Ley cried out: "Ow, now I've had it!" I lay in a trench about seven metres away. The others asked whether I'd caught it too. Heini called out a couple of times to the Oberfeldwebel. The Oberfeld asked where the bullet had gone in, but he only said very faintly "in the stomach." It was actually half a hand span below his heart. When he passed away, I leapt up out of the ditch, over the road and screamed to the others. I wasn't fired on, which was lucky, but I still had my overcoat and pullover on, and had to crawl through the cornfield in sweltering heat. I then had a little breather with the others. We then had to work our way about 800 metres through the cornfield. About half way back I came face to face with a Russian who was standing about five metres in front of me. Because I didn't have a rifle, I shouted out "Halt!" really loudly at him, mostly in fright. This attracted my comrades. The Russian wanted to capture me but when he saw the Oberfeldwebel approaching with a levelled rifle, he put his hands up but was still gunned down. There were still anxious minutes in the cornfield.

By abandoning the motorcycles, the patrol was able to disengage from the enemy without further loss, thanks mainly to Oberfeldwebel Locher who overpowered several Russians in a hand-to-hand struggle within the steamy confines of the vast cornfield. Grimm and his four men made it back to Ilinka:

> Reaching the village exit, Oberfeldwebel Locher commandeered a machine-gun from the Radfahrschwadron that was securing there and fired from the ridgeline of a roof, keeping the enemy down to prevent the destruction of our motorcycles. In the meantime, I brought forward 1st and 2nd Platoons and got in contact with Regiment 578 to organise heavy weapons support. On my orders, Leutnant Beigel prepared the company for a counterattack. First and 2nd Platoons were fired on at really close range from the cornfield while moving into assembly positions, whereupon Feldwebel Grassl pushed into the field with his platoon[4] to destroy the Russians and avoid casualties to his platoon. Beigel led his platoon to the north by striking out for the railway embankment. He hit the enemy in the flank and then made contact with Grassl's platoon. Because it was not possible to bring up heavy weapons immediately, 1st Platoon pushed forward.

> The attack quickly gained ground. While 1st Platoon and myself were moving forward to 2nd Platoon, we were fired on from the front and rear by Russians left behind in the cornfield, as well as from the railway embankment. Through the swift advance of 1st Platoon with flanking fire and the enveloping attack of 2nd Platoon by Feldwebel Grassl – in front and behind the railway embankment – the Russians discontinued their fire. Gefreiter Ley and the vehicles were retrieved without enemy interference. The three well-constructed wooden bunkers and fortified command post in the railway embankment were destroyed with explosives. A heavily reinforced bunker, presumably for an anti-tank gun, was unoccupied and had not been occupied earlier. Found in the cornfield were numerous open positions for heavy weapons that had apparently not been occupied. Captured were two heavy machine-guns, a light mortar and numerous rifles.

The number of Russian dead was not recorded but Gefreiter Rinck noted in his diary that twenty-five prisoners were taken. Vorherr describes what happened to them:

> Later, when we combed through the buildings, there were over thirty Russians, they

---

4   Feldwebel Grassl was temporarily leading 2nd Platoon in the absence of Leutnant Hepp, who had been transferred to 3rd Company.

were all shot down. We then got our bikes back. I'll never forget 3 and 5 July. He is
now our tenth man who has fallen in front of the enemy... What is a man in war?
Nothing!

Gefreiter Ley was buried in Ilinka and a sketch of his resting place forwarded to the
graves registration officer. The death of his friend was still foremost in Vorherr's mind
a few days later:

It's a shame about Heini Ley. I still haven't come to terms with it. It'll be a little
while until his mother receives the news because at the moment there is no-one to
make the report.

The counterattack on the signalman's shack was acknowledged as an assault day –
the company's fourth – and announced in the battalion's order-of-the-day on 11 August.
The most interesting aspect of this action, however, is contained in Beigel's rendition of
the attack:

My company commander ran into some Russians during a reconnaissance with a
solo motorcycle and two motorcycle-sidecars. He received murderous fire and came
to me, totally beat. I had just arrived in the village after a thirty kilometre march.
While the company commander fetched heavy weapons, I threw forward two of the
company's platoons in lorries and attacked. We swamped the Russians, approached
the motorcycles and were able to get them back. A pillion passenger was
unfortunately killed by a shot to the heart. We then got twenty-four Russians out
of the positions. I had them all shot. The company commander was astonished
when he saw his motorcycles roar up.

The coldblooded killing of prisoners aside, this account allows a glimpse into
Beigel's mindset. It may not be apparent at first glance, but the slight differences in
accounts by Grimm and Beigel reveal signs of an unspoken rivalry between the two. In
Beigel's mind, it was he who had saved the day, as Grimm was "totally beat" and later
"astonished," but Grimm's report purports to show him in full control and dictating
Beigel's actions. Did Grimm tailor his official report to incorporate Beigel's initiative as
his own, or was Beigel just big-noting himself? Beigel's words make it clear that he felt
he was responsible for the victory and Grimm played no active role in the counterattack,
yet Grimm clearly states that he did ("while 1st Platoon and myself were moving
forward..."). The unit mentioned was Beigel's 1st Platoon. Both men cannot be right. It
is impossible to determine who is telling the truth, but it is unimportant because the

critical point is their relationship with each other. Grimm was Beigel's superior, so no obvious evidence about the tension between them exists, but undertones and occasional snide remarks in Beigel's letters show that he did not have the highest respect for Grimm and rated himself a better leader and soldier. As this point it would be wise to know more about Grimm. His cherubic face belies a strict authoritarian and he had at least one trait that was not particularly endearing, which he mentioned in letters home. Here he talks about his batman, Gefreiter Paul Nuoffer:

> He's only too happy to forage [for food for me] but sometimes I have to tell him off because he is missing when I want to drive off. In general I scold a lot but everything is then ship-shape and my men soon come to know me. Order must be maintained.

Grimm saw his "scolding" as an essential part of his job, the best method for keeping his men in line and performing properly: "I sometimes have to – or must – scold. I also bare my teeth sometimes, just no mellowness. The tone of communication is much rougher than on the barracks yard." At 5'5" (168 cm), Grimm was one of the shortest men in the unit, and when a short man is being assertive, let alone aggressive, many are likely to think it is due to his size, a social stereotype derisively known as a "Napoleon complex". "Just no mellowness," wrote Grimm, and he obviously thought that letting a minor infraction slip through would be a sign of weakness. A leader cannot control his men effectively without respect: some command it simply by their physical stature, others earn it by leading by example, while others think it can be gained by yelling like a drill instructor. The commander is the father of a unit, and this is a good analogy to apply to Grimm: he loved his soldiers like a father does his children, but like any father, some stern words are occasionally needed. Long after the war, Grimm wrote the following about his style of discipline:

> Since detraining in Merefa, I – as company commander – had never forced one member of the company to fulfil my instructions, had not verbalised a punishment and definitely did not carry one out. If someone occasionally got out of line, for example, during a march, then they received a reprimand and were given sentry duty during the night.

In essence, he was much the same as Oberstleutnant Hertel, whom he disliked because of his authoritative manner, but Grimm could not see this quality in himself.

– – –

The German offensive was in full swing by 5 July. Forty-eighth Panzer Corps had formed three bridgeheads over the Don, one reaching to within four kilometres of

Voronezh. Moving up from the south and approaching the Don near Korotoyak was 40th Panzer Corps. Field Marshall von Bock issued Directive No. 2 for Operation "Braunschweig," (Brunswick) the new name for "Blue":

> *The enemy has not succeeded in organising a new defence anywhere. Wherever he is attacked, his resistance quickly collapses and he flees. It has been impossible to discern any purpose or plan to his retreats. At no point thus far in the Eastern campaign has such strong evidence of disintegration been observed on the enemy side.*

Paulus' 6th Army was ordered to "stay on the enemy's heels" while Army Group von Weichs was to release Hoth's 4th Panzer Army and put it at the disposal of Army Group B.

Motorised Chase (6-9 July 1942)

Wearied by the previous day's effort, the infantrymen were permitted to move off at a more leisurely hour. Leutnant Beigel and his men utilised the unexpected reprieve to savour the provisions scavenged the previous day: a lorry full of tinned meat and fish, sugar and a lot of tobacco. Midway through a letter to his girlfriend, he was ordered to get his men together and move out. The division's task was to pursue retreating enemy forces. Krüder's and Winzer's regiments formed dual spearheads. The main body of the pioneer battalion trudged along behind Krüder on the left. The pioneers of Feldwebel Grassl's 2nd Platoon were mounted on captured trucks and attached to the advance guard: if no mines were encountered, they would be able to sit back and enjoy the ride. Opposition ahead of the main regimental columns was not noteworthy, in fact, some Russians seemed to have completely lost their heads. Lorries drove directly into the advance guard. In this way, two fuel tankers, as well as other vehicles, were collared, and the petrol tanks of the captured trucks topped up. Gefreiter Rinck wrote in his diary: "1000-1700 hours, with the advance detachment on lorries! Many prisoners." Beigel was thrilled by the lack of resistance and used their flight to assuage his girlfriend's fears: "If you could see how the Russians run when we turn up, then you'd get a hell of a lot of joy out of it. If they keep running, then I can kiss you sooner."

Recon planes buzzed about over the retreating enemy. Keen eyes spotted one of the planes heading towards them. It swooped down and dropped a message canister, its location marked by an expanding pillar of purple smoke. The "eye in the sky" reported that the Russians were in full retreat. Vanquished columns flooded back either side of the German spearheads, particularly along the valley road running parallel to the Ivani Creek in the right neighbour's sector. German ground troops adored the recon planes: "The 'duty sergeant'" Beigel told his girlfriend, "is a twin-fuselage reconnaissance plane that continually watches over the Russkis like a jealous woman without a man. The Russians tell me that they really hate them. You know, if they see a transport or an artillery battery or a tank, then they always get some support and then it's 'bon appetit, Russki!' If danger threatens us somewhere, then he'll fly down to us and drop a message. He is a great guy, he's always there and offers us a vital service."

Stubborn resistance briefly flickered just short of the day's objective, but it was quickly broken by the advance guard. The division directed the infantry regiments to occupy an advanced line based on a series of hillocks in order to continue the pursuit at 0200 hours on 7 July. They did so, but the green light to attack only came towards 0600 hours. Beismann was ordered to assemble his pioneers in Oleinikovo by 0330 hours and have bridging equipment handy.

Once the attack was underway, General Oppenländer issued fresh instructions based on the latest intelligence. Nothing but weak rearguards were in front, so both regiments were ordered to gain crossings over the Kalitva and Olkhovatka as quickly as possible. As a consequence, Hauptmann Beismann was spurred to send Hauptmann Häntzka's company to Regiment 578 and follow with the main body of his battalion so that its river crossing equipment could be deployed in a timely manner. Grimm's company was kept busy in the rear. In Krivaya Beresa his men extinguished a flour warehouse that had been doused with petrol and set on fire. Other pioneers were detailed to search the village for stragglers. They brought in twenty-five prisoners. Beismann ordered Grimm to send more pioneers forward. He, in turn, instructed Leutnant Beigel to load a reinforced platoon on two lorries and reach the advance detachment in all haste. Racing along dusty, pot-holed roads, Beigel's platoon joined the advanced detachment in Mariyevka at quarter to twelve, just in time to check a bridge in Kalitva and secure it for Krüder's Regiment 576. The mad dash thrilled Beigel:

> My platoon and I were with a small motorised advance detachment. That was a superb thing. We drove fifteen kilometres behind enemy lines with our group and formed a minuscule bridgehead. Resistance was very light. We didn't lose a man.

Elements of Feldwebel Grassl's 2nd Platoon had been attached to Krüder's regiment prior to the assault. This from the diary of Gefreiter Rinck:

> 0330 departure with advance detachment. Reached the village of Mariyevka, however, the [assault guns] turned off to the right. Bridgehead formed over three bridges. Enemy transport column with about fifty vehicles scattered. Many prisoners.

Grimm and the rest of the company joined Beigel's detachment at 1700 hours, combed through the western part of the village and took forty prisoners. Two commissars were killed in a hand-to-hand scuffle and their papers passed to battalion staff. Beigel's platoon captured a Russian captain. In all, his platoon took sixty-five prisoners and captured a medium Russian lorry, two MG-34s – their origins unknown – plus various types of Russian equipment and weapons. Gefreiter Vorherr chauffeured Grimm around in his motorcycle-sidecar and was witness to the swift conquest of Mariyevka:

My company took a village, conquered everything there, several cars, tractors and took a number of prisoners, also a lot of horses. There were two commissars amongst them. Some of them were able to escape… We captured a lot of conserves and cheese, a few hundredweight, also some tea… Things are moving quickly in Russia!

The bridgehead was broadened south-west and south of Olkhovatka. Oberst Winzer pushed his units closer to the daily objective without regard for Russian units still fighting on the flanks. Enemy resistance ahead of his regiment was very weak and the day's objective was reached at 1900 hours. According to captured orders, the division was pushing into the reserves of an enemy line and in doing so had unhinged the defences of the Soviet 28th Army.

The pursuit continued at 0230 hours on 8 July and swiftly drove a deep wedge into the softening Russian front. All three infantry regiments were in the line. Out in front was an advance detachment. As usual, 305th Pioneer Battalion was split up: Häntzka's 3rd was the advance detachment, Klein's 1st was with Regiment 578 and Grimm's 2nd with Regiment 576. On the basis of reports that panzers were advancing from the south ahead of the division, all stops were pulled out to hinder the Russian withdrawal. Feldwebel Platzer's 3rd Platoon, reinforced with extra men, climbed aboard a pair of lorries and joined the advance detachment of Regiment 576, which took the bridge near Lisinovka into its hands undamaged. Platzer's men checked it for explosive charges and mines, improved the road surface, and put out covering parties and bridge sentries. One of Platzer's Soviet lorries broke down; another was quickly found. Gefreiter Hans Russ was wounded on the left upper arm by a grazing shot while securing the bridge but remained on duty. The rest of the company reached Lisinovka at 2000 hours and took up quarters. In the process, another assault day was accrued by Grimm's 2nd Company: "Occupation of bridge over the Svinika, near Lisinovka, by advance detachment."

The story was the same on the division's right wing. Regiment 578 pushed through to its daily objective. The Russians streamed back on lorries, horseback and foot. Supported by assault guns, Winzer's battalions occupied the crucial high ground. The Russians noticed that a ring was closing around them and tried to break out.

The pace of the Russian retreat was causing all sorts of problems for the German leadership. On the afternoon of 8 July, von Bock advised OKH that the second phase of the offensive was useless because his armies would "most likely strike into thin air" if they implemented the existing plan, and then suggested the objectives of the armoured forces be reconsidered. The bulk of German armour had been concentrated in the north for the thrust to Voronezh, but with the rapid fall of this city on 6 July, the panzers were now awkwardly positioned to exploit the Russian retreat further south. Signs of a flexible Soviet defence were evident during "Wilhelm" and Fridericus", but the speed of the withdrawal during "Blue" caught the German command off-guard. The entire concept for the summer offensive had rested on the assumption that Soviet forces would act like they did in 1941, that is, hold their positions at all costs. If that had been the case, "Blue's" small envelopments would certainly have netted the vast majority of Russian forces opposing them, instead, only 70,000 prisoners were taken. Enemy forces fleeing hell-for-leather were unlikely to be caught by horse-drawn infantry divisions; this was a job for mobile units, and unfortunately for the Germans, most of theirs were piled up near Voronezh.

To switch the main effort from north to south, Hoth's 4th Panzer Army HQ, 48th Panzer Corps (with 24th Panzer Division and Grossdeutschland) and 24th Panzer Corps (with 3rd and 16th Motorised Divisions) were ordered south from Voronezh. Upon arrival in the Rossosh – Novaya Kalitva area, Hoth also took command of 40th Panzer Corps and 8th Army Corps. Both corps – 305th Infantry Division belonged to the latter – were already surging south, while Hoth's other two corps began the 175km journey from Voronezh. When the second phase commenced on 9 July, the offensive was a good two weeks ahead of its original timetable but preparations for it were half-baked at best: the fuel tanks of 24th Panzer Division and Grossdeutschland ran dry midway between Voronezh and Novaya Kalitva, while 3rd and 16th Motorised Divisions were stuck in Voronezh until infantry divisions arrived to relieve them. The second phase also signalled the end of Army Group B's hegemony over the operation: Field Marshall List's Army Group A assumed control over the southern sector.

The division was granted a rest day on 9 July while the left neighbour, 40th Panzer Corps, tried to get weakening Russian forces out of the northern flank by swiftly pushing south-south-east from Rossosh. Oppenländer's regiments dug in on the dominating high ground and awaited the arrival of the left and right neighbours. The pioneers had an easy day. Grimm's company remained in Lisinovka to improve its dual bridges. The only other task was sentry duty. Lack of resistance was greatly appreciated but the relentless pursuit was taxing nonetheless. The sweltering heat rapidly drained energy and generated cascades of sweat that attracted dust and flies. Hygiene was a low priority in a fast-paced advance and it showed, as Beigel attempted to convey to his girl:

> *After several days of strenuous marching, I finally have some time to write to you. You can imagine how swiftly the advance is progressing. For us foot troops, however, it is not as simple as it is for the motorised units. Our horses are slowly but surely dying. Our infantry have lice and full beards because they have not been able to wash. We, that is, our company, have gradually motorised ourselves. We've already been driving for four days. That is a massive relief for our men. Despite the fact that we're able to drive, we hardly ever arrive before night falls.*

General Oppenländer no longer felt like he was on his own. The division's right flank was shielded when 376th Infantry Division moved level. Panzer divisions passed by on either side. By the evening of 9 July, Paulus' armoured pincer – the three mobile divisions of Stumme's 40th Panzer Corps – had carved a broad corridor through Timoshenko's defences to the Don River near Boguchar. The right flank of 1st Panzer Army moved off on the morning of 9 July with orders to link up with 4th Panzer Army, the assumption being that 6th Army would tie down the enemy, but it was hard pressed keeping up with fleeing enemy troops, let alone pinning them down, so the potential trap would soon be empty. Moreover, Paulus' units were advancing so quickly that they would be beyond the anticipated link-up site long before 1st Panzer Army arrived. As there was no point in having 1st Panzer Army continue on its assigned course, OKH ordered it to head due east toward Millerovo while Hoth's 4th Panzer Army took a bridgehead on the Don at Boguchar as a springboard for a subsequent thrust towards Stalingrad. The Bodensee Division would help secure 6th Army's northern flank along the Don. As such, it was directed to gain a bridgehead over the Krinitsa Creek.

Hot Pursuit (10-12 July 1942)

The division's spearhead regiments (576 on the right and 577 on the left), with pioneers incorporated into both formations,[5] moved off at 0200 hours with orders to establish bridgeheads as quickly as possible. Grimm's men once again climbed aboard lorries. Division advised that "the enemy only has command of weak forces in the division's sector [but] it's possible that reinforcements (tanks and motorised infantry) may be moved up." Marching in the sticky heat and cloying dust was demanding, but progress was rapid, "barely no resistance, numerous prisoners and booty" noted the

---

5   Häntzka's 3rd Company into Inf.Rgt.576, Klein's 1st and Grimm's 2nd Company into Inf.Rgt.577.

divisional war diary. The giant banners of dust thrown up by the columns attracted roving Russian planes, though their attacks had little effect. A low-flying plane strafed Grimm's company near Kolbinsk. A ricocheting round lacerated Unteroffizier Fritz Dietzsch's forehead and he was admitted to hospital.

The Krinitsa Creek, a miserable runnel hemmed in by eroded banks, was rapidly reached. The two leading infantry battalions crossed easily with assistance from the pioneers but their vehicles required a proper crossing point, so Grimm put two of his squads into a captured lorry and sent them to the road bridge near Krinichnaya in the neighbour's sector to check for mines and explosive charges. Oberst Voigt's entire regiment advanced another seven kilometres and pushed into weak resistance in Novaya Kalitva. Two mine-detection and clearance teams from Grimm's company came under fire as they entered the village. Leutnant Beigel went in to look for mines, found none, so the teams pulled back. Artillery fire howled in from the northern bank of the Don the whole time.

The division moved out at 0200 hours on 11 July. Two of Grimm's squads were attached to the advance guard of Regiment 577. At Gedyuche they began the restoration of the bridge and regulated traffic flow. Enemy contact up front was non-existent but Soviet planes carried out nuisance raids, sometimes with devastating effect: toward 1530 hours, a bomb landed right on one of Regiment 578's columns, putting thirty men and thirty horses out of action. The lack of combat was appreciated but conditions were unbearable. One of 2nd Company's horses died of exhaustion. The men were also fatigued to the point of collapse, but they kept marching, as Beigel describes:

A two hour rest is coming to an end. The hottest time of day is past and we'll shortly be driving to the front again. Our infantry have to perform amazingly. For almost the entire week they've had to march forty to fifty kilometres per day, and often more. On top of all that is the sweltering heat.

The rest of Grimm's company reached Gedyuche at 1930 hours and constructed a fascine crossing for tractors and heavy vehicles in order to preserve the wooden bridge. Sentries were put out and quarters taken up for the night. All officers in the pioneer battalion were blessed with industrious orderlies who cared for their superiors like well-trained servants, but Gefreiter Nuoffer aspired to be the best. As driver of Grimm's car, he had access to wheels, so he would disappear for hours, searching and scrounging for any sort of foodstuff to vary his commander's diet. Grimm constantly reminded him that he shouldn't become blasé about foraging on his own because a nasty surprise might pop up at any time. While on the hunt for some honey, the inevitable happened:

Nuoffer was caught unawares by a Soviet tank, though fortunately for him, the tank was out of high-explosive rounds. Instead, it took potshots with armour-piercing shells, which had little effect other than to scare Nuoffer out of his wits and reduce his comrades to laughter. The tank was later knocked out by an anti-tank gun.

The division's task on 12 July was to defend a sixty kilometre front along the Don River. Soviet troops with dug-in tanks were still on the southern bank, in a river loop, and held the villages of Galiyevka and Grushevo as bridgeheads. Bands of Red Army soldiers also lurked in the forests between the river loop and Novaya Kalitva, emerging in the mornings and evenings to ambush lone vehicles. Soviet forces on the northern bank of the Don were weak and showed little propensity for combat. Villages spread out along the river were combed through and the southern riverbank reported enemy-free on 14 July. The main tasks for the pioneers were bridge construction and ascertaining conditions along the Don for an eventual crossing. One consideration for the summer offensive was an advance towards Stalingrad along the northern bank of the Don, and the sharp kink in the river near Novaya Kalitva was considered the best site for a bridgehead. While Grimm's company prepared inflatable boats and wood for two boat ferries, he headed off to the river with a recon team to explore crossing possibilities. The area was still an active combat zone. Grimm's team moved along the road to Grushevo and met up with an infantry company from Regiment 577 tasked with combing through the village and occupying it. As soon as the infantrymen did their job, Grimm would do his. However, flanking fire from heavy machine-guns, heavy mortars and crash-bangs[6] halted the infantry. Leutnant Beigel was in the thick of the action: "We met heavy enemy resistance… I personally had a shoot-out with several Russians in a village but I don't mind." The recon team returned and reported their failure.

Defence Along the Don (13-16 July 1942)

Oberst Voigt's Regiment 577 attacked Grushevo in the morning with sturdy support from artillery and assault guns. The village was captured after a brief struggle. The assault guns were shifted to Regiment 576, which then moved against Galiyevka, though it was only conquered after a protracted battle that extracted quite a few casualties. The division then set itself up for defence throughout 13 and 14 July. After the exertions of the last few weeks, some rest and recovery was exactly what was needed. The division had covered more than four hundred kilometres from Volchansk to the Don, almost three hundred of those in attack and pursuit.

---

6   Crash-bang = German nickname ("ratsch-boom") for the Soviet 7.62cm anti-tank gun, so-called because it was a high velocity weapon and the sound of the gun firing was heard just before the explosion of the round.

Establishment of static positions and river crossings required the skill of the pioneers. Grimm set out on 13 July to complete his investigation of the Don, while Grushevo and Galiyevka were still under attack. Accompanying him were Leutnant Beigel and a small team. Once well-constructed positions on Hill 159.0 near Galiyevka were stormed by Regiment 576 just before noon, Grimm's patrol entered Grushevo. The commander of an infantry company positioned at the village exit promised Grimm that heavy weapons would keep an eye on the meadows north-east of the village and gave him two squads of infantry. The infantrymen would cover both sides of the kilometre-wide forest cutting during the recon team's infiltration to the Don. Grimm's men reached the riverbank unnoticed. The sparse woods had not been checked for enemy stragglers, so everyone was on edge as they investigated the river and its banks. The width of the river was about 120 metres at this point. A crossing site with a ferry was located; the ferry itself, camouflaged with branches, was berthed tight against the riverbank. Spotted fifty metres upstream was a concrete bunker and an expertly camouflaged wooden pillbox whose firing slit leered at the Germans. Well-constructed field positions occupied by Red Army soldiers were identified both sides of the ferrying site. The enemy, however, did not see Grimm and his men. Using binoculars from atop small rises, Beigel intently scanned the sectors upstream and downstream and gained valuable intelligence. Satisfied that he had all he needed, Grimm signalled his men and they headed back. After moving to the forest cutting eight-hundred metres downriver of the recon site, a violent scuffle broke out in the forest, but the team was able to extricate itself and return to Grushevo without loss. Grimm's conclusion was that a forced river crossing in the area north-east of Grushevo was viable.

An addendum to a divisional order issued at 1915 hours on 13 July laid out special dispositions for construction and consolidation of positions. Once a main combat line was established along the chain of combat outposts, construction of powerful defensive positions was to begin right away. Although generally over a hundred metres wide, the Don River flowed at a leisurely three kilometres-per-hour, so the infantry regiments were reminded that the river was not an insurmountable obstacle due to its numerous fords. The elevated bank of soft, white chalk cliffs in some sections offered excellent possibilities for strongholds and observation, so the sector was tailor-made for a defensive line. The great breadth of the sector, however, forced the adoption of a strongpoint-based defence with the intermediate ground being covered against enemy infiltration, particularly, at night, by constantly manned nests. The troops were ordered to dig in deep and begin the installation of dug-outs. Civilians were cleared from riverside villages in order to prevent contact with the foe on the opposite bank, though this measure was partly negated by Russian soldiers escaping over the Don. Most were caught, some as they were crossing the river, but a few made it.

Construction of positions was the concern of the troops but pioneer detachments were allocated for special work and technical guidance. The pioneers inspected defences on behalf of the division. A pioneer company commander was also sent to every regiment as an adviser. Grimm and his 2nd Company were dispatched to Regiment 578 on 14 July. Grimm reconnoitred available crossings, fords and crossing possibilities near Novaya Kalitva, as well as the condition of the riverbank. He then went to regiment staff to begin work as a technical consultant and immediately established contact with the officer responsible for constructing positions. In the meantime, his company set up camp in a cherry orchard in Orobinskiy and had the afternoon off. The lighter workload was appreciated, as noted by Leutnant Beigel:

> You cannot imagine what the land looks like. Infinitely wide, long and drawn out, elevations with cultivated fields, every now and then small patches of oak forest and dried-up rivers in the valleys. From time to time it's possible to see twenty kilometres into the distance. When one sees these endless expanses and the long villages straddling the rivers, then you can become melancholic. It's dreadfully hot at the moment. The wells are soon empty and everyone's looking for a shady spot to sleep in… A rest is doing us all some good. Tiredness gradually leaves our bones. I pitched my tent in a fruit garden in which there are many cherry trees that'll be covered in ripe fruit at any time. I had a table built for me, as well as a handsome bench. You can't imagine what it means to be able to eat and write at a table again. Most times these things are done either standing up or lying down.

The leisurely pace continued on 15 July. Grimm scouted favourable sites for artillery observations posts. His company only moved out at 1700 hours to construct some positions.

While the infantry divisions of 8th Army Corps secured the Don River, the main battle was being waged to the south. The panzer and motorised divisions from Voronezh tried to cut off the Russian retreat at Millerovo but only partially succeeded because the cordon was too porous. The haul of prisoners was once again disappointing. With the current Russian penchant for withdrawals, it would have been better to doggedly chase them to prevent the establishment of solid defences west of the Volga. An eastward pursuit would achieve the original aim of "Blue" of striking the enemy in the Don bend and then gaining the Stalingrad area. Generaloberst Halder, the Chief of the Army General Staff, favoured this approach, but Hitler thought otherwise. The small number of prisoners and booty confirmed his preconceived opinion that the Russians had nothing left and were at the end of their strength. He maintained that the infantry power of 6th Army was sufficient for the drive to the Volga, regardless of the fact that it also still had to protect the northern flank along the Don. Hitler therefore stripped Army Group B of 4th Panzer Army and almost half of 6th Army – including the panzer and motorised divisions – and sent them south to Army Group A. He issued a new order to Army Group A on 13 July: encircle the enemy in the Donets bend north of Rostov. Army Group B was left to guard the flank and rear of Army Group A as it attacked southward. The difference between the original plan and this new one was startling. Army Groups A and B were supposed to punch through with a powerful armoured fist, but the new plan entailed a slender finger of infantry slowly poking towards Stalingrad while the other fingers seized Rostov. The forces aimed at Stalingrad were greatly affected by another ramification of Hitler's tinkering: the lion's share of the fuel went south with the armoured forces and the petrol tanks of 6th Army ran dry. Fuel shortages had plagued the army's units for weeks, but now they became so serious that the advance threatened to stall. Supply lines stretched taut and then broke. A few days after Hitler diverted the fuel, Gefreiter Vorherr wrote:

My company is about forty to sixty kilometres further ahead. We can't follow up with our motorised vehicles because we still haven't received any petrol. This has gone on for two days now... no rations, no mail, nothing.

Little could Vorherr know that this situation would last a lot longer. On 28 July he wrote that "I don't even know where the company is at the moment, only that it's probably about one hundred and eighty kilometres away from us here. God only knows when we'll get there. The company cannot be very far from Stalingrad."

This fuel-induced disarticulation of the company was in the future. On 16 July, the pioneers continued fortifying the defensive positions at night, a task not without dangers. They left at 1700 hours and during the next twelve hours made ready five heavy machine-gun nests, two anti-tank gun positions and one artillery observation post. Five more heavy machine-gun positions were half-finished. However, while stringing barbed wire along the riverbank near Derezovka, 2nd Platoon was ambushed from behind by Russians streaming back towards the Don because at this point the river was only 1.5 metres deep, so it could be waded. The men were blocking a gully that opened onto the Don when grenades and gunfire crashed down upon them from out of the darkness. Gefreiter Jakob Taschner was killed and Pionier Rudolf Mutter injured.

Mutter was taken to a dressing station with a bullet lodged in his right thigh. Gefreiter Taschner was buried in Derezovka.

Late in the afternoon of 16 July, the division was ordered to halt its positional fortification. Hitler had instructed Paulus to expand his operations eastward to prevent the Soviets from forming a stronger defence line in the large Don bend. General Oppenländer was to hand over his sector and resume the advance. All units had been well rested over the previous four days. The rigours of another long march stood before them.

Endless Marching (17-29 July 1942)

The neighbouring divisions would each take over half of the divisional sector. As units were relieved, they immediately began marching east. Ahead of them were the three other divisions of 8th Army Corps. Advance detachments had already crossed the Chir River near Bokovskaya, one hundred and thirty kilometres further east.

The pioneer companies were relieved in place early in the day. All maps of the divisional sector and river surveys were handed over to their replacements. Grimm's platoons returned from their nocturnal work at 0800 hours and received their marching orders. All vehicles remained behind because of petrol shortages. In the late afternoon 2nd Company reached Tverdokhlebova and awaited Oberst Winzer's Regiment 578, to which it was attached for the march. The men were warned that if they bivouacked in the open, they must minimise the risk from aerial attacks by widely spacing their tents, digging numerous air-raid trenches and putting out aircraft spotters so that the troops could get into cover in good time during the approach of every plane, be it friend or foe.

Night-time drizzle strengthened during the morning of 18 July and turned roads into quagmires. Grimm's objective was Radchenskoye. En route, he was ordered to improve the surface of the bridge near Poltavskaya. Traffic was interrupted. In pouring rain the pioneers fixed about twelve metres of the bridge decking, then picked up their kit and slogged on. Soviet planes swooped in low and strafed the column twice, but no-one was hit. The infantrymen and pioneers were glad when they entered Radchenskoye early in the afternoon. The village's huts were soon tightly packed, though the men were at least able to get their things in order and dry out. The downpour continued. It was hoped that 19 July would be a rest day as the roads were swamps, the men and horses totally exhausted and no more petrol was available.

The rain did not let up. Despite abysmal conditions, the troops started the mud-slog at 0330 hours. The daily objective was Pozdnyakoff but at dawn, shortly after setting out, a new order directed them to Makaroff. The glutinous roads increased losses amongst the horses. Supplies only came forward in trifling quantities. Fuel was no longer delivered, so the motorised vehicles were left behind as fuel tanks ran dry, until finally, even ambulances were forsaken by the side of the road. The rain ceased in the afternoon and the roads started to dry out but the doughy mud placed a huge strain on the horses. Even though the final stretch of road was decent, the forty kilometre march claimed the lives of two horses in Grimm's company, one from exhaustion. The long days of marching made Leutnant Beigel introspective:

> We have been on the march again for the last couple of days, the entire day yesterday in pouring rain and deep mud. Today, however, the sun is drying things up beautifully. You know, there is a lot of time to ponder and think while marching. How is it going with you? You've got it much worse than I have. You are always at home, you always have your daily routine and almost always worry about me. We often don't have time to think of home. When we are under fire, then there is only fighting and for that one requires all thoughts to be in order. The Russians no longer oppose us very often. They run! But we have lice and fleas, that is, I "only" have fleas. They often make fine hunting. If you ever get some good insect powder, then send it to me. Until then, I'll trap and kill.

Beigel also mentions a certain city for the first time, a name that haunts the German psyche to this day:

> I'm in my tent. Tomorrow, as usual, we'll wake up at 0200 hours and then march until about 1700 hours and later. If petrol comes forward again, then we'll be able to drive. If we keep going, we'll soon be on the Volga, general direction Stalingrad...

The march continued early on 20 July in tropical heat. Roads dried out. The last drops of fuel were consumed: the only means of propulsion now were feet and hooves. Units endeavoured to compensate for the loss of horses by getting hold of local ponies. The men sought refuge from the midday heat in Meshkoff and only recommenced the march at 1400 hours, destination Alekseyevskiy. This village, which was reached four hours later, proved to be a beneficent place. Milk, eggs and potatoes helped compensate for inadequate rations and the thirty-seven kilometre march.

The rest of the division was also on the move. Klein's 1st Company, attached to Voigt's Regiment 577, reached Meshkoff late in the afternoon. On a divergent course was Krüder's Regiment 576, which aimed to join up with the rest of the division near Bokovskaya. Häntzka's company, as part of a motorised group containing 305th Panzerjäger Battalion and Radfahrschwadron 305, was idle near Novy Byt awaiting petrol. Once its tanks were full, it would push through to Bokovskaya in one go.

Heavy rain virtually halted offensive operations from 18 through 20 July, but on 21 July the infantry divisions of 8th and 51st Army Corps, supported by 14th Panzer Corps, slowly pushed east in extreme heat and powdery dust towards Serafimovich-on-Don and the Chir, crossing the latter river in three places by nightfall.

The sun glowered down remorselessly on 22 July. Grimm's pioneers fabricated a fifteen-metre long fascine causeway to overcome a boggy depression, then marched to Verkhne Grushki. To avoid the heat, Grimm's company set off at 0200 hours on 23 July, destination Yagodniy on the eastern bank of the Kriusha Creek. Conditions became more gruelling with each passing day. Intense heat (30° C in the shade, 43° C in the sun), poor roads and few watering spots were compounded by marathon-length marches. The blazing sun caused a few cases of heatstroke, hardly a surprise given the woollen uniforms, leather boots and full marching kit. Quite a few men, slightly stooped and clutching their stomachs, scurried out of formation to squat beside the road: diarrhoea was beginning to run rampant through the ranks. Eighteen hours was needed to cover the distance. As ration deliveries were infrequent, the troops were told to live off the land. Livestock was rustled. There was very little fodder for the horses: it could only be procured in diminishing amounts, so large quantities of fruit stored in the collective farms was used, but this led to some deaths. By some estimates, the division still only had half of its original horses, the shortfall being made up by approximately double the number of hardy local ponies.

While the bulk of the division pushed east on 24 July, Regiment 578 veered north-eastward to seal off Russian forces in the Serafimovich area. Grimm's pioneers accompanied III./578, which relieved the recon battalion of 100th Jäger Division in the hills north of Verkhnaya Tsaritsinskaya. Grimm's men set up camp in a forest cutting and attended to personal matters. Beigel explained the situation to his girlfriend:

> *You've had to wait a long time for mail. I'm sorry about that but I can't help it: we've been marching for eight days with an infantry regiment behind a panzer division. We've marched over two hundred kilometres.[7] Our battalion hasn't brought along food or mail. They've been held up because of fuel shortages. As a result, we receive no mail, but also can't post any... You know, we don't grumble, mainly because we're much too tired for that, the main thing is that our panzers have enough fuel to keep them rolling forward... Do you know how far we've*

---

7   According to Gefreiter Rinck's diary, 2nd Company had actually marched 260 km over the previous eight days. Beginning on 17 July: 20 + 30 + 38 + 37 + 32 + 36 + 42 + 25 = 260 km.

already marched since coming to Russia? Surely a thousand kilometres. Only an infantryman can comprehend what that means. And my men have held tough…

Incidentally, our company has been living off the land for days. One day there'll be nothing at all, the next there's pancakes, chicken and fried potatoes. Today we're living in a forest. There's absolutely no villages in the surrounding area where we can organise something. Tomorrow we'll probably be somewhere where we can get some gherkins, radishes and eggs. Well, that's war. Nice now and then, hard now and then. We haven't seen the Russians for a long time. They seem to have gone MIA.

Grimm's company left their forest camp-site at 1300 hours on 25 July in drenching thundershowers and plodded along twenty-five kilomteres of sloppy roads to their new billeting area on high ground between Verkhnaya Tsaritsinskaya and Serafimovich. After nine days of marching, the company was granted a rest day, which allowed Grimm to turn his attention to the dire supply situation. After the rain-soaked roads had dried out a little, Grimm and his Rechnungsführer[8] drove back to search for the battalion staff in order to address the growing supply problem. Eight days had passed since the company had received rations. Food was procured in villages while on the march. Some early season potatoes were also scrounged from gardens. In the previous two days, however, the company had not passed through any villages and rations, including water, were extremely scarce. Grimm recorded more details in a letter home:

I'm sorry that I haven't got a message to you since 21 July. Even if I had written to you while on the march, the letter would not have been on its way anyway. So, last night I arrived at my staff a hundred kilometres behind my company front so that I could personally take mail, rations, market goods, etc. back to the front. On a sixty kilometres stretch of road there is neither a house nor a village for as far as the eye can see. I also celebrated a jubilee, one thousand kilometres linear distance according to the map, mostly on foot, a tiny part on horseback, but all on Russian roads. Don't ask about other distances. We have marched in over thirty degree heat in the shade for the last ten days. On our feet every day from 0200 in the morning until 2100 in the evening, if not longer. I do not want to and cannot report details, only that we are fit and healthy and without loss. That is a quite a feat.

Grimm was suffering from mild diarrhoea: "Many comrades also have it and almost

---

8   Rechnungsführer = accountant and pay NCO.

no-one escapes." The anguish of completing day-long marches carrying all equipment and weapons in blazing heat on parched roads – while being tortured by liquid bowels – can barely be imagined. However, it seems Grimm took it in his stride: "I am proud of everything I have done. Hopefully I am granted the strength to be able to serve my Fatherland even more. For now, that is the most beautiful and greatest task of my life."

Leutnant Beigel also used the opportunity to attend to correspondence:

We long for mail just like we do for rations. I've tried to mail a few letters lately. I tried to hand them in to a panzer unit because they have fuel and can therefore get them to the rear quicker... We have bivouacked in a field and now we wait until we move on again. The rain, thank God, has stopped, though the wind is still shaking our tents.

The lack of mail may seem trivial but it deprived the soldiers of their only link with home. When Grimm returned from the rear with food and bulging mail sacks, like Santa Claus in a Feldgrau uniform, his men felt like Christmas had come early. Even officers were excited. "After fourteen days, the very first mail arrived from home," wrote Beigel. "It was a happy day for all of us. I thank you for your mail. Three letters and a double letter with cigarettes reached me."[9] For fathers, contact with home was vital because their children were growing up with little paternal input. Advice and encouragement – and if need be, chastising words – were passed through the postal system. Only through this route and on the rare stints of leave were fathers able to head up their families. Many soldiers became dads while at the front after earlier furloughs bore fruit. Such was the case with Grimm. His first child was born while he was in France, and he was fortunate to receive a few weeks to visit his young daughter, but he had not seen her since, not even in photos as he had taken his camera with him to Russia. He was therefore delighted when Lotte sent a photo:

That was very nice of our friends to have taken the photo. They must surely know how happy a father is to receive a photo of his child. As I see in the snapshot, she looks like a North Frisian Grimm type, particularly like my grandfather, as far as I can tell from the photo. Yes, I am very happy that our dear child is a great distraction to you and that you are not alone.

A few days later Grimm wrote: "I keep pulling out the small photo of Gisela from

---

9   In an earlier letter, Beigel wrote: "That you have sent me cigarettes has made me happy. You obviously know that I'm smoking again. It's because of the war. I'll give it up after the war if you have something against this vice."

my wallet to have a look." Beigel was overjoyed when his girlfriend sent a photo of herself, but he protected it as if it was her:

A brief glimpse in my wallet is enough to bring me great joy because of your picture inside. I don't carry my wallet with me. I have my reasons for that. I only want to look at you when I don't have the Russians in front of me, when the war isn't being too hard on us. I also don't want your photo to fall into the hands of a Russian.

Later in the campaign, during a quiet spell with little to distract their homesick minds, the soldiers fretted upon mail call and were disappointed when they missed out, as Beigel describes:

I was expecting a letter from you today. Nothing came, not even the Mühldorfer newspaper. Mail will come again tomorrow and there'll surely be something then. Mail call is almost as important as food deliveries. That's because there's nothing else pleasant for us.

The importance of Feldpost cannot be underestimated. Beigel expressed the feeling of many soldiers swallowed up in the vastness of the East:

Do you know that mail has a completely different significance here than in, say, France? When you're totally cut off from culture, a letter transports you back into the faraway homeland, into the most beautiful land of culture, and gives you a lift.

Feldpost contributed greatly to positive morale, and lack of it had the opposite effect. Rations nourished the body but mail from home sustained each man's spirit and morale.

The division set itself in motion early on 27 July, following the light infantry of 100th Jäger Division in order to cover the northern Don flank from Kletskaya to Yelanskoye. Oberst Winzer, whose Regiment 578 continued to screen the Serafimovich bridgehead, was ordered to set Grimm's company in march to rendezvous with Regiment 577 near Izbushinskiy. The pioneers, led by Leutnant Beigel, left their billets with growling stomachs; Grimm was still in the rear, almost a hundred kilometres from his men, attempting to get them some food. He loaded up a Studebaker lorry with rations and market goods (chocolate, alcohol and cigarettes) and, together with a motorcycle-sidecar, headed back to his company at 1500 hours.

On 28 July, in searing temperatures and choking clouds of dust, Hauptmann Klein and Leutnant Beigel led their companies along the main road to Orekhovskiy, a trek of forty-two exhausting kilometres. Not a drop of water could be found for the field-kitchens or horses. Fortunately, Grimm arrived with the truckful of food. His company remained in the village on 29 July to recover from their epic march. He put out a platoon-strength picket to secure the billeting area, which proved to be a wise move as two armed deserters were seized a short while later. Throughout the day the clamour of heavy fighting resonated from Ventsy to the south-east and Kletskaya to the north-west.

One Death is a Tragedy (30 July 1942)

The division's performance received supreme recognition on this day when General der Artillerie Heitz, Commanding General of 8th Army Corps, presented the Knight's Cross to Generalmajor Oppenländer at the divisional command post in Tsymlov. Bestowal of this medal was proof that the Bodensee Division, a former occupation division, had matured into a full-value Ostdivision.

The operational breather allowed Grimm to carry out one of the pleasurable tasks of being a commander. On the morning on 30 July, he handed out five Iron Crosses and a War Service Cross. The decorations had been approved on 18 July, but the pace of operations, and the fact that Grimm had only picked up the medals during his visit to the rear, meant they were only awarded now. Seven Iron Crosses (one First Class and six Second Class) were officially awarded but only five were bestowed.[10] Awarded the Iron Cross Second Class were Feldwebel Emil Grassl, Unteroffizier Hermann Bielefeld, Obergefreiter Josef Lachermeier, Gefreiter Johann Russ, Gefreiter Ludwig Rützel and Gefreiter Konrad Weid. A milestone was the presentation of the company's inaugural Iron Cross First Class to Unteroffizier Erwin Bub for his role in destroying the T-34 on 24 June. The recipient of the War Service Cross was Unteroffizier Georg Jäckel. The award winners accepted the congratulations of their comrades, shared a drink and then it was back to business as usual.

Grimm ensured that pickets and security outposts were changed throughout the day. He was called to a conference in the afternoon and informed that the infantry would take over security protection when they arrived the following day and his company would become a regimental reserve. Meanwhile, as assault guns were soon expected to arrive, Grimm's men had to establish a four-metre wide ford next to Orekhovskiy's north-west bridge as it was only rated for a maximum of sixteen tonnes. While shielding their fellow pioneers, a security picket north of the village was ambushed by an enemy patrol. Behind a rise in the ground and out of sight of his

---

10 Two of the seven recipients were absent: Feldwebel Grassl, who was transferred to 3rd Company at some point in July, and Gefreiter Johann Russ, who suffered a grazing wound to his left bicep on 8 July and was undergoing treatment.

buddies, Pionier Heinrich Erath was shot in the face and his twitching corpse callously rifled. Despite an immediate response, the fleeing Russians escaped retribution. His death angered Beigel:

A sentry of mine was shot by a Russian patrol. I've lost another good man. I rode up as quickly as possible but only saw them running off. You can probably imagine how I feel. I'll certainly not be taking any more prisoners.

The isolated death of 35-year-old Erath was insignificant in the larger scheme of military operations but devastating for his family. An initially sad duty for Oberleutnant Grimm evolved into a tragedy that weighed upon him for the rest of his life as Erath's death set in train a sequence of events that would baffle and stun him. He revealed the details to Lotte:

Now to the sad case of the man who was killed on 30 July. I have written to his wife, who is childless. I will only write about the precise cause of death if the next-of-kin specifically asks about it… When you go to the indoor market, visit Stand 42… The owner is apparently [Erath's] sister-in-law, and [Frau Erath] helps out… Go there and tell them that you are Frau Grimm and have heard that they are related to Herr Erath and that you give your sincerest condolences for their great loss. You don't know the precise facts [about his death] but if they want to know more about the circumstances, write to me immediately. He was actually a brave guy and it's a great shame about him. The Russian dogs completely plundered him, but please don't say anything.

Decades after the war, the details of Erath's death and subsequent ramifications were still razor sharp in Grimm's mind:

The order was conscientiously observed that company commanders should inform next-of-kin about deaths the following day at the latest, by hand – written in ink – with a sketch of the grave enclosed. We were mostly advancing, the supply train often lagged far behind, and I can clearly recall how I often, for example, in the steppe, lay on my stomach, using my map case as a base, writing the sad message. The queries from next-of-kin were often devastating, particularly from those whose loved ones were killed in a mine accident and there were no effects – like rings or wallets and other personal items – that could be handed back. I always gave the next-of-kin my home address and if possible the home addresses of witnesses. I have noted I did this because of our heavy casualties and the possibility existed that wounded and sick men would still be able to give information to next-of-kin. Here, I want to relate a tragic case. Pionier Heinrich Erath was killed on 30 July 1942 and the enemy took everything off him before a nearby strongpoint could intervene. As far as I recall, I had not disclosed my home address [...] but his wife must have known that I worked in Stuttgart and was living in Degerloch. She asked whether she would be allowed to visit me in Degerloch during my next furlough. I agreed... I was home the beginning of December under outpatient treatment in hospital. Frau Erath helped out her sister who ran a fruit and vegetable stand in the indoor market, so we went there first of all and the sister told us that they were worried about their Trudel. She no longer laughed and moved about so strangely. Frau Erath arrived and put her inquisitive questions to me. I had to repeatedly confirm to her that I knew him and what he had once told me about his home and his large gardens, and so on. Then I described precisely that I had him laid to rest in the presence of Oberfeldwebel Jakob, our Spieß [company sergeant major], and I delivered a eulogy, which unfortunately was only possible in a few cases. A precise sketch of his grave and other details were handed over. Indeed, how he was hit. A partisan in the cornfield hit him in the lower part of the face with a submachine-gun round from close range and death was instantaneous. With that, Frau Erath became much calmer. We had complied with her request for a visit and she seemed relaxed and resolved, almost joyful. She wanted the circumstances of burial described in great detail... [Before I returned to the front] I thought I would meet her at the fruit and vegetable stand in the indoor market but that unfortunately did not happen. Her sister thanked us heartily, their Trudel was doing well...

Grimm returned to Russia in late December. The dire situation compelled men returning from leave to be formed into ad hoc formations and placed under an officer; Grimm was given command of such a unit. The makeshift arrangement prevented normal delivery of mail, so the first letter that he received from Lotte was in February 1943. It contained news that hit him like a tonne of bricks: Erath's wife had taken her own life with gas. Beforehand, she had put all her affairs in order, right down to the littlest detail. Grimm was shellshocked. "What had I done wrong?" he asked himself. The blame did not lie with Grimm. The catalyst lay within Frau Erath herself. All joy had been drained from her life. Happy memories became painful. She lost her future.

She lost her children. All their birthdays. Their weddings. Grandchildren to spoil. To go on living would be just treading water, living life without any purpose, and she saw just one way out of her grief-stricken limbo. How many other wives went down the same dark path? When reading military history, it is easy to become blasé about the deaths of soldiers as that is part and parcel of warfare, but a spouse taking her own life serves to illustrate the fact that each man was not a faceless, nameless cog in the military machine: he had a wife, children, parents, siblings, people who cared about him, and his death left a gaping hole.

Dogfight (31 July 1942)

While Regiment 578 successfully captured half of the Serafimovich bridgehead in a joint operation with the Italian Bersaglieri, the rest of 305th Infantry Division held defensive positions out in the steppe. Gefreiter Rinck noted in his diary that it was a "quiet day." Grimm used the opportunity to attend to some office work:1

> I have done nothing else but complete a mountain of paperwork, mainly recommendations for the Iron Cross, Wound Badge, Assault Badge, and so on. The sad messages will be done tomorrow.

The company remained in Orekhovskiy. A few men were out on picket duty but everyone else lounged in the shade. The sound of aircraft engines filled the pale blue sky; a squadron of heavy Russian bombers flew overhead and dropped bombs on the Rollbahn (main route of advance). On the return flight over the village, the squadron – protected by fighters – was bounced by Messerschmitts. Shielding their eyes, the pioneers gawped at the spectacle, cheering and whistling as German aces slashed through the enemy formation. Leutnant Beigel was still shaking with excitement when he wrote home half an hour later:

> Today we saw a fantastic aerial battle. Twelve Russian bombers, escorted by three fighters, drew over our village at about 200 metres altitude. We thought they were just about to unload. Suddenly, four German fighters roared in like madmen and cleared the sky! The fighters really went at it hard. The Russians no longer knew what was what. Two Russian bombers were quickly shot down. We had a good view of one of our fighters pumping a full burst into a Russian from behind so that it emitted a shower of sparks. The air battle quickly shifted toward enemy lines, so we weren't able to see any more of it. Later on, the fighters came flying back, all four of them.

Grimm was entertaining Oberzahlmeister Keppler when the racket drew his gaze:

I saw a glorious dogfight for the first time... Thirteen new Russian bombers calmly flew over our village but were not malicious at all. A short while later we heard them drop their eggs behind us. Then there was a commotion and my men shouted with joy. Our fighters are here! Right above us, they positioned themselves behind the Russian bombers, which flew a little higher. The Russian escort fighters zoomed away just over our heads. Our fighters attacked and we were able to confirm the downing of three Russian bombers.

The company's security pickets were not relieved by the infantry as promised because Voigt's regiment was being shifted to another sector. As darkness was falling, the advance commando of 100th Jäger Division began relieving Regiment 577. Grimm greeted the arrival of reinforcements with relief because his men felt the steppe had swallowed them up. If a large-scale Russian attack came their way, they would be ground into the dirt and no-one would ever know. "At first, we were completely on our own together with an infantry battalion," wrote Leutnant Beigel. "Since yesterday, the heavy weapons have also arrived. In addition, assault guns, anti-tank guns and an entire regiment of infantry. Nothing can happen to us now." Grimm was allowed to pull back his pickets to the edge of the village during the night. Freed of the burden of worrying about his men standing guard out in the vast steppe, Grimm settled down for a decent night's sleep. His command post was in the courtyard of a derelict farmstead, but he still slept in a covered foxhole complete with an inflatable mattress and blanket:

This hole offered good protection from shell fragments and I did not need to disturb myself during bombing attacks or mortar barrages. Of course I received reports from time to time, even during the night, but I've grown accustomed to that, I simply roll over and go back to sleep.

– – –

The men were now acclimatised to the Eastern Front. Combat-lean and marching fit, hundreds of dusty kilometres passed under their boots in conditions impossible to imagine just a year earlier. Long gone were the soft garrison soldiers from France. Grimm and his brother soldiers were equal to any other infantry division in Paulus' army. The Knight's Cross around Oppenländer's neck was not just a personal decoration for him: it represented the efforts and sacrifices of the entire division. It symbolised the fulfilment of orders, the attainment of objectives. Summer was at its height and the men knew there was still some way to go until the campaign had achieved its goals, but the end seemed to be in sight. All that stood between them and the Volga was a vast expanse of parched steppe – perfect country for mobile operations – the Don River and more steppe. Red Army formations were still fleeing east. Victory was within grasp.

CHAPTER 5

Bodensee

Mobile units overtook the division during its long march from the Boguchar area, proof that Hitler had finally been persuaded that 6th Army needed them if the land-bridge between the Don and the Volga was to be won. This wise decision, however, was offset by others. Hitler felt certain of success in the southern sector and therefore ordered von Manstein's 11th Army, currently recuperating from its conquest of Sevastopol, to be transferred to Leningrad. And when Rostov fell on 23 July and Red Army forces fled south over the Don, Hitler decided to dispense with a unified advance to the Volga by both army groups and sent Heeresgruppe A south with the new objectives of Batum on the Black Sea and Baku on the Caspian Sea, therefore not just into the Caucasus, but over the mountains, into the Transcaucasus. Hitler broke a golden rule. Instead of concentrating strength for consecutive attainment of two objectives, he split his forces and demanded that Stalingrad and the Caucasus be gained concurrently. All that was left to Army Group B for its advance on Stalingrad was Paulus' weakened army, now drained of most of its fuel. Worse still, 6th Army's mission was expanded: after capturing Stalingrad, it would push along the Volga to Astrakhan and block river traffic there. Hitler even voiced the ludicrous idea of advancing through the Middle East to link up with Rommel in North Africa. The Chief of the Army General Staff, General Franz Halder, told him that he was grossly underestimating Soviet strength, but Hitler countered that the Russians were nearly broken. The Führer should have paid heed because the Red Army was in no way beaten. Its units were pulling back to the Volga ahead of Paulus' spearhead, occasionally standing their ground in order to delay the Germans and gain time for the preparation of defences west of Stalingrad. Hitler's amateurish dabbling in grand strategy granted the Russian leadership an unexpected time extension and they exploited this by assembling an entire army west of the Don.

The increasing resistance faced by 6th Army from 23 July onwards demonstrated that it was too weak to reach the Volga on its own. Hitler therefore revised his directive on 30 July and attached Hoth's 4th Panzer Army (two German corps with four divisions and a Romanian corps, likewise with four divisions) to Army Group B so that it could support 6th Army's frontal attack on Stalingrad by thrusting towards the metropolis from the south-west. The summer offensive was irrevocably set on diverging paths as the two powerful army groups aspired to reach objectives seven hundred kilometres apart.

Boredom (1-14 August 1942)

While Regiment 578 remained behind to capture the remaining half of the Serafimovich bridgehead, the rest of the division pushed east, but as so often happened in difficult situations, it had been split up. Only four of its nine infantry battalions were within the divisional framework; other divisional elements were attached to 100th Jäger Division for a southward attack. In return, Jäger-Regiment 227 was subordinated to the Bodensee Division. Every man was needed because worrying signs of Soviet resurgence had appeared and the division reckoned on powerful attacks in the coming days. Oppenländer shifted his units to capitalise on advantageous terrain features, though with too few men on the ground, the pioneers were deployed as infantry. Hauptmann Häntzka and his 3rd Company secured the right flank of Regiment 227 by blocking the narrow ridge near Hill 243.3. Initially, Klein's and Grimm's companies were held at the division's disposal but the dire need to fill gaps forced their commitment, so both companies were ordered to reach Hill 243.3 as quickly as possible. Grimm pulled in his pickets and marched fifteen kilometres south-eastward to the hill. Klein did the same. The forty-five degree heat was unbearable. Now subordinated to Jäger-Regiment 227, Major Beismann[1] was given the task of holding the hill with his battalion and closing a large gap. Grimm was required to occupy a thousand-metre section of front, normally a job for a battalion:

> The terrain here is very monotonous. The fields are not cultivated and the wheatfields are full and grown through with grass. There is nothing but these fields as far as the eye can see, perhaps a small row of bushes about ten metres long after twenty kilometres. Deep, almost impassable gullies cut through the terrain and they can only be spotted when they're right in front of you. Of course, these often prove to be obstacles and lengthy detours are needed. Otherwise you can march a hundred kilometres without running into an obstacle.

As soon as they arrived, the pioneers excavated foxholes and set themselves up for defence. A few artillery shells shrieked in, punctuated occasionally by mortars rounds, but there was no immediate threat. More ominous was the delivery of forty extra anti-tank mines. The pioneers remained in these positions on 2 August. Temperatures once again reached the mid-forties (around 110 F). Grimm pushed two squads forward about four hundred metres as combat outposts. A small copse nearby was combed

---

1   Hauptmann Beismann was promoted to Major with effect from 1 August 1942. The actual date of his promotion is not known but his new rank will be reflected in the text from this point forward.

through and four prisoners wheedled out. Their main activity, however, was spectating. From their hilltop position, the pioneers had box seats for a frenetic clash in front of them: six kilometres away, heavy tank and infantry battles played out the entire day. Grimm's men lounged in their foxholes and watched vicariously. From time to time an enemy tank fired on the hill. Pionier Josef Geiger was wounded by a shell fragment. Grimm, normally reticent in describing combat activity to his wife, could not resist mentioning the scene: "From my position I see at some distance continual tank battles and ceaseless infantry attacks for days on end. The burning Russian tanks light up the night and one could believe that they were heliotropes on the slopes of our mountain." Leutnant Beigel was also enthralled by the spectacle:

> The entire battalion lies in position on a hill. We won't be here long because as far as we've been able to observe, the Russians have been thrown back. Today, I have, that is, we have, experienced a splendid battle. About four or five kilometres ahead of us is a large, flat plateau on which a mighty tank battle raged. To the left we saw the Russians attacking with many tanks and on the right we saw German anti-aircraft guns, panzers and infantry in position. As the Russian attack got underway, sixteen German Stukas intervened in the battle. They flew [four][2] sorties. Through the magnificent cooperation between army and air force, all attacks were smashed back. Many Russian tanks are burning over there. The clamour of battle has just died down. Every now and then our artillery fires overhead and every now and then several Russian mortars fire. Tomorrow morning, when we crawl out of our foxholes to have a good stretch, the Russkis will be done for. Then we'll have to leave our beautiful positions that were shovelled out with a lot of sweat.

The monotony of the following days was broken only by continual artillery harassment and griping about the water shortage. "You must excuse me if the white paper bears ugly stains," Beigel wrote to his girlfriend. "We hardly ever wash because there is not a drop of water here." Grimm had a solution to dirt-stained letters: he headed back to the supply train and used a typewriter set up in the shade of a wagon, a privilege reserved for company commanders. The lack of potable water was serious. All watercourses had dried out. Here and there were small reservoirs, once used to water cattle, but the stagnant melt water from early spring was only suitable for the horses. Initially, drinking water had to be fetched from twelve kilometres away but the need was

---

2   Beigel claimed it was three sorties but Grimm, who kept more accurate records, stated it was four.

so great that teams went searching for it – successfully, as it turns out: "Water for our coffee [comes] from a newly discovered source," wrote Grimm. "Shaving, washing, and such, naturally goes by the board. If someone wants to wash his hands, then it'll be taken from his precious tea or coffee. However, we're very glad that it's not raining." With poor personal hygiene came pestilence. Grimm took care of a few fleas but his flea powder was far away, safely locked away in his officer's chest, two hundred kilometres further back in the rear.

The garrison of Hill 243.3 quickly settled into a routine. Sentries were relieved every few hours. Harassment by Soviet artillery was ongoing. Nevertheless, each day offered something different, an event of note – even some precipitation – that Gefreiter Rinck recorded in his diary:

3 August: Got ready in the morning. 1200-1800 march into the forward positions. Construction of tank-proof foxholes. A lot of rain during the march (20 km).

4 August: Stuka attack from 0400-0800. Tank assembly ahead of our line was smashed. Thirty-eight tanks destroyed. Russian artillery bombardment.

5 August: The Russians attack every night with tanks. To the right of us, sixteen tanks were destroyed by rocket-launchers. Mortar barrages. Pionier Geyer wounded. Increased outpost duty at night.

The overwhelming sensation remained heat, thirst and dirt, and all soldiers bellyached to their families about the trying conditions. Beigel tried to explain the conditions to his girl:

It's dead quiet today. We've been in the same position for the past three days. The sun beats down brutally, almost like in Africa. At 0500 in the morning, when the sun starts burning down on the tent-quarters that cover our foxholes, then we have to get out of our incubators. Imagine the water in your canteen staying hot the entire morning because it stood in the sun. Now we generally receive butter and fat as liquids. We spend the entire day under stretched-out tent sheets, which are our only protection against the sun. Our main occupations are reading the paper, rubbing the filth off our mitts, cutting off our beards with scissors and sleeping. If only there was a little bit of water nearby. The closest well is eight kilometres away.

The pioneers were fortunate that the climate was the only thing with which they had to contend. After their drubbing on 2 August, the Soviets had abandoned their attacks against the division, but recon planes spotted tanks assembling, mainly in the river bend north of the Sirotinskaya–Perekopskaya line. The divisions of 8th Army Corps were warned to expect attacks from the south-east, east and north-west at any time. The pioneers were directed to thicken the defensive line with mines. Beismann ordered Hauptmann Klein to install trip-wire mines on likely approach routes while Grimm's men covered them. Grimm took two squads from 2nd Platoon and was allocated one platoon from Klein's company. As the sun gloriously painted the western sky, the night-shift of 2nd Company assembled their gear and double-checked weapons. All was quiet. They moved off towards Klein's command post in the purple twilight. Accompanied by the chirping of crickets and distant gunfire, the delicate job was carried out without any appreciable enemy interference. Grimm then received the task of personally occupying an artillery observation post with one pioneer squad for the night. Throughout the day the observation post was secured by a quarter of the company and lay under constant heavy mortar fire. While Grimm spent the night in the vulnerable post, his deputy, Leutnant Beigel, had a relaxing evening with another platoon leader:

> I spent a long time with Willi [Platzer], a Feldwebel from Linz, who has always been in our company and with whom I'm good friends. We prattled on to each other until late into the night about happy times from home. In our minds' eyes we have quickly finished off the Russians, been driven to France and have then knocked out the English. That's what we thought about…

Grimm and his squad returned from their security task at 0600 hours on 6 August just as Stukas droned overhead and dived on a Soviet tank assembly. His company remained in defensive positions throughout the day under the ever-present artillery and mortar fire. Dive-bombers and mail call were the sole distractions on another incandescent summer day.

Random artillery rounds shrieked in during the morning of 7 August, one shell arching straight into the supply train of Klein's company. Gefreiter Füssinger was there: no human fatalities, but eight horses were killed. Klein ordered his train to shift outside the fire-zone, which made Füssinger very happy. When word reached Grimm that 1st Company's supply train had been hit, he shifted his own to Mukovnin, six kilometres to the south. By a stroke of fortune, good quality drinking water was found there. Grimm was able to use this discovery to provide some personal comfort:

> I had my hair cut (don't ask how) and washed my whole body. The water was fetched from far away. I sent my laundry back with Nuoffer to be boiled up, which killed the vermin. I now wear fresh beautiful clothes that will make me stand out

for a few days. They will be quickly blackened by the black earth because we are currently in foxholes.

If the pioneers thought previous temperatures had been high, they were shocked by the stifling heat on 7 August. The mercury reached fifty degrees Celsius. Shirts were off all day. The sameness of the daily routine caused the days to blur. Beigel wrote "Today is Monday, as I have discovered in a calendar. One loses track of time." He obviously had because it was a Friday. He continued:

We've been in the same foxholes for the past five days. The Russian has become considerably decent. He no longer shoots so much with his artillery and no longer attacks wildly. Today I learned that the Russian divisions and tank brigades are surrounded. They're now sitting in the Don bend and have no more supplies. That's great, isn't it! We've not seen a single cloud since last Monday. The only good thing is that a wind is blowing, even if it is warm, or sometimes even hot. From time to time exploding shells kindle the dry grass of the pastures, resulting in small grass fires.

There was little relief from the heat during the night when Grimm led another two squads back to the artillery observation post. "It's early in the morning, when the sun is not beating down too fiercely," wrote Leutnant Beigel on 8 August, "during the day it is impossible to do anything. At most one can still read, but that's it. The heat makes us all so listless that we just spend all day under our tent-quarters like stuffed pigs." The ongoing battle against the elements was taxing but at least the only threat to life was swelling artillery and mortar fire. The situation was not so carefree in other sectors.

The mission for General Heitz's 8th Army Corps, to which Oppenländer's division belonged, was to protect the northern flank of 6th Army while a new offensive was being prepared. Soviet relief attacks originating from the small Don bend climaxed on 6 August and tangibly slackened in the division's sector from the 7th, although they began increasing in strength further south. Heitz shuffled his forces southward to counter this. Two regiments of the neighbouring 113th Infantry Division were extracted from the line and sent south, with the resulting gap filled by the 305th Division. This was accomplished by organising all units into three sectors controlled by the most trustworthy officers. In the north was Oberst Winzer, the central sector was held by Oberstleutnant Neibecker's Jäger-Regiment 227, and in the south was Major Brandt of Regiment 576. Häntzka's company was attached to Winzer, while Beismann and the other two companies remained in the central sector subordinated to Regiment 227. The lengthy sector and low headcount compelled a strongpoint-based defence with particular attention paid to sites where experience suggested the Soviets would attack. Positions were to be held unconditionally. "Elastic defence does not come into question at all," stated the divisional order, "because the recapture of lost positions will cost more blood than a rigid defence." With new sectors came new tasks for the pioneers, as Grimm reports:

> At 1700 hours [I] received the task of taking over fifteen minefields with two hundred mines from 113th Pioneer Battalion in the defensive sector south-east of Mukovnin. Laying pattern: ring-shaped with five-metre separation, camouflaged with grass, in front of heavy weapons. The handover took place under heavy enemy fire because the Russians launched a disruptive attack on the right sector at nightfall.

These new minefields would provide a lot of work for Grimm over the coming days.

On 9 August Grimm and Oberfeldwebel Locher clambered aboard Gefreiter Vorherr's motorcycle-sidecar and set about the quotidian task of discussing pioneer operations with infantry officers. As they drove from one place to another, bullets whizzed by their heads as Russian snipers took potshots. Grimm urged Vorherr to speed up. Mortar rounds straddled their route and threw up dirt and dry sods. "Go Vorherr, go!" The trio went hell-for-leather, dodging and weaving through clouds of dust and smoke. "You've got to be lucky," Grimm wrote to Lotte. "We scrammed as fast as an express train and the salutations whistled about our ears…"

– – –

The enforced hiatus and shared adversity enabled the pioneers to bond even more tightly. Friendships formed in France were sealed forever in Russia. Each soldier knew he could rely on the man next to him. Acquaintances met up to talk about old times. Beigel visited his best friend whenever he could:

> I met Peter [Buchner] yesterday. He is irrepressible as always. In 1st Company, where he has been now for a year, they believe he's indestructible. I believe that the two of us will make it out of this war.

Buchner was a tough nut. He was the only platoon leader in Klein's company to survive the Kharkov battle and emerged unscathed from countless engagements. Grimm wrote an interesting appraisal him:

> A determined and strong-willed officer who understands how to lead and inspire his men in a foresighted and purposeful way. Ambitious, hard-working, tenacious and strong, almost reckless towards himself. In battle he distinguished himself

through exceptional nerves and acts of daredevilry, leading his platoon prudently and skilfully in battle. His military manner is good. His background (north Bavarian) makes him relatively robust and he is very popular and highly regarded by his subordinates because of his thoughtfulness and cool character. His company commander, Hauptmann Klein, did not know how to encourage him in a way befitting an officer, with the result that Oberleutnant Buchner remained somewhat strikingly true to his Lower Bavarian ways. He is honest, open and honourable, respected by his superiors, popular and esteemed by his comrades.

What turned ordinary men like Buchner into first-rate soldiers? Most of the division's men were not prime specimens of manhood. The youngest, fittest men in Germany were more likely to be in a panzer or a plane than slogging across Russia on foot with a re-purposed bodenständige-Division. Some young blood had been pumped in during the unit's transformation into an assault division, but overall, the calibre of manpower was – to put it nicely – average. Very few were active soldiers. Almost all had been drafted: they were teachers, farmers, salesmen, led by officers in their thirties and forties, yet here they were, full-fledged combat soldiers, able to accomplish any mission set for them, equal if not better than neighbouring divisions with illustrious histories. National traits like thoroughness, loyalty and dutifulness had been harnessed to convert ordinary citizens into first-class soldiers, but it was the superior understanding of warfare by the Germans at the time that is key. Most of the credit for the performance of the pioneer battalion goes to Major Beismann. His influence and aggressive attitude cascaded to all command levels in his battalion: those who aspired to be combat heroes had a role model, those who simply wanted to survive feared disappointing him. The German knack for producing so many combat-capable units is undeniable. The German soldier was respected on the battlefield, so much so that the imaginations of their opponents conjured up images of steely-jawed Aryans, yet, just like in the Bodensee Division, they were more likely to be scrawny kids, world-weary fathers or glass-wearing students like Buchner.

– – –

Only after ammunition and fuel stocks were replenished and more forces brought up was 6th Army able to crush Soviet resistance west of the Don. In hot, bloody fighting from 7 to 11 August, Paulus' units encircled nine rifle divisions and seven tank brigades west of Kalach and then crushed them. Although ending in victory, this battle proved that the Soviet leadership had capitalised on the opportunity presented to them when Hitler shifted the offensive emphasis south: the Soviet command had paid a huge price, but it gained a further two weeks for defensive preparations on the land-bridge west of Stalingrad. Paulus immediately began regrouping for an attack into the small Don bend, another set-piece operation that may have been avoided if Hitler had not temporarily abandoned the thrust directed east to the Volga knee.

Grimm's company remained in position on 11 August under a glowering sun and moderate artillery-mortar fire. Thickening of the main resistance line with mines continued. He pulled two squads of Feldwebel Kurt Ritter's 2nd Platoon out of the defensive line and gave them the hazardous task of disinterring minefields and relaying them. Of fourteen minefields taken over from the 113th Pioneer Battalion, eight were lifted and six relaid with a total of 110 mines. Laying pattern was ring-shaped around the gun positions with five pace intervals. Two adjoining minefields in the left and right sectors were detected by a Soviet patrol, so Grimm cunningly removed all mines from both fields and replanted them during the night about fifty metres from enemy lines. The following day (12 August) Grimm, his HQ section and a squad from 3rd Platoon booby-trapped a gully with ten anti-personnel mines connected to a pair of hundred-metre long tripwires, one lying behind the other inside the gully. If a wire was tripped anywhere along its length, all ten mines would go up. The pioneers were always devising cunning traps to outwit their opponent.

The pioneers' only tasks consisted of mine-laying and sentry duty. Leutnant Beigel took his turn up at the artillery observation post but everything was so calm he just lay on the ground, gazing up at the Milky Way, pointing out a shooting star to a chum and wishing he was at home kissing his girlfriend. His letter from 13 August conveys the general mood:

> We're still in the same spot as we were fourteen days ago... Not much has changed
> in that time. The paths are now well-trodden, we have to replace the camouflage on
> our positions often and the number of bullets, as well as hand grenades, are available
> in half quantities. The good Russki wouldn't dream of disturbing our peace. He
> actually has more artillery than before but has basically given up his futile attacks.
> As always, the sun shines down impudently, so that we are burnt brown.

After almost two weeks of fretting over defences and worrying about being overrun, aerial recon and patrols ascertained that the direct opponent (the Soviet 205th Rifle Division) was definitely on the defensive. Paulus had decided to clear up the potentially threatening situation in this north-eastern corner of the Don Bend before proceeding with a river crossing and advance on Stalingrad. Armoured units of 14th Panzer Corps arrived. Several more infantry divisions followed on 11-12 August. Eventually, eleven divisions from five corps were concentrated on a 55-km wide sector. The mission of this formidable grouping was to push north-eastward to destroy the weakened 4th Tank Army and then gain some bridgeheads over the Don to facilitate the subsequent drive toward Stalingrad. The role of 305th Infantry Division in this offensive was simple: it was to protect the growing left flank of 14th Panzer Corps. First, it would throw back opposing forces and then, trailing behind the panzers, take the high ground south and south-east of Kamyshinskiy and thus cover the northern flank of the panzer corps.

After almost a fortnight of seeming isolation in thinly held defensive lines, the dirt-encrusted pioneers were thrilled to see endless columns of tanks and trucks converging on their sector. One officer wryly noted: "An unfamiliar but delightful sight – our panzers and motorised vehicles in magical numbers." General Oppenländer issued his attack orders a few hours before sunset on 13 August: the division would advance along a line of hills (229.3 – 226.0 – 236.9), reach the area south and south-east of Kamyshinskiy, then pivot to secure the northern flank of the panzer corps. Armoured units would roll straight through the division's positions and stun the opposing Soviet forces with a left hook inside the river bend and form a sack open towards the west.

General Paulus arrived mid-morning to inspect the battleground. After a basic orientation at corps headquarters, he travelled by car to view the attack zone and the best location for that was an artillery observation post on Hill 243.3, the very hill being held by the 305th Pioneer Battalion. One of Oppenländer's orderly officers briefed Paulus on the terrain and talked about the division's attack objectives, all while Soviet artillery rounds intermittently screamed in and landed on the hill and the busy road behind it. The pioneers in foxholes dotted around the hill had no idea that an illustrious visitor was in their midst. A messenger informed Major Beismann that he was wanted, so he tidied himself up and strode over to the observation post. He saluted smartly, shook Paulus' proffered hand and briefly described his battalion's activities during the previous few weeks, including heavy losses in dead and wounded, particularly during the lifting of minefields. A few minutes later, Paulus climbed into his car and headed back down the arid road to Verkhnaya Buzinovka.

As a company leader, Grimm did not receive the divisional attack order until 1600 hours on 14 August, the day before the attack. Lotte, his wife, celebrated her birthday on 15 August and he was uncertain if this was a good or bad omen. An hour later, Beismann ordered him to lift the minefields and boobytraps before the attack began at 0400 hours the next morning. Feldwebel Ritter's 2nd Platoon (minus one squad) and Feldwebel Platzer's 3rd Platoon reached the minefields at nightfall. Those lying directly in front of the enemy positions were cleared without incident before midnight.

Terror (15-18 August 1942)

Stukas and fighter-bombers howled overhead at 0200 hours and unloaded upon the Soviet positions. Grimm's men, who had just started clearing another minefield, tried to not let the awesome spectacle distract them from the delicate job at hand, but they occasionally craned their necks to watch a low-flying plane career over their heads. This was no time to watch the sky show: a hundred mines needed to be lifted. Consulting a sketch, Grimm set his clearance teams to work unearthing the deadly disks. Eventually, he faced a dilemma: 99 of 100 T-mines marked on the sketch were lifted, but the last one could not be found despite the most meticulous search. Time was pressing. Grimm ran over to some nearby infantrymen and questioned them. It turned out that the rogue mine had detonated. The minefield was declared clear. Panzers that had assembled on the rear slope moved out at 0300 hours, right on schedule. This armoured spearhead from 16th Panzer Division drove through the lines of 305th Division and moved north-eastward astride a saddle of high ground flanked on both sides by impassable ravines. Two motorised divisions followed in their wake.

As the panzers rolled past, Grimm set about his next task: dismantling the tripwire boobytrap laid two days earlier. One mine had been triggered without setting off the others. The remaining nine mines were de-activated. The work was completed in the nick of time: at 0400 hours, a reinforced infantry battalion launched an attack and took Hill 231.3 in order to cover the southern flank of the advance. Häntzka's 3rd Company trailed behind the battalion's left wing, ready to protect its flank should the need arise. The infantry battalion continued to move in an eastern direction to capture a few more hills and shield the main force before linking up with its parent regiment. The bulk of the division began advancing from of its assembly areas at 0500 hours.

Grimm cleared minefield 'Schneck' in the right sector under heavy artillery and mortar fire. Eighty-six anti-tank mines were lifted, thirteen were set off by artillery impacts and one had been deactivated and reburied by another pioneer unit. Towards 0700 hours, after a short lorry ride, they rejoined the rest of the company in Perekopskiy. An order was awaiting Grimm: "Make contact with the staff of Infanterie-Regiment 576."

In the meantime, the panzers had burst through and the attack was in full swing. Large numbers of enemy troops were encircled in small pockets behind the front-line. Open expanses were empty: the Russians defended themselves bitterly from every gully. Dust filled the air. Pillars of smoke snaked skyward. The pale blue sky now looked tea-stained, the sun an amber orb, somehow smaller than usual. Once again the heat was intense, the clamour of battle constant. Rocket-launchers wailed, planes swarmed overhead, artillery thundered. At midday, the panzers of 16. Panzer-Division overran the headquarters of 4th Tank Army but its commander, General Kriuchenkin, eluded capture. Far behind the spearhead, Russian resistance intensified. Vehicles hauling back wounded soldiers were fired upon and small columns ambushed. The quick divisions were making good progress but lacked infantry to eliminate all of the enemy groups caught behind the line. Von Wietersheim's motorised troops were too valuable to recall for such a task, so Paulus ordered 8th Army Corps to do it. With admirable alacrity, the corps arranged for two battalions to mop up the pockets. Meanwhile, Grimm's and Klein's companies marched forward five kilometres. At 1400 hours, at Krüder's command post, both officers received the order to comb through a small copse:

> Fighter-bombers had already prepared this small forest in the morning, panzers and infantry did their destructive work when pushing past the forest and the ground was covered with masses of dead Russians, often strung together or in clusters in their foxholes. You could tell at first glance that the Russians fighting here were elite troops, young fellows, with the best weapons and exceptionally well equipped. In any case, we were ready for a tough nut to crack.

Grimm subsequently learned that a few panzers and riflemen from 60th Motorised Division had pushed through the copse in the morning, then Krüder's regiment approached the small woodland a few hours later, expecting little resistance. One battalion went in to clear it but suffered heavy casualties, and when it became known that two pioneer companies were on their way, the battalion pulled back from the woods. Despite this, Grimm and Klein were informed by Regiment 576 that they would only have to deal with weak enemy forces.

Both companies assembled on the western edge of the forest. Surveying the terrain through binoculars, Grimm saw rolling hills, ravines hidden in folds in the ground and an unimposing stand of trees. The copse, roughly eight hundred metres long by four hundred wide, had not been thoroughly searched, so an unknown enemy force lurked inside. Gunfire crackled in the distance but the tree-line ahead was silent. The attack began at 1500 hours, 1st Company to the north, 2nd Company to the south, both companies crossing open grassland and pressing into the woods without opposition. As they pushed uphill into the copse, every man noticed that it was criss-crossed by deep gullies and ominous fire-lanes. After a third of the forest was in their hands, Grimm's men bumped into the enemy. Machine-guns hammered, followed by a brief flurry of single shots. The leading teams went to ground and the advance immediately stopped. Grimm ran from platoon leader to platoon leader and told them to keep moving. Bitter fighting was triggered as they pushed forward again. Gunfire swelled and mortar rounds began falling. Shouts echoed through the steamy forest. The broadening of the attack by 1st Company kicked a hornet's nest. Volleys of heavy mortar fire crashed down. Shell fragments fizzed through the air, slicing off small branches and leaves, and occasionally thudding into human flesh. The assault teams pushed on through the undergrowth, avoiding the sunlit fire-lanes and exposed gullies. Bursts of fire erupted from nowhere. Where was the enemy? There! Sprayed with fire and pummelled by grenades, enemy positions were charged. Soviet riflemen fought to the bitter end. Only then did the pioneers see their enemy's solid positions sunk deep into the forested slope. The fire discipline of the Red Army soldiers was excellent; they held their fire until the pioneers were directly in front of them, while others let the Germans pass before blasting away from behind. This treacherous fighting terrified the pioneers. As men were wounded, they were dragged back by friends eager to escape the nightmare. "We had two wounded men at the command post, then three, and many fighters had to bring in the wounded, thus leaving the forest!" recalled Grimm. He tried to put on a brave face but was scared witless, confessing decades later that "the situation was such that we all wanted to run away, but didn't." The lifeless bodies of Obergefreiter Willi Weller (bullet in the head) and Gefreiter Heinrich Prüll (hit in the abdomen) were dragged back. Grimm watched in stunned silence as a man, clinging to consciousness, was hauled in and laid on the ground three metres away, his back to him. Gefreiter Eduard Schürer had been shot from behind and the bullet was still lodged in a lung. His body was so torn up that he was breathing through the bullet hole. Bloody froth built up on his back. Grimm was aghast that he was still alive and knew his chances of survival were slim. Oberfeldwebel Locher bandaged the gaping wound with first-aid packets but they quickly soaked through. Grimm and Locher frisked the pockets of the two dead men and found more, Locher finally staunching the flow with a large dressing. Medics carried Schürer to an ambulance waiting on the outskirts of the copse.[3]

Soviet troops holed up in fortified positions in the gullies could not be destroyed due to a lack of heavy weapons, but the pioneers pressed on regardless. Ferocious fighting surged through the trees. Hand-to-hand scuffles broke out. Several mortar pits were overrun and their crews gunned down because they continued to resist, even when staring up the barrels of German guns. Mortar fire lessened but volleys still landed about the area. Grimm's and Klein's command posts were right next to each

---

3   Incredibly, Schürer survived his grievous wound. About one and a quarter years later, in Italy, Schürer returned to service but only as a driver with the supply train and rear area services. In 1950, Schürer wrote to Grimm in order to obtain a sworn declaration about the severity of his wounds. In post-war Germany, a disability pension was the only source of income for incapacitated veterans.

other to enable coordinated operations, so Grimm witnessed Klein being dropped by a shell splinter. A messenger dashed through the trees to inform Klein's deputy, Oberleutnant Buchner, that he was now in charge of 1st Company, but minutes later a bullet thwacked into his left hip. Morale is sustained by seemingly invincible leaders like Buchner, so his wounding had a devastating effect on the company's will to fight. His men watched, stunned, as he too was carried out of the deadly woods. Grimm felt the situation spiralling out of control when his reliable junior leaders started coming back to his command point, tattered and bleeding: first was Feldwebel Ritter, leader of 2nd Platoon, followed by the company's only winner of the Iron Cross First Class, Unteroffizier Bub, with a nasty arm wound.[4] Grimm took command of both companies and assumed total control of the battle. He was in little doubt that his men were seriously outgunned. Definitely identified in the forest were six heavy mortars, at least six heavy machine-guns, a large number of automatic weapons and two 76.2mm anti-tank guns. The high casualty toll eventually saved the pioneer companies. "It was actually a ghastly situation and thank God there were so many wounded in the forest, otherwise, most of us would be dead!" Grimm recalled. When Major Beismann was informed of casualties and enemy superiority in heavy weapons, he ordered Grimm to break off the attack, disengage from the enemy and pull the companies out of the woods. Grimm did so towards 1830 hours, after all the dead and wounded had been recovered. The pioneers brought back six heavy mortars with a dozen crates of ammunition, one light mortar and twenty-seven prisoners. A large number of semi-automatic rifles and a few machine-guns were destroyed during the retreat. Both companies then set up defences along the northern and western edges of the copse in order to prevent the enemy from ambushing vehicles on the nearby supply road.

Everyone was shellshocked by the day's fighting. Eastern Front veterans, those from 1941, knew both sides of the Russian soldier's psyche: one day he would surrender meekly, the next he fought like a cornered lion. Grimm's men were familiar with the first, but only on 15 August did they become acquainted with the alter-ego. "I will not so quickly forget this, your 26th birthday," Grimm wrote to his wife, "because it was by far the toughest day – I mean the hardest day of fighting – that I have so far experienced. It was full-on, from midnight to midnight, and despite everything that was going on, I kept thinking of you." Even Beigel could not hide his feelings:

---

4   Bub was transported back to Germany for specialist treatment. Weeks later, he updated Grimm on his wounds. Grimm wrote to his wife on 3 October: "Unteroffizier Bub, who visited you that time from France, was wounded on 15 August, and it is possible that he may visit you. He is the only one in the company with the Iron Cross First Class, for destroying a tank. I don't think he'll be able to come back to me [...] He must have an operation on his arm."

We had our most difficult day of the entire Russian campaign. It was the big attack after a long period of waiting. I thank God that I'm still alive. I cannot tell you what it was like. In any case, the Russians fought until the last drop of blood, didn't yield an inch and didn't surrender even when the muzzle of a gun was pointed at their chests... We are without many a good comrade. Peter [Buchner], who attacked with us, was wounded, as was his company commander and another platoon leader. Now, that is more than enough...

Casualties in Grimm's 2nd Company were two dead (Obergefreiter Willi Weller and Gefreiter Heinrich Prüll) and five wounded (Feldwebel Kurt Ritter, Unteroffizier Erwin Bub, Gefreiter Friedrich Steiner, Gefreiter Friedrich Küstner and Gefreiter Eduard Schürer). Casualties in Klein's 1st Company were four dead (Obergefreiter Karl Bulling, Josef Kunzelmann, Obergefreiter Alfons Reistenbach and Gefreiter Otto Fischer), wounded were two officers (Hauptmann Klein and Oberleutnant Buchner), one NCO (Feldwebel August Erbacher) and at least three men (Gefreiter Alfons Huber, a man called Göbel and an unnamed soldier). Not surprisingly, the day was later counted as a Sturmtag ("15 August 1942, attack on the copse near Ventsy") for both companies. This was the seventh such day for Grimm's company but only the third for Klein's.

While the pioneers were enduring their forest nightmare, the battle unfolded around them. After breaking strong resistance, 16th Panzer Division took the dominant high ground north-east of Sakolovskiy and sent its advance detachment down to the Don crossing near Ostrovskiy. The intention was for the division to advance further south, seek contact with the spearhead of 24th Panzer Corps and form a ring around all enemy forces west of this line. As the front extended eastward, it was the job of 305th Infantry Division to shield the new northern flank. Throughout the day, all three of its infantry regiments slotted into their designated sectors in this new line. The assault had cleft 4th Tank Army in two, forcing three rifle division to withdraw north towards Kremenskaya and the remainder to fall back to the Don. German front-lines were relatively solid but Soviet groups cut off in the rear continued to stubbornly resist.

The night passed quietly at first. Emotionally and physically drained by their traumatic encounter, Grimm's pioneers tried to snatch some rest, but were constantly called to arms by an enemy unwittingly conducting subtle yet effective psychological warfare. Anything out of the ordinary was interpreted as Soviet deviousness. The pioneers were spooked. Rifle shots occasionally cracked in the darkness. The wind carried Russian voices. The wily Soviets also used the prevailing breeze to their advantage by setting fire to the steppe grass so the wind could carry it toward German positions. Grimm was unnerved: "It was a ghostly sight at night." The wall of flames advanced upon them. Everyone remained in their foxholes, ready to repel the wave of Soviet soldiers presumed to be following in its scorched wake. As the fire swept around the German burrows, the soldiers crouched below the lip, waiting for the heat to lessen. The critical moment had arrived. They were in trouble if the Soviets were within grenade range. Grimm ordered a machine-gun to fire through the flames. One by one the pioneers popped up, eyes stinging from the smoke, ready to engage their enemy. Nothing. Just a wasteland of smouldering grass stumps.

Sixth Army continued its attack before dawn on 16 August, crushing one part of Kriuchenkin's army and pressing the rest into diminishing footholds south of the Don River. Mixed combat groups of 16th Panzer Division reached the Don in several spots and even seized a scorched bridge at Trekhostrovskaya.

Back at the copse, deep in the rear, Grimm's men had a much quieter day. Single shots occasionally zipped out of the forest, whereupon the pioneers retaliated with heavy fire, more a release of nervous tension and frustration than anything else. Grimm sent three prisoners back into the forest to ask their comrades to give up their senseless resistance. By 2100 hours the number of prisoners had risen to 130. Beismann was ordered to free up both pioneer companies and send them forward to the divisional command post, but the dangerous copse still needed containment, so only 1st Company was pulled out. Beismann also assigned it a new leader: his orderly officer, Oberleutnant Schaate. Gefreiter Füssinger voiced the feelings of everyone in the company: "The new commander was certainly no stranger to us because he had served as an Oberleutnant in our battalion for a long time, so it would work out very well with him."

Grimm was forced to considerably broaden his company's front on 17 August after the withdrawal of 1st Company. At 1300 hours he dispatched a squad-strength patrol into the copse. Its leader had orders to pull back if he met strong enemy resistance. Shortly after penetrating the forest, fierce gunfire exploded. Cracking rifles and burping PPSHs were overlaid by stuttering light machine-guns. The Russians threw such vast quantities of hand grenades that it was initially thought the impacts were from heavy mortars. The patrol bolted out of the sinister woods. The true strength of its garrison remained unknown. The Russians knew they were surrounded and had almost run out of rations and water. At 2045 hours, for the first time, they tried to break through Grimm's positions by employing heavy machine-guns. The attack was repulsed. Several Russian patrols were able to work their way across the scorched plains to within a few metres of Grimm's machine-gun posts but they were recognised in time and destroyed. The Russians repeated their attacks throughout the night and suffered dearly. The company had one dead (Gefreiter Anton Rüb) and one seriously wounded (Gefreiter Eugen Kusterer). Kusterer died in a field hospital. As his company gained the upper hand, Grimm's attitude changed and the shock of recent events dissipated:

We destroyed a vast quantity of fancy self-loading rifles and other things in the forest. The Russians were destroyed in the surrounding area and a huge amount of booty brought in. The fighting was tough and unforgiving. Stalin can issue instructions as he wants, the German Wehrmacht will break any resistance. How pointlessly the Russians can fight: whoever has not experienced it simply would not believe it.

The number of prisoners taken by Grimm's company rose to 153 by day's end on 17 August. Many were tough young officer cadets, defiance still glimmering in their eyes. Found amongst the prisoners was a German-speaker with a Franconian-Swabian dialect. His grandparents had come back from America and he lived in a German-speaking colony near Odessa. He provided an interesting impression of the Red Army, particularly emphasising the uncomradely behaviour amongst the enlisted men. Grimm sensed an opportunity and asked him if he was prepared to go into the copse and look around. He agreed and soon encountered a commissar with the rank of major and a senior-lieutenant, whom he promised he would fetch bread and water. Instead, he brought back to Grimm a few deserters. As senseless resistance was being sustained by the commissar and officer, Grimm hatched a devious plan to eliminate them. The new collaborator and two other prisoners, who had already been with the company for weeks, made their way back into the forest on 18 August, coolly assassinated both leaders and secured their copious records. Further reconnaissance by the turncoat revealed that only isolated resistance nests were left in the forest and they were swiftly mopped-up. Grimm was justifiably proud of his innovative scheme that entailed zero risk for his own men. He bragged to his wife: "My company smoked out the copse, a commissar and a first-lieutenant were shot. I sent important documents, identification cards and papers to division, some of their soldiers were taken prisoner." The number of prisoners rose to 173 men. Russian corpses lay everywhere.

The offensive to eliminate Soviet forces in the north-eastern corner of the Don Bend was a complete success. The two motorised divisions and 16th Panzer Division assembled in the rear to rest and refit. A six-page order titled "Army order for the attack on Stalingrad" had been issued on 16 August and everything was falling into place accordingly. With his flanks secure and German units ensconced on the towering western bank of the Don, Paulus was free to contemplate the drive to the Volga and subsequent conquest of Stalingrad.

Don River Duty (19-31 August 1942)

General Oppenländer was tasked with taking over security of the Don bend from 3rd Motorised Division with two infantry regiments while Winzer's Regiment 578 went into reserve. As the infantry regiments tramped towards their allotted sectors, the 305th Pioneer Battalion was unexpectedly called away on a new assignment. Bridge construction, the officers were told. Beismann's units were assigned to the engineering leader of 8th Army Corps. "We were abruptly shipped down to the Don River on vehicles to build a bridge," recorded Grimm. "I had barely driven onto the site when we spurted from the vehicles like a flash." The wooden bridge had been saved a few days earlier after a few Soviet soldiers set the middle of the bridge on fire. Acting quickly, the burning sections had been detached and the western end saved. The significance of an intact Don bridge so close to Stalingrad was not lost on the Soviet leadership. Russian bombers roared in at all altitudes, strafing working parties and scattering bombs. Soon after arrival, Grimm's company watched anti-aircraft guns shoot down four bombers. Fighters also chimed in and vigorously harried every plane bearing a red star.

The partly demolished bridge had been repaired to handle light traffic. Towards 1600 hours on 19 August the bridge was opened for vehicles weighing less than four tonnes. The bridgehead was only two hundred metres deep, so the crossing lay under direct fire from infantry weapons. Grimm's task was to chart the river bottom and accurately sketch the bridge. Assisted by his HQ section, Grimm worked his way along the bridge, dropping a weighted line into the water every metre to measure the depth. Then a swarm of bombers cruised in. "Take cover!" Grimm's team bolted along the exposed deck but, realising they wouldn't make it to the riverbank, jumped into the shallows to seek cover in water-filled craters. Flak intimidated the bombers, causing them to jink and strew their bombs wildly. As the planes turned for home, the pioneers, sandy and sopping wet, clambered back onto the bridge to complete their task. Grimm gathered all data by 1700 hours. Later at night, by candlelight in a darkened passenger car, Grimm analysed and evaluated the results and prepared a sketch of the bridge with profile graph. The bridge was 251 metres long. The sketch was ready at 0400 hours the next morning and conveyed to Battalion. Grimm and his assistants were completely exhausted. The rest of 2nd Company had busied themselves preparing secure billets upriver of the bridge. "We are on the Don since noon," wrote Beigel."

We've dug ourselves into a gully and are therefore fairly secure against all the steel sent over to us by the Russkis. Otherwise, we're going well. When the infantry move further forward over there, then we will strengthen the bridges over the broad river.

Thanks to our anti-aircraft guns and our fighters that circle above us, the Soviet bombers don't get too close.

Throughout the hours of darkness, however, the Soviets laid on the heaviest fire imaginable, with bombs, artillery and mortar fire as well as Stalin organs. Engines of Soviet bombers reverberated through the blackness. Explosions flashed. Tracers snaked back and forth across the sky. A few of Grimm's men were buried alive in collapsed dug-outs but were rescued in time. One man was lightly wounded by a bomb. The sole captured lorry, an American Studebaker used to haul mines, explosives and bridging equipment, was blown to pieces by a direct hit. Despite these repeated nocturnal bombing raids, the bridge was not damaged.

Oberst Winzer had been ordered to take over the northern half of the Ostrovskiy bridgehead during the night of 19-20 August by relieving a regiment of 389th Infantry Division. His battalions took cover in gullies not far from Grimm's pioneers and waited for the signal to cross the flare-lit bridge.

The small bridgehead in the lowlands was a tricky spot to hold as the forward line ran through alluvial forests that offered countless opportunities for stealthy Soviet units to infiltrate the thin German screen. Winzer knew his regiment would have to fight for a tenable position. The relief in the right sector was carried out without problems but on the left they needed to attack to occupy their designated positions. Even so, the line ran along an impossible course: the Russians sat deep in the left flank and fired on the bridge and ferries with machine-guns. An attack devised by Winzer and carried out at dawn on 21 August rectified the situation, though at heavy cost. Grimm thanked his lucky star that he was not involved in the fighting. As he walked down to the bridge, his ears were assaulted by the furious cannonade rolling across the river. His men gazed at the boiling bridgehead with a mix of dread and relief. Their task revolved around the bridge, but they knew full well that their knowledge of explosives and flamethrowers could be called upon at any moment. Not that the river was a safe haven: their construction work the previous day was carried out under heavy fire and continual bombing runs that killed one man (Gefreiter Xaver Seidl) and wounded six more. On 21 August they repaired parts of the wooden bridge and replaced others with metal B-Gerät so that the structure could bear eight tonnes. One of his platoons improved the exit ramp while Schaate's and Häntzka's companies carried out ferrying operations for heavy loads. Three bombing attacks came in against the bridge but caused no damage. The pioneers watched in awe as three aircraft were shot down by anti-aircraft guns. German fighters then swooped in to clean up the scraps, as Grimm recalls:

The Russian attacked with twelve new bombers and I saw the bombs released high overhead, though they caused us no harm because everyone found a foxhole between dead horses. Flak fired, a bomber burned brightly, two others further back went down in flames after being hit by flak. Then our fighters hunted them from behind and I reckon barely one of them would have made it home safe and sound.

The bridgehead and its vital bridge was a magnet for Russian planes. Aerial bombardments increased once darkness fell. One of the many bombs hit the bridge during the night of 21-22 August. Traffic was halted so that repairs could be carried out. Grimm's company played their part: two new piles were rammed in, two piles reinforced, six spans of the superstructure – about thirty metres of it – dismantled and relaid with available material. The company worked in squads in forty-five minute shifts. The bridge was back in commission by midday. His company lost two dead (Obergefreiter Franz Bergmeier and Pionier Josef Benz) and nine wounded. A small measure of revenge was taken when a bomber was ripped apart over the bridge by a direct hit. Grimm admired the steadfastness of the gunners manning their Flak guns while everyone else sought shelter:

Spaced a few hours apart, the Russians attacked again and again, entire flights were ripped apart by us, but they kept attacking, sacrificing bomber after bomber. The bridge, however, was only hit once. The heavens smiled upon me! The Russians fired with every weapon and my company had regrettable losses. But victory was ours, the bridge stood. Whoever saw the enemy planes downed here would change his opinion of the Flak.[5] At night we could in most cases move back a bit from the bridge because of the non-stop air strikes. They dropped their eggs about the area and everyone lay in his foxhole. Then the whistling of falling bombs and their detonations do not disturb us.

The situation along the division's front was relatively stable. Although Winzer's attack improved the defensibility of the bridgehead, Soviet units still occupied the large figleaf-shaped island to the north. The southern half of the Ostrovskiy bridgehead was to be handed over to 305th Division during the night of 22-23 August. Under cover of night, two battalions of Regiment 577 took their place in the bridgehead without any problems and the Bodensee Division assumed full responsibility for its defence.

---

5   This is almost certainly a reference to Flak units back in the homeland that were unable to prevent Allied bombers from attacking German cities.

Red flares hissed upward at dawn on 23 August, presaging a Soviet attack against the bridgehead perimeter. Defensive fire abruptly erupted and it collapsed. Soviet ground attacks halted completely during daylight hours but artillery and heavy mortars steadily pounded the bridgehead and its lifeline. German fighters were omnipresent and swept the sky clear of Soviet planes. Grimm's pioneers appreciated the aerial cover, yet that mattered little because they were constantly harassed by artillery and mortars. While repairing the exposed bridge, an artillery shell killed two men (Gefreiter Julius Klein and Gefreiter Georg Fischer) and two draught-horses with a direct hit. Job done, Grimm and his men were shifted into a rear position for a rest. Losses during their stint at the Ostrovskiy bridge were heavy: five dead and seventeen wounded, some of them severely. Beigel was glad to be out of there. He wrote to his girlfriend for the first time in four days:

We had a quiet day today. It's doing us all good because the last three days were nothing special. They were dreadful hours for us pioneers. Now it's over. We've been in a small forest since midday and are resting. It's a blissful relief when you take your filthy gear off. I slept in the afternoon, and now it's 2000 hours and I am lying comfortably in a cosy tent writing you this letter... When I sleep tonight, I'll probably dream of you. It is already very cold in the tent. However, I'll be able to bear it with my two blankets. Will we have to remain in Russia this winter? Maybe even only to recover somewhere in Poland? Our infantry look dreadful. If only the war was over soon... With darkness we await the "customary" visit by planes, which arrive promptly. The attacks were more powerful and fiercer than the previous nights, as was the artillery and heavy mortar fire. He probably suspects new preparations and attacks... German panzers are on the Volga north of Stalingrad. Our soldiers heard this news with deep pleasure and satisfaction. We helped with this. Our heavy losses have not been for nothing!

In a bold strike across the land-bridge between the Don and Volga Rivers, the armoured spearhead of 14th Panzer Corps reached Mother Volga north of Stalingrad and halted river traffic. News of this stunning blow swept through the ranks of the Bodensee Division. Soldiers fight for pieces of land, see friends die in the process and often know nothing about their role in the overall plan; now, the soldiers received feedback on why they were struggling to hold a small blood-soaked bridgehead. They had attracted Soviet attention upon themselves, thus dispersing enemy strength and enabling the panzers to punch through to the Volga in one go. For the moment, even to the higher commands, it appeared a decisive blow had been struck.

On his second anniversary as an officer (1 July), Beigel was officially promoted to Oberleutnant, although he did not receive notification until mid-August. Months earlier, he had decided to become a careerist, as he explained to his girlfriend:

The regulation has now finally appeared that reserve officers can register for training as an active officer. I'll register because I've already sacrificed five years of my youth to the military. Should I simply throw away those years? My seniority-in-rank will be set so that my Leutnant's commission is moved back half a year. If everything goes to plan, I'll soon be an Oberleutnant. I think you'll agree if I register myself now. I'll then at least have a secure future and enjoy being in the army.

All relevant paperwork was filled out, character checks made by officials back in Germany, and Beigel signed his life away: "In the event of my acceptance as an active officer, I am willing to commit myself to an indefinite period of service." His best friend in the battalion, Leutnant Buchner, lodged his application the same day. When Beigel was notified of his promotion to reserve Oberleutnant, he was ill-equipped to display his new rank, so he asked his sweetheart for help:

Would you mind supplying me with some pips? I only received two pips from the commander[6] (who wore them earlier as an Oberleutnant himself) and that is naturally too few. I still need eight more. You know what they look like: gilded and square.

Gefreiter Harald Baumann from 2nd Company also signed up for twelve years of service at this time, a process that inadvertently led to Grimm experiencing another run-in with Major Beismann. Decades later, when Baumann sought confirmation of his career, Grimm had no problem recalling the details as the day was still clearly burned into his mind due to the trouble with the battalion commander:

I had an unforgettable bout of resentment towards him on this day. When I went to the battalion command post to receive orders, I was accused of ignoring his instructions that I should lead the company back to our supply train in the Don hills.

Circumstances had once again conspired to cast Grimm in an unfavourable light: he never received the order, but Beismann did not believe him:

I was unable to lead the company back to the supply train before nightfall on the evening of 22 August because I had to submit Baumann's personnel files – as well as an appraisal of him – to Battalion. Because of enemy action, delivery to my command

---

6   Beigel is referring to Beismann, not Grimm.

post (a foxhole) was delayed. I therefore got the platoon leaders to spread out their men and echelon them deeply up in the Don hills, which caused no casualties. I had to wait a long time and was able to finish my paperwork by candlelight.

Such administrative matters could not be avoided, even in primitive field conditions. With his papers in order, Gefreiter Baumann was called away from his comrades:

I was summoned to the company command post, where Grimm informed me that I would head to Dessau-Rosslau for an Unteroffizier training course and that my request to become a career soldier – which I had submitted to Ludwig Beigel – had come into effect. I then had to go at short notice directly to the battalion command post to see Major Beismann, who had arranged it. After a brief talk with Major Beismann and the preparation of a permit, I went back to the company for the last time in Russia and handed in my things – I was a Gefreiter and machine-gunner at the time. That evening I was again ordered to battalion staff and promoted to Unteroffizier by Major Beismann and then handed my marching orders and the detachment to Dessau-Rosslau. I then went directly from there in a car, etc., via Kharkov, Berlin to Dessau-Rosslau and was there until 31 January 1943.

By signing on to become a full-time soldier, Baumann would miss the dramatic events ahead. At one point in the coming months, the Battalion would request his return in writing, but the pioneer school flatly refused.

– – –

The hotly contested Ostrovskiy bridgehead was a sideshow now that the panzers had punched a thin corridor through to the Volga. Ferrying operations for heavy loads were discontinued, so on 24 August Major Beismann was ordered to dismantle them and the landing stages on both sides of the river. Grimm was responsible for leading forward the empty trucks to retrieve the metal bridging components and pontoons. Working under heavy harassment fire, the pioneers swiftly broke down the ferry and docks, heaved them aboard the trucks and dispersed into cover before a random artillery round hit home. At 1600 hours Grimm and his company received the mission to reconnoitre a bridge over the Panchinka Creek in the bridgehead while simultaneously making the necessary preparations to throw a bridge over it. A run-of-the-mill pioneering task in a dangerous sector.

Grimm and his company crossed the fog-shrouded river and marched to their worksite at the Panchinka Creek. A B-Gerät bridge capable of bearing ten tonnes was already in situ but its two valuable trestles were urgently needed elsewhere. Grimm's men built a completely new bridge from wood, fifteen metres long, supported by two piers and able to bear eight tonnes. A fifteen-metre long fascine causeway was also relaid. Then the trestle bridge was taken apart and driven away. The work was interrupted a few times by haphazard mortar and artillery fire, but all in all, it took ten hours. The company returned to their billets on the high ground west of Ostrovskiy, always a relief after being in the target zone down near the river. Oberleutnant Beigel almost looked forward to seeing an impressive show from his box seat:

> My tent is sitting in a forest. The view is fabulous. If bombers come during the night and our light anti-aircraft guns open up with tracer ammunition, then there'll be a spectacular fireworks display, but it's even more amazing when the Stalin organs chime in with their projectiles. Then the sparks really fly. It is agreeable and fantastic to watch, but far less so to be caught under it.

Beigel tried to view the scene through his girlfriend's eyes. What would she think? Many German soldiers believed they were in Russia to safeguard their country and their womenfolk: men are to protect, women to be protected. Beigel followed this train of thought for a moment before pulling the emergency brake:

> I want you to be with us for a few days so that you can see us in action. But no, it would not be for you because there are certain moments and deaths; the fighting is so dreadful that a German woman should never be permitted to see it. So please remain at home and imagine that I'm always cheerfully lying in my tent and nowhere else. It's not always bad. Sure, every now and then it is, but mostly it's easily bearable. We don't have much further to march. This year we'll certainly see the Volga, the large river that is often sung about. Then we'll go into winter positions [unless] we are pulled out (this rumour does the rounds when there is a quiet day)...

Gefreiter Füssinger from 1st Company spent most of his time behind the lines but he occasionally ventured into hellish Ostrovskiy and heard talk about the campaign finishing once they reached the Volga:

> In recent days we had to build a bridge over the Don. Unfortunately, that didn't happen without casualties. The Don is very, very wide [...] and the Russians fired mortars, artillery and planes all night and for days on end. I was down there, in the village, for one night and one day. Then all hell broke loose. There was no talk of sleep because every moment you thought: you're in it now. I've certainly already

seen a lot since coming to Russia, but not a night like that. It should be at an end, but I see it coming, that we will ride sleighs on the Volga this winter...

Today we must give up everything. Since May, no bed, no roof over our heads, often only the bare ground as a place to lie, not even straw. No music, no movies, nothing that brings joy, only every day in mortal danger. The struggle here in the East last year was certainly hard, but other comrades who were there at the time say that it's even worse now because Stalin gave the order to not move back another inch. And so the Russians are very stubborn and you have to drag each one out of his hole. And then the advance continues.

Divisional staff had the impression that the Russians were no longer concentrating all forces against the bridgehead, and in fact, were already sending units south. They were correct: the land-bridge between the Don and Volga was attracting Red Army forces and as a consequence the positional fighting on the Don began to quieten down. A holiday atmosphere prevailed on the Don. Off-duty soldiers stripped off sweat-starched uniforms, relishing the sensation of the sun and fresh air on their bodies as they washed clothes and bathed. Eager vermin hunts were carried out and countless victories reported. Some went fishing with inflatable boats and hand-grenades (they brought in an admirable haul). Every now and then, however, a shell crashed into the blissful setting. Grimm's pioneers rested in the forest-capped hills. Small teams brought back fresh water from the river. Everyone caught up on personal hygiene and housekeeping. Grimm wrote out recommendations for the Iron Cross. His HQ section processed claims for bestowal of the Assault Badge, eventually submitting forty-two names to Division. One of those was Grimm himself: "Naturally I have earned the Assault Badge and in the next few days I will lodge the corresponding proposal, so that it can be awarded to me." Paperwork completed, he penned a letter to Lotte:

Last night I slept really well. Nuoffer dug me a beautiful hole and put straw in. On top of that went my rubber groundsheet, blanket and sleeping bag, and I covered myself with my coat. I closed the hole with the tent-quarter. At night the earth trembled after bombs were dropped and I went straight back to sleep. A foxhole makes you feel safe. Hopefully I can also sleep again tonight. It's now noon and I had a lot to attend to this morning. You know, it's honestly good for the nerves when the whistling of bombs and shells is not heard every minute and you don't have to look for cover on dead-flat ground.

As soon as the sun disappeared behind the high western bank, the peaceful scene on the Don changed. Groups dispersed. Everyone moved closer to their foxholes. When the first Soviet plane droned overheard, there was not a soul to be seen. Soon there was crashing from all sides. Bombs came in thickly. Interspersed amongst the whistling was the howl of Stalin organs and the earth-jolting impacts of heavy artillery. The pioneers in the hills had a grandstand view.

On the evening of 28 August, the Bodensee-Dvision was ordered to abandon its foothold east of the Don as the overall situation demanded it: with a viable Don crossing further south, the Ostrovskiy bridgehead no longer had any significance and every available unit was needed for the advance on Stalingrad. It was a bitter pill to swallow. Evacuation of the bridgehead was to be kept secret from the troops for as long as possible. Units deployed in the forward line would only be notified at the last moment. There was little time to waste. Vehicles, horse-drawn wagons and everything no longer absolutely necessary needed to be back over the river by 0500 hours the following morning (29 August) because that was when pioneers would start dismantling the bridge. Anything left in the bridgehead would be ferried across or sent on a lengthy detour via the Panchinka bridge to the Don crossing at Akatov. Deadline for the complete evacuation of the bridgehead was dawn on 30 August. Oberleutnant Grimm was one of the first officers in the division to be informed about the abandonment. Just after noon on 28 August, his company took over ferrying operations with two inflatable rafts and an assault boat. The pioneers were destined to play a major role in the final act of the Ostrovskiy bridgehead.

The serenity of the night of 28-29 August was unnerving. Soviet forces were absolutely quiet. A steady stream of traffic flowed west across the bridge. Grimm's men moved in the opposite direction. They crossed the Don before dawn and entered the bridgehead with a long "to do" list. As the division's new forward line would run along the towering western bank, it was decided to convert the soon-to-be-Russian lowlands into a killing zone. Together with other pioneers, Grimm's company thinned out the trees, manufactured fire lanes, laid devious minefields and installed a dozen tripwire anti-personnel mines. An S-mine detonated while its trip-line was being set but nobody was injured. Two booby traps were also rigged. Other pioneers laid all sorts of nasty surprises about the area. Nothing of use was to be left to the enemy. All dug-outs were levelled and filled in. Russian bunkers along the Don were blown up. The Ostrovskiy bridge was completely dismantled, its lumber and bridging equipment hauled away to supply depots. The entire position was thoroughly cleaned out and large quantities of ammunition and Russian weapons dumped in the river. Red Army soldiers offered no interference, but dense fire from big guns and heavy mortars sporadically thumped down. Grimm lost two men to severe wounds, another was hospitalised with light injuries. The company then returned to their billets and were told a long rest was on the cards.

Over in the bridgehead, the next phase of the evacuation was implemented. A few hours before sunset, the last designated infantry battalion pulled back from the front in waves and filed down to the river, where pioneers crossed them over. Rainfall steadily increased. It had been presumed that the Russians would follow up aggressively, but German masking techniques appeared to be working. Apart from occasional artillery fire, nothing else stirred. Each regiment left a reinforced company in the bridgehead to maintain contact with the enemy. Their orders were to pull back as late as possible, the earliest time being dawn on 30 August. At 0300 hours, enemy pressure forced Winzer's reinforced company to fall back to the river. Their evacuation proceeded smoothly. With that, the bridgehead was completely deflated and the operation deemed a huge success.

Grimm's final task for August was to lay mine barriers near the village of Ostrovskiy. As mighty riverside cliffs south of there formed first-class defensive ramparts, the area at greatest risk of Soviet incursion was the village and forested region to the north. At nightfall, Oberleutnant Beigel carried out the relevant reconnaissance and the company prepared the material required for the operation.

To garner more forces for the push towards Stalingrad, 389th Infantry Division would be released by evacuating its Don bridgehead and 305th Division would take over the river front, adjoining 384th Infantry Division on the right. The Bodensee Division had to occupy its new line by the evening of 31 August.

The overall situation seemed positive from the German perspective, and in fact, the end was in sight. Once Stalingrad was taken, the army could go into winter positions and simply hold the line while other armies went after grander objectives in the Caucasus. The capture of Stalingrad had got off to a cracking start when 6th Army's fast troops reached the Volga on 23 August and blocked it, but the initial exuberance rapidly disappeared in the face of bitter Russian resistance. The units of 14th Panzer Corps held the breakthrough wedge against powerful enemy armoured attacks from the north and break-out attempts from the south. The stage was set for an advance by 4th Panzer Army from the south, which was able to link up with 6th Army west of Stalingrad on 2 September.

From General Oppenländer's point-of-view, his division had completed its offensive phase of operations. It was now time to settle down into defence. Stalingrad would be seized by other units while his men moved into winter positions. And not before time.

The pioneers' impression of their foe had drastically altered in the second half of August. No longer was the Red Army soldier scared of the German bogeyman and inclined to retreat at a moment's notice. Instead, Grimm's men came face to face with the Soviet soldier feared and respected for his bloody-minded determination to keep fighting to the death. They hoped the past fortnight was an anomaly.

Final Days West of the Don (1-4 September 1942)

Awake, fed and kitted up while the moon was high, Grimm moved at the head of his men down to the river south of Ostrovskiy. It was unexpectedly cold. The mighty riverbank loomed up almost a hundred metres behind them. Defences atop the cliff were practically impregnable, but it paid to be careful, so the riverbank was being mined. Grimm's men installed twenty tripwire S-mines. Soviet troops over the river did not interfere. The pioneers left the exposed site when dawn began to blanch the darkness. While Grimm and his small staff drew up mine plans during the day, the rest of the company relaxed. Instead of heading out into the increasingly nippy nights, they were able to snuggle up in their tents, have a tipple and a good chat.

"It is now 0815 hours," wrote Gefreiter Vorherr on 2 September, "but still so cool that my fingers are freezing, even inside the tent. It's now already quite cool in the tent at night. No straw can be found in the area and so we lie on the bare ground. What will happen in winter? [...] Hopefully we move into a decent village or to Stalingrad for the winter." The first sampling of cool weather caused many to pin their hopes on warmer billets. Everyone knew that Stalingrad was a major metropolis, but few fathomed the extent of the devastation being wreaked upon it by the Luftwaffe. Nevertheless, any sort of permanent structure, even a partially destroyed one, was preferable to wintry months on the bleak steppe. Meanwhile, the senselessness of holding the Ostrovskiy bridgehead had been spun into a brilliant ploy that completely fooled the Red Army, attracting them like bees to honey while the genuine attack unfolded further south. Even Vorherr, a cynical and shrewd blue-collar worker, was convinced by the line being fed to the rank-and-file:

> [The] bridgehead [...] was a deception for the Russians, to which they committed themselves. They moved up all their forces against it. Then, when they had gone too far to turn back, our troops moved over the Don supported by panzers, pushed forward to the Volga and now stand on the river to the north of Stalingrad. In the meantime, we have given up the bridgehead and now have the Russians in a pincer from which no-one can escape because it has also been blocked off from above. The

bridgehead was only formed so that the Russians threw their main body at it. Now they're right in the soup! The pocket will be cleaned out in a few days. Then it'll be determined where we take up winter positions.

On 2 September Grimm was called to a conference at battalion HQ. Häntzka and Schaate were also present. Beismann did not keep them in suspense: no attacks, nothing too dangerous, just general guidelines for planning and constructing winter positions, a long-awaited moment because it signified an end to their role in the summer campaign. Beismann ordered Grimm to shift his company into the patch of forest north of the battalion command post. The downside of being in close proximity to his commander was compensated in other ways: a night off, access to the canteen – he bought some writing paper, envelopes and a bottle of brandy – and hearing that one of his best friends was nearby:

I learned that Dr. Scheffel was just a few metres away from us. In the evening I invited him over and, together with Leutnant Beigel – who has just been promoted to Oberleutnant – we sat in my darkened car and had a drink. But you know, after three shot glasses of booze, we had to stop because we could not hold our liquor. It was a very nice evening and we talked vividly. We reminisced a lot about our days in France.

The following day, the procurement of materials needed for the winter began in earnest, not an easy task in the sparsely wooded Don bend. Grimm contacted Regiment 577 to discuss construction of positions and sinking of wells. "We started digging wells, in case we are here a few more days," Grimm wrote. "Our successors will be glad about that. I have a dowser in my company, the rods turned down and I looked for water veins. It worked well." In this arid region, potable water was a precious commodity, as were vessels in which to store it, so Grimm was chuffed with a recent acquisition: "A 20-litre jerrican fell off a heavy tank on 15 August… such things are priceless."

The pioneers looked forward to their new assignment. Digging trenches and fabricating shelters was backbreaking work but losing sweat was preferable to blood. Transitioning to the defensive also offered better prospects of furlough or wholesale transfer to somewhere, anywhere, in western Europe. Beigel put his thoughts to paper:

We will construct beautiful positions and install accommodation under the earth… Times have become significantly quieter. Our dream is to be pulled out and sent to France. We'll remain in Russia over winter… The main affair is now behind us. The only question is where we'll move into winter positions. The Russian is no longer very strong. It's doubtful whether he can hold out over the winter. In any case, he'll go hungry.

Beismann recognised Beigel's potential as leader of a unit bigger than a platoon, so his recent promotion to Oberleutnant combined with an expected lengthy period of positional warfare provided the perfect timing to withdraw him from 2nd Company and attach him to battalion staff, first for some on-the-job training, then a stint back in Germany at an officer's school. Beigel accepted the transfer with mixed feelings:

I'm no longer a platoon leader, but am training as company leader for special purposes! I'm very sorry that I can no longer lead my group of men. I like them. We've had a lot of joy together, have buried many comrades and become really tight. We lived very well together. All for one. They were my boys, like a family. Even though they are now led by an Oberfeldwebel, I'll always be with them if something is happening.

Oberfeldwebel Lorenz Locher, Grimm's trusted HQ section leader, took over Beigel's 1st Platoon. All three platoons were now led by seasoned senior NCOs. Assuming Locher's old job was Unteroffizier Baumhauer.

The future looked comfortably sanguine for the pioneers. Alas, it was not to be. The general situation forced 6th Army to enact a major reorganisation. A patchwork of battalions from three divisions was manning the Don defences and buttressing the western end of the long corridor stretching across to the Volga. In addition to extracting one of these divisions for the advance of Stalingrad, the army's staff fully recognised the value of unit integrity, so it decided to return detached units to their parent formations. The Bodensee Division was ordered to relieve 384th Infantry Division, which in turn would take over the Don defences. This new move would place the division firmly on the defensive line strung across the Don-Volga landbridge. Procrastination was not a feature of German war-making, so the reshuffle began immediately.

Divisional staff issued a pamphlet to its units to help them prepare for a prolonged period of defensive combat. The opening paragraph reads: "Defence rests on the repellent effect of automatic weapons. When these weapons are correctly deployed, the effect is insurmountable. Not one case is known where infantry have overpowered a firing machine-gun with a frontal attack." The exposed steppe was the perfect location to implement this advice. Nevertheless, the fact that they were holding position was no excuse for the troops to be confined to their positions. "While on the defensive, reconnaissance is frequently neglected. It is just as important as during an attack," reminded the leaflet. Patrols must monitor possible approach routes and maintain contact with the enemy while snipers are kept at bay by German crackshots equipped with scoped rifles. "The more active the troops are with reconnaissance and fire behaviour, the more hesitant the enemy to attack at that spot." The Bodensee Division had never been on the strategic defensive: now, on the Don-Volga isthmus, it was being asked to construct a solid line of defences to be held indefinitely, so-called "winter positions". In this vein, the pamphlet emphasised the importance of sweat equity:

> Every spadeful of earth saves blood. This principle is to be hammered into the troops. The spadework of the Russians shall be used as an example. Most troops believe that their role in positional construction is done after they excavate a rifle pit or tank-proof foxhole. On tactical, but also psychological grounds, the expansion of nests (rifle pits connected by short trenches) is required.

Positions must be continually reinforced, trench walls braced, dug-outs and shelters created, approach- and connection trenches dug to simplify traffic and prevent pointless casualties, and finally, wire obstacles erected in front. Wire entanglements were acknowledged as "vital" because they "provide moral support" but it was "necessary to inculcate the troops that barbed wire does not protect absolutely." All of these tasks fell within the remit of the pioneers, so the future weeks looked set to be a busy time for Beismann's men. For the moment, they had no idea what the future held. They were just glad to spend a quiet day in their rear billets. Grimm summed up the mood:

> It's been pretty quiet in recent days and we have remained in the same place. Enemy activity is very low and a blessing for us. Things will liven up very quickly, however. I write when it's bright and early, so as not to be disturbed. Soup is dished out at 1800 hours and a quarter of an hour later you can no longer read a newspaper. So we get to bed early and recoup the many hours we missed during our forced marches… Since we have a radio in the company, we are connected with the world. Sometimes we listen to music and then spirits are buoyed. Soon we'll find a permanent place where we can set up our winter quarters. Then we'll go into Mother Earth. We've already worked out the plans and are waiting to start. Hopefully it happens soon.

During the day (4 September) Grimm handed over the Ostrovskiy minefields, collected all of the company's pioneer equipment and had the vehicles loaded. He was then called to another conference at battalion HQ. Beismann briefly outlined the plan and assigned each company to a specific regiment. "Grimm, you'll be supporting Regiment 578."

On 4 September, 6th Army laid down its "aims for continuation of operations", the crux being the attainment of a winter position between the Don and Volga after the seizure of Stalingrad. A line deemed most favourable yet attainable was set down. All Soviet forces south of this line were to be smashed by an attack with two panzer and two motorised divisions setting off from the area south-west of Yerzovka and moving westward with the right wing following the proposed "winter" line. In this way, all Soviet units east of the Don would be rolled up from the east. Then, four to five infantry divisions would shift northward and occupy the new winter position. This operation was scheduled to begin ten days after Stalingrad was taken.

Warm Welcome (5-10 September 1942)

The finely synchronised exchange of 305th and 384th Infantry Division's units continued during the night of 4-5 September. The old sector of 384th Division now held two of the Bodensee Division's regiments, and vice versa, so at 0800 hours both divisions assumed command of their new sectors. And not a moment too soon because the Soviets gave the newcomers a warm welcome. Mid-morning, with the aid of tanks, they charged both wings. More assaults rolled in against other sectors further east. The attacks against Winzer's line repeatedly struck the Kulturstation (cultural centre) and Point 73.2, where Winzer had set up his HQ in the old command post of Regiment 535, but the thrusts were repelled. Grimm's company missed these attacks but had a clear view as they marched along the Don High Road to Vertyachii. Under fine skies occasionally swirled with light cloud, his soldiers trudged along the dusty plateau, watching the battle down on the other side of the river. The racket had died down by the time they descended to the river crossing. Dusk was falling when they tramped across the bridge. Hooves clip-clopped on the wooden planks. They reached Vertyachii at nightfall. This large village – soon to become the division's nerve centre – reinforced the soldiers' prejudices. Gefreiter Füssinger's expressed what many thought:

> One thing Russia has taught us: better to be just a worker in Germany than a landowner or any other citizen here. Our cottage is a castle compared to the houses here. If only the people knew how poor they are! It is a populace in rags and with dirt and lice, no question of roads. There's hundreds and thousands of people who barely saw a car before the war and have never been in a city. Everything is at least a hundred years behind.

While resting in the village, Grimm was called to the phone. The situation up front demanded immediate deployment of his company. The sounds of battle heard all day had been emanating from Oberst Winzer's sector and a few holes needed to be plugged. Grimm was ordered to reach the Point 73.2 area but leave his supply train in Vertyachii. The night march was strenuous and difficult because the route lacked signposts for Regiment 578 and at times passed through calf-deep sand and across open fields. Twelve gruelling kilometres later, the totally exhausted men moved into position to close a gap in the line and began digging themselves in under the Russians' noses. Grimm was just happy to have arrived:

> I was incensed by the guide on the motorcycle and I certainly did not wish him well. I knew that the main combat line was not one solid line and there was a danger that we might slip straight between two strongpoints directly into enemy hands. The sand near the Don caused me grief. Would our machine-guns jam if we bumped into the enemy? There were no roads to lead us to our objective. In any case, we made it and I was able to report to Oberst Winzer and his adjutant, Hauptmann Schwarz. It was refreshing to know that you and your men are valued. When I reported our combat strength with twelve operational light machine-guns, Oberst Winzer reckoned that there would be no breakthrough. From Hauptmann Schwarz I learned that the guide who failed us was 'sleepyhead' M. and I was absolutely stunned to discover that he was a fellow Kirchheimer whom I had known in my youth, Edie M.

> We laid out our foxholes and light machine-gun positions about two hundred metres in front of the regimental command post. Before my foxhole was ready, I already had a telephone connection.

During the night, word filtered through to Grimm that Hauptmann Häntzka had been severely wounded during the Soviet attacks. The news also reached Beigel at battalion staff:

> A report came in from the front that the commander of 3rd Company was wounded and will not be returning to his company. Now I might rejoin the ranks as a company leader. It's up to my commander if he'll let me go or not because a Hauptmann has got away from the staff, so that means there are only two pioneer officers on the staff. Well, it's all the same to me where I go. I'd make a good company leader, even if I don't have a specialised commander's school behind me.

Fritz Häntzka died of his injuries later that night. His wife Alice and 12-year-old daughter Ursula were absolutely devastated. Worse was to come: British bombers attacked Duisburg in mid-May 1943 and levelled large parts of the old city, including the Häntzka home. Alice's parents were also killed. Almost 100,000 people were left homeless. In September 1943 Alice and Ursula moved across the country to Bockwitz near Dresden, Fritz's birthplace, for familial support, moving in with Fritz's sister and her family. They tried to continue their lives but grief inexorably crushed them and contributed to a tragic domino effect. On 24 April 1944 Häntzka's wife and daughter, sister Frieda, her husband and three children all committed suicide. Their bodies were found the following day.

For the moment, leadership of 3rd Company was entrusted to its senior platoon leader, Leutnant Hepp. The company was pulled back into reserve positions behind Regiment 578, Schaate's 1st Company was still west of the Don, so only Grimm's 2nd remained up front deployed as infantrymen:

Taking up a defensive positions in a few hours at night in unknown territory, in addition to awaiting an attack at dawn, mobilised our last reserves of strength and caused sleep to be forgotten. I was able to locate a few foxholes in the defensive sector allocated to us. Who had dug them was beyond my knowledge. In any case, the company had to be dug in before dawn. The expected dawn attack did not eventuate. In the first rays of light, every man looked for the enemy out in front of his defensive sector. Eternity was seen when looking into the steppe, like looking from a beach at the flat open ocean. The distant ground appeared to be dead flat. Not a tree or a bush to be seen, the steppe grass was withered and almost dried up. As a result, it was not possible to adjust target and range data by eye because the enemy wasn't visible. During our approach march in the night, we noticed that we crossed several shallow depressions when the sandy soil of the Don valley was left behind us. Running through the steppe grass like short stretches of earth was dried-out knee-high scrub, similar to our 'wild carrots'. Such dark brown coloured patches were perceived as small islands in the forefield. Once you left an [elevated] observation post, however, these dark patches disappeared from view. The enemy must therefore have been preparing his attack in the shallow depression lying opposite us. We were later able to establish that such shallow depressions meandered their way through the arid steppe like river courses.

As the sun rose on 6 September, it burned away the milky haze, revealing a beautiful day with occasional light cloud. The hours ticked away. No sign of the Russians. Then, shortly before 1000 hours, the noise of battle swelled in the neighbour's sector and the pioneers heard the rattle of tank tracks from enemy lines. Right on 1000 hours the Soviets hit their positions with artillery and heavy mortars. An artillery observer called from the company's left wing: "The Russians are attacking!" Grimm used one hand to press the phone against his ear and the other to keep binoculars up to his eyes. The battleground was laid out before him:

As the field-telephones were interconnected, I overheard the reports to and orders from Oberst Winzer, and also the orders to the heavy infantry weapons. I immediately recognised the impact – that is, the detonations – of the shells from our heavy infantry guns because these burst about twenty or thirty metres above the ground and produced a massive shrapnel effect.[1] (I understood nothing about artillery, I think someone said they were firing with a raised focal point). In order to save ammunition, the order to fire would only be given to my company when the mass infantry attack was clearly discernible to the entire company. Unbelievable anxiety gripped my men because it took a long time for the enemy – working his way towards us in dense waves in the gently undulating terrain and under cover of the knee-high shrubbery – to be visible to the entire company.

The pioneers had never faced such an attack before. The inexorable approach of the earth-brown figures was unnerving. Soldiers on the left end of the line watched the Soviets grow larger in their sights. Tension knotted their stomachs. When could they open fire? Those on the right sector could see nothing because of the cross-slope. As their heavy weapons fired into German lines, the Red Army soldiers worked their way through the shrubby grass, eventually coming to within four hundred metres of the company's right wing. Only now were they visible. Grimm gave the order: "Open fire!":

The concentrated fire of the entire company had an undreamt-of effect because, in my opinion, the enemy was not expecting it. I was in shouting contact with Unteroffizier Trost, I want to say clearly here. Fairly early during our defence, I received the call that Unteroffizier Trost was wounded, then shortly after that he

---

1   Although the Germans did not have proximity fuses (an Allied invention first used by field artillery in late 1944), they were able to create low airbursts by using "double action fuses", a combined variable time fuse plus impact fuse (the impact fuse was a back-up if the timed fuse did not work).

was dead. Our regiment commander, Oberst Winzer, ordered me to bring the fire
of our heavy infantry guns closer to our position because I was now the one best
able to observe the battlefield.

The heavy artillery barrage combined with the company's dozen machine-guns reaped a fearful toll and the Russian attack disintegrated. At nightfall three deserters confirmed severe casualties. Grimm lost one dead (Unteroffizier Karl Trost), one severely wounded and one lightly wounded. "After the successful defence," recalled Grimm, "Oberst Winzer expressed his complete respect to our company for our infantry performance. I learned that a battery of 8.8 cm Flak on self-propelled chassis – far to the right of us – had shot up three tanks, setting two of them on fire." Grimm was not quite correct in his description: the guns, known as "Sturer Emil" (Stubborn Emil), were fearsome 12.8 cm tank killers mounted on a fully-tracked and armoured chassis.

General Oppenländer ordered his regiments to hold the line. Due to the lack of natural defences, the 305th Pioneer Battalion was to consult sector commanders and explore the possibility of laying mines in front of especially vulnerable positions. Subsequent requisition of the necessary mines was to be expedited so that the two pioneer companies subordinated to Regiment 578 could begin promptly laying mines and then be released. Bunkers also needed to be built. Sector commanders would direct requests for construction material to the pioneer battalion, which was responsible for procuring material for the entire divisional sector. With that, the battalion's mission was set for the next two weeks.

The last infantry battalions of the Bodensee Division arrived east of the Don during the night of 6-7 September. The ensuing day was very quiet for Oppenländer's men. Not so for the right neighbour. Powerful Soviet attacks rolled towards the defences of 76th Infantry Division throughout 7 September, but all attacks, including those supported by tanks, were thrown back in costly fighting. Grimm's company remained in its positions, sniping at individual Russians who approached the front-line. Oberst Winzer visited Grimm and the pair crawled on alone to a forward post. As they scanned the terrain, Grimm pointed out certain features, emphasising the hidden depression where Soviet troops had formed up and were now positioned. Both officers returned safely from their reconnaissance. Grimm curled up in his foxhole "like a small hare sitting in its burrow" to write a letter: "The Russians have just flown over us, a number of bombers with strong fighter cover. The bombs fell elsewhere. A Russian sniper is harassing us at the moment." Just then, a messenger poked in his head to tell him that he was wanted at the regimental command post. Winzer called a meeting at 1700 hours to discuss an attack planned for the following day – and Grimm's company was to play the lead role. The previous day's unexpected mass attack had unsettled the division's senior leaders because Soviet troops had assembled without being detected, and if the unseen depressions were not cleaned out smartly, a fresh assault was sure to follow. Grimm describes how this nocturnal reconnaissance-in-force would be implemented:

> We had to advance about 400-500 metres towards the enemy on the right wing of the *Radfahrschwadron*, then move parallel to our position across to the company's left wing. There, turn to the left along the entire width and head back to our foxholes. We therefore had to push through a long drawn-out depression in which the enemy had assembled for his attack. It was assumed that we would bump into tanks, so we took along magnetic hollow charges for the first time.

This operation was the battalion's inaugural deployment of the "Hafthohlladung," a powerful magnetic anti-tank mine. It gave the pioneers a weapon with which to combat tanks, as Grimm explains:

> Prior to the Don, the only means available to take out tanks "in close combat" were hand grenades, pioneer explosives – which could be combined into a demolition charge – and anti-tank mines. We had heard about this close combat weapon a short while before and it promised a great deal. As a forerunner to the Panzerfaust, the hollow charge mine would soon fade into obscurity. I want to describe it from memory. It looked like a funnel with an outer rim diameter of 20cm and a round pipe of about 5cm diameter, similar to the hand grenade shaft, in which the fuse was encased. Attached to the outer rim of the funnel were magnets, which kept the charge attached to the steel [of a tank]. Someone had discovered that detonating explosives in a certain hollow configuration achieved a weight reduction on one hand and a focusing effect on the other (like a magnifying glass) which pierced strong steel plate and had an unexpectedly destructive effect on the interior of a tank.

After being briefed by Winzer, Grimm returned to his company, summoned his platoon leaders and told them to get their men ready. There would be no sleep for 2nd Company during the night of 7-8 September:

> At 0100 hours details were ordered, no speaking, no shooting at fleeing enemy soldiers, no lights, no cigarettes, count your steps, march compass, note information. The North Star was clearly visible and came to our aid yet again. The likelihood that we would be back in our positions before dawn was extremely good.

The company left its foxholes at 0200 hours and assembled in the dark. The aim was simple: outflank the Russians from the east and comb through the shallow depression. Grimm had overall control of the attack. Light and heavy infantry guns, as well as a battalion of heavy artillery, were on standby, ready to lay down fire if needed. A signal was given and the troops moved out. Dry grass crinkled underfoot. Twigs snapped. Several halts were called but the night remained silent. At the pre-determined point, they turned west and began moving parallel to their front-line. Everyone was on edge. Darkness posterised everything into a few shades of black: only the sky and ground could be clearly differentiated. In the lightless steppe, hearing was the dominant sense, although the nose sometimes came into play. Non-smokers caught occasional whiffs of machorka, the pungent Russian tobacco whose odour lingered for hours, but enemy soldiers were nowhere to be found. It soon became clear that they had swiftly pulled back to avoid being outflanked. Grimm's company reached the attack objective but came away empty-handed:

While pivoting back towards our position, day broke and the sun was up in a few minutes. We had to move through a very heavy Russian barrage of mortar and machine-gun fire. About eighty metres from our positions I bumped into a heavy Russian machine-gun, ready to fire with crates of ammunition. A moment later – it was unbelievable, we were still under heavy fire and had not yet reached our positions – Edelhäuser crawled over to me and showed me his right hand with the index finger shot off at the first knuckle with an almost elated cry that I could hear despite the ear-piercing sounds of battle: "But Herr Oberleutnant, now I can go home and get married!"[2]

More casualties were suffered despite a fierce German counter-barrage. Gefreiter Hans Sparrer recalls:

Karl Simmel and Georg Schmitt from seventh squad of 3rd Platoon were hit shortly after the surprise artillery barrage on 1st Platoon. I believe Schmitt was able to move himself out of the firing area. Karl Simmel, who was moving behind me, had an injury to his thigh. Initially I carried him on my back, then a second man helped carry him on a rifle, then he was bandaged by Sanitäts-Gefreiter Michael Stampfer once we found cover.

Simmel remembers his wounding:

---

2   Gefreiter Adam Edelhäuser did not return to the unit as a convalescent.

I was shot in the right thigh during the night attack and Hans [Sparrer] and another
comrade carried me back to the dressing station. Georg Schmitt, also from 3rd
Platoon, was likewise wounded (shot through the calf). Georg and I stayed together.
We went into hospital in Kharkov on 16 September and from there further back
where we landed in a reserve hospital in Proskurov on 22 September. At last a bed.
We felt like we were in heaven. We'd both had dysentery for months and after eight
days, because of our fevers, the doctors diagnosed malaria. Now we would survive.

The most serious case was Obergefreiter Georg Mark,[3] leader of the third squad in
1st Platoon, who was struck mid-thigh by a bullet. Blood pumped furiously from his
torn right leg until a tourniquet was tightened and the flow staunched with dressings.
He was stabilised at a dressing station, but once in a major hospital in the rear, the
decision was made to amputate the leg in order to save his life. Just under a year later,
sufficiently recovered from his wound, he sent a letter to Grimm:

Healthwise, I'm going well, although I'm having a lot of trouble running, namely
because I received my prosthetic leg a few days ago. Jedes Pionier ein Behelfssoldat![4]
Although it has hit me hard, leaving me with no more than a 30cm stump, I'm very
disappointed that I can no longer be with my comrades of 2nd Company.

Once back in their defensive positions, a headcount showed that a man from 1st
Platoon, Gefreiter Josef März, was missing. Subsequent enquiries revealed that he had
died: his comrades saw him taken down by a huge explosion, presumably a direct
mortar hit. Casualties from the nocturnal sortie were therefore one dead (März), three
heavily wounded (Mark, Simmel and Schmitt) and one lightly wounded (Edelhäuser).
Oberfeldwebel Locher informed Grimm that Gefreiter März was carrying one of the
brand-new magnetic mines and it had not been accounted for. The use and loss of every
hollow charge was to be immediately reported to Battalion. Grimm felt ill: entrusted
with a new secret weapon and one was already missing, perhaps even in enemy hands.
He was anxious about reporting it. "Locher, get back out there, find März and the
hollow charge mine." The patrol found neither. At 1700 hours a second patrol, again
commanded by Oberfeldwebel Locher, headed out under heavy artillery cover. The
Russians returned fierce fire and laid down a smoke screen. Locher's investigation
showed that the magnetic mine had detonated and taken März's life. This news sent
shivers down all their spines:

---

3   No relation to the author.
4   Translates as something like "Every pioneer is a makeshift / substitute soldier!"

The explosives carried by us, including hand grenades, demolition satchels, magnetic charges and mines, all consisted of so-called "safe to handle" explosive. This could only be caused to explode with a fuse or a blasting cap that reached a temperature of about 300° C upon ignition. Well, we had already suffered very heavy losses during previous engagements on the advance because when the explosives were struck by a shell or explosive projectile, indeed, even a piece of shrapnel, they detonated.

The effort to recover März's remains and investigate the missing charge resulted in four more men being wounded, one mortally. "Oberfeldwebel Locher, my courageous Lorenz Locher," Grimm remembered, "was himself wounded by shrapnel, and the loss of such a platoon leader hit the entire company hard." Even though he had taken over Beigel's group just four days earlier, Grimm considered Locher to be his best platoon leader. After learning that Locher was injured,[5] the company's Spieß,[6] Hauptfeldwebel Jakob, reported to Grimm and asked to lead the platoon. Jakob was a valuable member of Grimm's staff because he was "able to carry on the paper war so that I could lead the men." Nevertheless, Grimm acceded to Jakob's request and gave him immediate leadership of 1st Platoon. Assuming responsibility as Spieß, "mother of the company," was Unteroffizier Franz Dittenhöfer, a mountain of a man with the hulking presence of a heavyweight boxer. He was no rear-office denizen: he came directly from the front-line to take over the job, wore the Iron Cross tri-colour ribbon in his buttonhole and was owed the Assault Badge, but possessed administrative skills stemming from years of overseeing the family manufacturing business in Nürnberg.

The company also lost its medic, a man who went into action unarmed, and defied gunfire and shrapnel countless times to bind wounds:

Brave Sanitäts-Gefreiter Michael Stampfer was wounded together with the recon team leader during the second mission. He lay in an ambulance and died of his wounds before it departed.

The vehicle bearing the red cross, a symbol of salvation and safety to which he had carried many a comrade, was where Stampfer took his last breath. Total casualties for 8 September were two dead, six heavily and one lightly wounded. Oberleutnant Beigel

---

5   Locher recovered from his wounds and returned to action the following year. He was wounded for a third time in Italy. Ultimately, he survived the war, a grizzled veteran who had seen too much. He died in 1972.

6   The "Spieß" was a senior NCO, who took care of the administrative matters of a company, similar to a Sergeant Major. He was wryly known as the "mother of the company" (in contrast to the commander, who was considered "father of the company").

was dismayed when he heard about the grim tally because he had led many of them, and was especially close to one of the dead:

> Second Company was deployed and suffered a few casualties. My successor, an Oberfeldwebel [Locher], was wounded, as were several men from my platoon. My old orderly, Gefreiter Josef März, was killed. There are always painful losses. My old platoon was hit particularly hard. I am still on the staff and will probably get 3rd Company in the next few days. Their old commander succumbed to his wounds in the meantime. It's an eternal shame about our good old Hauptmann Häntzka. I'll be quite happy if I'm allowed to take over the company. It has a really excellent reputation within the division. If we didn't have so few officers in the battalion at the moment, then there'd be a possibility that I'd go to the Kompanieführerschule [company leader's school] in Rosslau and then to France... My battalion commander would have certainly sent me because I want to remain a soldier.

Military operations marched on, so Grimm had little time to lament the latest fatalities. One platoon was dispatched at nightfall to lay fifty anti-tank mines while the rest of the company assembled behind Winzer's command post as a reserve. Grimm was then confronted by a task that filled him with dread: informing Beismann about the lost magnetic mine:

> The report went to Battalion on 9 September and our commander, Major Beismann, called me, raging, and I was called to account because of the loss. Because I had this conversation in the presence of Hauptmann Schwarz at the regimental command post, I reported the same thing to the regiment commander, Oberst Winzer. It must have backfired, as they say, because Major Beismann never raised this matter again.

The tension between Beismann and Grimm is exposed by this post-war report which, together with occasional asides and a "between-the-lines" undertone in other correspondence, reveals the deep rift that existed between the two men. Grimm saw definite proof of Beismann's attitude in 1944:

> During high mountain operations in southern Italy with the new battalion, various letters from Frau Beismann arrived that did not please me personally. Nothing could be done to change her false point of view, which she could only have got from her brave husband.

These uncomplimentary sentiments about Grimm have unfortunately not survived. It is not known for certain when their relationship deteriorated but by September 1942 it was at rock bottom. Beismann lacked complete trust in his senior company commander, while Grimm felt his boss had it in for him, even though he did everything asked of him. It was little incidents that tainted an otherwise exemplary performance. In addition to problems with a superior, Grimm also had issues with a subordinate, although he was unaware of it. Beigel's unspoken rivalry with Grimm flourished once he was no longer under his command. Ever the keen socialiser, Beigel spent most spare moments during this hiatus chatting with fellow officers:

> Yesterday I had a chat with Hepp (Leutnant Hepp is a platoon leader in the company). We sat in a darkened car for a long time. Because the yearning for home was awakened, the big desire returned, as so often happens, to be allowed to go home, even if only for a few days. I don't know what's happening with furlough. Our adjutant is away.[7] Then comes Grimm who has not been home for over 14 months. Who comes next, I don't know exactly. Probably the commander. But at least there is furlough and that's the main thing...

The only officer he did not fraternise with was Grimm. All their interactions were business-related, as is clear from the way he signed off a letter:

> Now I'll go see Schaate. You know him. He's been leading 1st Company since the earlier commander was wounded. He's in the same gully, not far from me. And then I'll go to Grimm, but only because of the mail. I haven't received any for a long time. They're still going to the old address!

– – –

Paulus' army had gained operational freedom to move forces between the Stalingrad front and the Don-Volga front and secured suitable starting positions for the assault on the city proper. Ongoing attacks against the northern defensive front continued to gain strength but plans were in place to rectify the situation. Once Stalingrad was taken, the army would recuperate for ten days before launching a limited offensive north into the land-bridge. Two proposals were on the table: a "medium solution" and a "small solution," both with the aim of forming a solid blocking position on favourable terrain

---

7   The adjutant, Max Fritz, would be away for a long time, at least two months, because he did not return until 7 November. Such a long furlough was very unusual in the German army. He lived with his parents in the centre of Stuttgart, a town frequently bombed by Allied planes, so it is possible he may have received extended leave to assist with repairing some sort of bomb damage.

north of Stalingrad. The differences between the two plans were the depth of thrust, the final position of the line and the number of divisions involved. Everything depended on developments in Stalingrad. When Gefreiter Vorherr finally rejoined the company on 9 September, it is evident that he was swiftly and accurately informed about the army's intentions, a sign that this plan was not kept secret from the rank-and-file: "I believe [our current] positions will be held until the troops around Stalingrad are freed. Then we will again move a bit further north."

By dawn of 9 September, Grimm's company had moved into a hollow 1200 metres south of the main combat line and began setting themselves up for a long stay. The five-metre deep "hollow" was in fact the Nizhny (Lower) Gerasimova Balka which meandered in whorls, roughly from east to west, like a slow flowing river with a flat road-width river bed and gently sloping walls. Oberst Winzer kept Grimm's men within easy reach – about two hundred metres behind his regimental command post – so he could deploy them instantly as a "fire brigade" during an enemy breakthrough. The main purpose of Grimm's long-term attachment to Regiment 578, however, was engineering tasks and Winzer wasted no time in putting the men to work. Grimm was summoned to a conference to discuss the installation of new minefields. Accompanying him to every meeting of this type was Unteroffizier August Baumhauer, his HQ section leader, who was responsible for creating the mine plans. Baumhauer was a structural engineer and architect by profession, so drawing accurate maps – upon which men's lives depended – was child's play. He accompanied the mine-laying teams into no-man's-land to record the precise position of each lethal disk. At nightfall, under heavy enemy fire, the pioneers laid four minefields with a total of fifty mines.

The 10th of September was a red letter day for Oberleutnant Beigel: he was finally given control of a company, an independent command and the smallest building block of the mighty Wehrmacht:

> I took over the bicycle company [3/305th Pioneer Battalion] at midday... My new group of men are really great together. The company is on bicycles. Apart from those, I also have seven lorries, three motorcycle-sidecars, two solo motorcycles and a passenger car. Leutnant Hepp, my earlier comrade in 2nd Company, is a platoon leader under me. We get along really well. I know most of the men from France. When the ball gets rolling again, when I have settled in, then it'll definitely be great.

Grimm was sorry to see Beigel go: "I've had to let go of my Oberleutnant Beigel because he will now lead a company. Thus I have no more officers."

Grimm was summoned to Vertyachii in the afternoon by Major Beismann for a company commanders' conference. He had not seen his commander for five days and was expecting a roasting over the magnetic mine incident, but it never eventuated. Relations were strained but cordial. Before the meeting, Schaate and Grimm congratulated Beigel on his new appointment. The three men knew each other intimately. Schaate and Grimm served together as new recruits in Pioneer Battalion Ulm in 1934-1935. They were of the same generation, grew up in similar conditions and rode the wave of the nation's resurgent power. The trio understood one another perfectly. Once the meeting was adjourned, Grimm used the opportunity to get his house in order by visiting all rear echelon elements encamped around Vertyachii:

> *The supply train with field-kitchen and MuMa (compressors, power saws, drills, plus flamethrowers, inflatable boats, ropes, etc.) had been brought up by Hauptfeldwebel Jakob, as was my petrol-driven British staff car, which had only been available to me for a few days during the advance. Petrol was so scarce that we were only able to fill up our motorbikes. The supply train was billeted under thick trees just north of Vertyachii in the Don valley. My brave and excellent riding horse, Friedel, with which I had carried out many a tricky reconnaissance, was able to recover there. I did not ride my hand horse (replacement horse) Peterle, with his short prancing strides, very much. Peterle was wounded twice but did not drop out. Futtermeister [fodder NCO] Borcher, with whom I had a particularly good understanding, really cared for our horses. My driver and orderly Nuoffer organised petrol for the Morris so that he could also come to Vertyachii. Thus Nuoffer was once again able to give me a haircut.*

The appearance of his long-absent orderly enabled Grimm to take care of other hygiene matters:

> *We suffered from vermin, clothing lice and fleas, one plaguing us, the other not really being noticeable, even when we were completely infested. There was no chemical means to use against them, only heat, which also destroyed the lice eggs. There was an iron with the supply train but not up front.*

Grimm only had a few hours in Vertyachii, so he was unable to get Nuoffer to scorch the lice out of his clothes, but he did rid himself of another pest:

> *There was water [in Vertyachii] and so I had a golden opportunity to rid myself of fleas. I placed a bucket of water right next to me, slowly rolled up my shirt and the first fleas were soon itching in the water. A change of underclothes alone did not*

eradicate this plague. A change of uniform had been unthinkable since the beginning of our advance.

While at the supply train, Grimm tried to address one of his men's biggest gripes – food. Vorherr was typically blunt:

You often get hungry and many are never full. The food is not astonishing, so it often makes you wonder how we've made such achievements. Actually, the food is very, very bad, more water than anything else. Everyone grumbles about it.

In the egalitarian Wehrmacht, officers suffered alongside their men. "The rations were not enough to fill us," Grimm remembers, "but the little that we did receive was at least warm. I recall that after a few weeks on the Don I was twice able to eat until I was full, and that was on fish from the Don fried in their own fat, only the garnish was missing and bread was scarce." Grimm had little luck in procuring more rations for his men because supply lines were severely strained and all available trucks were devoted to feeding the attack on Stalingrad. Units in secondary sectors just had to tighten their belts. It was at this point that the postal system gained even greater importance. As a result of being on the defensive and consequently using very little fuel and ammunition, the division was able to keep the mail flowing, including parcels up to one kilo in weight. Families usually sent small comforts, but now, hungry soldiers looked upon the parcels as an important supplement to their rations. Gefreiter Füssinger asked his parents to send something baked, like biscuits, rather than sausage because that was one of the foodstuffs still being doled out by the army. Vorherr's mother sent some white bread but the long trip and incredible heat throughout August had completely dried it out. "I urgently need some sweetener," he wrote on 11 September, "because then the coffee will taste better." Officers were not immune either. Grimm asked Lotte for some cakes or pastries "but something that won't spoil quickly." Beigel was ecstatic when a small collection of brown-paper parcels was waiting for him:

Four of your packages reached me yesterday. And nothing but good things inside. Every single little thing is a treat. The cigarettes of course please me greatly. This new drink will be fantastic. I will make some tea from it today. Our tea's not very good and the coffee is not much better. When I look at myself now, then I must say that I'm becoming damn thin. When I rub my hand over my chest from time to time, I feel nothing but bones. There's no longer any trace of my stomach.

While wandering around Vertyachii, Grimm recognised some familiar faces, but such meetings during war-time were always bittersweet, as Grimm related to his wife:

Occasionally I see people from my old unit [light pioneer column], and what they say fills me with pride and joy. Some sad news, however, because [Gefreiter Jakob] Bezler from Jesingen, whose wife wrote to you once, was the first man of the unit to be killed [on 4 September 1942]. You know, they are used mostly for supply and rarely have casualties. To me it hurts. You must not say anything to anyone because [...] his wife will get the news later.

Grimm also spotted Vorherr and ordered the motorcycle brought around so he could be driven back to the company. It was common knowledge that the open plains east of Vertyachii were like a shooting range during daylight hours: anyone foolish enough to leave the village while the sun was up was asking for trouble. Supply runs to the front only took place at night, yet Grimm needed to return before nightfall because his company was scheduled to lay more minefields. Vorherr zoomed out of the village with Grimm in the sidecar, trailing a pennant of dust that immediately drew fire. Artillery rounds impacted here and there. A few landed perilously close. The frantic pair gained no benefit from the wall of haze because they outpaced it. Four kilometres of dead flat ground needed to be covered. A shell burst in front of them and they zipped through the cloying dust cloud. After tense minutes, salvation was soon in sight and they disappeared into the gully a few hundred metres from the company billet. After washing away a week's worth of dirt in Vertyachii, Grimm's face had regathered it in the mad dash.

As darkness fell, Grimm's pioneers moved out, weighed down by cases of Tellermines. Accompanying them was Unteroffizier Baumhauer with his plotting equipment, compass and sketch pad. Instructions were to lay a minefield in front of some anti-tank guns. Unteroffizier Josef Lachermeier was the first to begin work. Nearby was Gefreiter Hans Sparrer, the second man to start digging a hole. Just then, a body-shattering detonation shook the ground and Lachermeier disintegrated. According to Vorherr, "nothing could be found of him other than the debris of his wallet." Gefreiter Sparrer was caught by the blast:

To this day it is still not clear to me why the mine detonated. I received shrapnel injuries to the left side of my head, coupled with speech loss and paralysis to the right side of my body.

Sparrer was both lucky and unfortunate: blessed to have survived the explosion but ill-fated as well because this was the second time he had been injured by German mines – back on 30 June, his entire squad was felled when an artillery round triggered secondary detonations in mines being carried.

Two other men were taken down by the mine accident, both soldiers Grimm could ill-afford to lose: platoon leader Hauptfeldwebel Jakob and Unteroffizier Baumhauer were peppered by fragments. Baumhauer's gaping leg wound would leave him trussed up in a hospital bed for a very long time. Casualties were therefore one dead (Lachermeier), one severely (Sparrer) and two moderately wounded (Jakob and Baumhauer). Work was immediately suspended and the dispirited pioneers returned to their billets. Grimm was stunned when a messenger brought the news: "Hauptfeldwebel Jakob was already wounded on his second day of operations and I immediately visited him at the main dressing station, where Unteroffizier Baumhauer also lay, though both were not severely wounded." Cause of the detonation was not determined.

With the wounding of Jakob, Grimm was left with just one senior NCO as a platoon leader, the indomitable and seemingly indestructible Feldwebel Platzer. The 2nd was led by Unteroffizier Pauli, while the 1st was in an even worse position, as Beigel discovered:

> My former platoon has been heavily bled in the meantime. After me, an Oberfeldwebel was platoon leader, then an Unteroffizier, and then a Gefreiter. I moved off with thirty-five men and now there are twelve, the rest are dead or wounded. One of the few visited me yesterday. They still want their old platoon leader back.

Winter Positions (11-30 September 1942)

The construction of permanent winter positions began and many pioneers, including the officers, thought their offensive role in the 1942 campaign had finally ended. Beigel's 3rd Company moved into a forest to prepare lumber for the division. The diary of 2nd Company details the type of activities that now consumed the attention of all the division's pioneers:

> Night of 11-12 September 1942: Laying of two minefields with a total of 48 T-mines under enemy influence (open placement pattern).

> Night of 12-13 September 1942: Installation of 90 Spanish riders under enemy influence.

> 13 September: Laying of a mine barrier with 40 T-mines (openly spaced). Connection of the Spanish riders set up the previous day with wire.

And so it continued for the rest of the month. The stark plains offered no natural defensive attributes, so it fell upon Beismann's troops to create artificial ones. Going out each night and working right under the enemy's nose and sometimes under fire, Grimm's company alone achieved incredible results in the period 14-30 September:

erection of 541 Spanish riders (total span length of over 1000 metres); installation of 126 trip-wire anti-tank mines covering a length of 1260 metres; laying of two anti-personnel mine barriers and two anti-tank minefields with 74 T-mines; clearance of three minefields and removal of 100 Spanish riders from a friendly sector due to be abandoned.

With advice from a pioneer officer, the infantry regiments determined where extra protection was needed and what type of defensive measures should be installed. The pioneers became night-shift workers because the exposed terrain prevented any kind of work in daylight. The heavy workload was only fulfilled because of extra hands, as Grimm explains:

> Wood, as well as all pioneer material and ammunition, had to be fetched by the company from the Don valley and from Vertyachii, almost exclusively at twilight or night-time. The Spanish riders had to be built up front. Along this broad front, energy-sapping foot marches had to be covered and ammunition and material needed for obstacle construction carried along. Prisoners who came along with us as Hiwis made it possible for us to master all of our tasks. We looked after them quite well and fed them, and they remained because they knew our behaviour was fair, in contrast to prisoners and deserters.

Four months of constant deployment had drastically weakened every unit so that some platoons were being led by enlisted men, so Beismann's three company commanders received a pleasant surprise on 12-13 September when reinforcements arrived. This field replacement battalion contained 857 replacements for the entire division. At first this infusion was warmly welcomed, but it soon became apparent that the men were undertrained. The division gave them a crash course in vital drills and front-line tactics before distributing them a week later. The pioneer battalion received a surprisingly large share; around forty men per company. Pioneers normally underwent six months of intensive training. These young men had far less. A few old-timers and convalescents were amongst the new arrivals but the remainder were woefully instructed in the craft of military engineering. Grimm recognised their deficiencies and immediately assigned a few of his best men to whip the newcomers into shape. Their first lesson was how to install tripwire mines. Training continued for several more days.

Not all of the reinforcements assimilated seamlessly. The replacement system in the German army generally supplied combat units with replacements drawn from the same military district. However, war-time pressures disrupted this sensible regulation. Maintenance of regional identity was an integral part of a formation's cohesion.

Germany's different regions engendered certain characteristics amongst its citizens, and like all nations, biases and rivalries – sometimes friendly, other times not – abounded amongst different provinces, especially neighbours. Men from certain districts never mixed because they were from different ends of the country. Such was the case with some of the newcomers: cold, disciplined Prussians, polar opposites of the battalion's south German majority, quickly rubbed Gefreiter Füssinger the wrong way:

> Half the company's old cadre is gone, new faces everywhere, very clever Prussians
> at that, and wherever they are, then us dumb Swabians don't have much to say. It's
> fortunate that we have an understanding Swabian as our new commander.

The men of 3rd Company were glad that Beigel, a Bavarian, was leading them. The feeling was mutual. "I've gradually settled down with my new group of men," wrote Beigel. "Of course, I don't know all of them, but I'm already somewhat trusted with the whole business. Believe me, it's a wonderful duty when you're allowed to do what you think is right as company commander and 'supreme leader' of such a group of men." Barely had he settled into the job when he was asked to incorporate the reinforcements. His opinion of them after two weeks is interesting:

> My forty replacements are completely lacking. They are not well-trained and above
> all lack spirit. Our instructors in the replacement units must go to the front so that
> we receive strong and assured soldiers.

– – –

Artillery and mortar shells fell sporadically during the night of 13-14 September, but at 0300 hours the fire strengthened and focused upon Oberst Winzer's sector, particularly around the Kulturstation (cultural centre). The racket roused Grimm's men from their bunkers. Poking heads out of their shelters, they saw dozens of rocket-tails glimmering through the overcast firmament before screeching to earth a kilometre or so to the north. After an hour of softening-up, a Soviet battalion attacked towards the Kulturstation. As the pioneers pulled on combat harnesses and checked weapons, they followed the course of the battle with their ears: after the bombardment ended, there was a brief pause, then scores of machine-guns opened up simultaneously, followed by several minutes of furious gunfire, and finally, as the fusillade died down, the diminishing crackle of small-arms. Word came back from the front: attack repulsed. The divisional sector then slumped into an all-day silence.

Life was easy for Beigel because unlike Grimm, he did not need to train replacements while on the front-line. Safely encamped in pine plantations north of Vertyachii, his mission was to generate lumber for dug-outs, and he controlled a swollen workforce to accomplish it. In addition to his own men, he was the boss of a small army of POWs:

I've got 150 Russians working for me now. The Russian company will be strengthened by about one hundred men in the coming days. There's simply loads of work.

A few unpleasant duties intruded upon his idyll: "Today I completed the final death notifications. The sad news goes into the homeland via air mail and I feel sorry for those that receive these letters." Overall, however, he was content. He even had a small Cossack horse simply for entertainment: "I ride it around without a saddle. It really tears along. I take it out when I have a crazy hour and want to blow off some steam."

The weather turned cool on 15 September. Those deployed out in the steppe were exposed to a cutting wind, but down in the Don valley, Beigel and his men were shielded by the trees. All of the company's vehicles, horses and tents were clustered into the narrow bands of pine plantations. They kept warm felling the trees as the breeze skirled through the tree-tops. The men knocked off at sunset:

The break is doing us all good. In the evening it's already dark at 1830 hours and the radio is turned up. Out of it comes music from home, which is so far away and which we all desperately want to see again. Because these wonderful songs contain the names of our girls, our thoughts are with you. It's then very silent in our circle. Every now and then the glimmer of a cigarette, but otherwise, nothing else stirs. I often wonder what the men are thinking when we sit around the radio in a circle, listening. Probably the same as me: I am at home, thinking of you and living with you. Homesickness is then often unbearable. When I crawl into my tent, there is only one great wish for me: that the Lord will guide me through these difficult hours so that I can see you and home again.

Beigel was surprised to see Major Beismann and his aide ride up on horses on 16 September, unannounced. Without much ado, Beismann indicated to Beigel and Hepp to stand next to each other so he could award both of them the Iron Cross First Class. Beigel barely had time to fetch his camera and hand it to his orderly. The first award presentations in Russia were carried out in front of entire companies, now they were just informal gatherings, and so Beismann pinned the prestigious medals on their chests. Beigel could not wait to tell his girlfriend:

My wish, to receive the Iron Cross First Class, was fulfilled... My commander presented this medal to me, which was awarded by the division commander. I was recommended for it as a platoon leader. At least that's what I believe because I've only been a company commander for a few days. My old company commander Grimm still only has the Iron Cross Second Class. If my good old Lutz was here, then he would have certainly received it. It is no coincidence that the Iron Cross First Class and the heart are so close together. Every enemy who falls from my bullets, and is killed, is revenge for my Lutz.

Two weeks later, Beigel received further recognition of his efforts:

The assault badge was awarded to me by the division. Now I've got a nice collection going. The assault badge particularly pleases me because the Iron Cross First Class is worth even more when one has looked the enemy in the eye.

While Beigel thought about avenging the death of his friend Leutnant Lutz, Grimm was notified about the demise of another pioneer officer: "Dr. Reiner Haan, who was with me in "maison rouge" in St. Georges as supply officer, was killed as a company leader. More and more holes are ripped in our former nice circle." Leutnant Dr. Haan had served as the battalion's commissary officer in France before being transferred to the infantry as a platoon leader in Regiment 578. Severely wounded on 14 September, an ambulance rushed him across the Don to an aid station but he succumbed to his injuries. Comrades buried him in a grave at the western entrance of Perepolnyi, a small village on the western bank of the Don.

– – –

Observers in the air and on the ground detected lively vehicular traffic and march movements, as well as tanks regrouping and preparing for an attack. The threat of a major attack against the neighbouring 76th Infantry Division prompted 8th Army Corps to condense the imperilled sector, so on 17 September General Oppenländer was order to expand his frontage by taking over part of 76th Division's front-line. As forecast, a major attack crashed into the German "Nordfront" defences on 18 September. Two Soviet tank corps with 150-200 tanks and parts of six rifle divisions aimed at the boundary between 8th Army Corps and 14th Panzer Corps, concentrating mainly on the front of 76th Infantry Division. Tanks busted through in several places and some pushed south to the main supply road. All available forces counterattacked to seal the gap and throw back the enemy. The men of the Bodensee Division were unaffected by these dramatic events, so much so that they warranted but passing comment in letters home. Gefreiter Vorherr wrote: "There was something happening to the east of us this morning. Our planes were there bright and early, but I don't know what's going on up front." By the afternoon, over 100 Russian tanks had been knocked out or immobilised. After counterattacks from the south-east, east and west, the Russian forces that had broken through were destroyed in fierce combat.

Clear skies enabled every available Luftwaffe plane to engage the penetration in non-stop sorties. German aces shot down 77 Soviet aircraft. Grimm's men could be forgiven for believing that God himself had smote two of those Russian planes from the sky. For the first time during the Eastern campaign, 2nd Company was being spiritually fortified by a church service carried out in the field. Everyone participated. Only one man kept watch, and that was Grimm. The company was about 1400 metres behind the front-line. Grimm had his back to the proceedings, scanning the cloud-covered sky for any sign of danger. As the chaplain's holy words fell upon fertile ears, two Soviet fighters zoomed overhead, collided mid-air and then crashed to the ground only 150 metres away, right on top of the company bunkers. All heads had swivelled to watch the spectacle. A murmur rippled through the solemn ranks. They had been sitting in clusters around those same bunkers just minutes earlier, hunting lice and writing letters home. The biggest buzz, however, was caused by the collision: it seemed the planes were magnetically attracted to each other, as if a higher power had personally intervened. Even Grimm, a less-than-pious Protestant, was affected: "You know, whoever does not come closer to his God in such moments is lacking something." Beigel's 3rd Company took part in a field service a few days earlier and it, too, made quite an impression:

Our divisional priest was with us in the forest. I had two beautiful altars set up for him. He then began the service. You cannot imagine how much effort the priest had put in to pack his suitcase: the candles were put up, his things that he needed to read for Mass were got out and he donned his robes. All my men sat around the altar. Then the Holy Communion began. We sang songs like we did when I was in school. I'm amazed that I still knew them. My pioneers and soldiers sat there, motionless, mute, not bowing and humble, but upright and straight, like their lives. Many were certainly there who had previously not wanted to know anything about the church. Now they were there with open minds. Our priest delivered his sermon in front of us as a soldier and a clergyman. He knew us all very well. He was himself up front during all of the attacks and has seen and experienced what it's like around us. He even found the right words for us. Then he gave us general absolution. Everything was quick and simple. Is it necessary that a soldier confess? It would only do him good, and his eternal commitment of his life for his fatherland would wipe away his guilt. Finally, we all took communion, as it has to be. That was also different than in church at home. Not kneeling, but upright, we stood in front of our priest. Many a man stood silent, as if in front of a superior officer. I will never forget these hours. Most kept quiet as we headed back to the front.

– – –

A foreboding breeze roamed the naked plains, forcing collars to be upturned and hands stuffed deep into pockets. And Stalingrad continued to grumble and smoulder on the eastern horizon. "Already the weather is no longer fine," wrote Gefreiter Vorherr, "there's a cold wind and you notice that it's autumn, particularly here in the steppe, where there's no shelter from the wind." Grimm was thrilled when his long-forgotten box full of personal belongings unexpectedly turned up: "Hurra, my officer's chest is here, with all the trappings!" He was even more pleased about its arrival when the weather turned cold: "In recent days it's been somewhat quieter, but colder... From my chest I took the toque and thick gloves. It's super that I have everything with me again. One thing is missing: mittens." The sinking mercury made it difficult to keep clean, as Vorherr explained:

> *The time will soon come when we have to quickly get out of our clothes because the lice have made themselves comfortable in them... Because it's now so cool, there's no desire to wash every day. I've had my shirt on for almost three weeks now because there's no opportunity to wash here. Oh, this beautiful Russia! If the Russians could ever see us, they'd know resistance was futile. Instead, he holds out – even though he has nothing to eat – until the very end. Every day we wait and reckon that Stalingrad will fall. Whoever has fought in this battle will know what he's achieved.*

Gefreiter Füssinger believed the capture of the metropolis would not end the war:

> *At the moment we are still in quiet places, even though events at Stalingrad can be observed well from here. I very much doubt that a decision will be reached this year, nevertheless, the imminent fall of Stalingrad will present a better chance for a decision. Stalin has already tried everything to stop this. But warm days are numbered and winter will soon show its rugged face. The hope that we'll return to Germany still persists, that is what we all want. Our division has had to carry out a lot of difficult tasks, but the Führer himself sent a personal letter of thanks to us. Now we hope for the best, we have always been lucky.*

Little did Vorherr and Füssinger know that the Moloch of war, now in residence at Stalingrad, would soon beckon the men from Swabia. For the moment, apart from harassment fire, the sounds of Soviet entrenchment work and the arrival of a dozen deserters, the 18th of September passed quietly.

Apart from taking fire while laying mines during the night of 18-19 September, Grimm's men had an easy couple of days. They finished construction of three dug-outs for Oberst Winzer (a command bunker, an orderly office and an accommodation bunker). Beigel's makeshift lumberyard provided the wood. At nightfall on 19 September, the company allocated some squads to an infantry battalion to act as guides for nearby minefields while another platoon cleared some other mines. At 1800 hours, well-placed fire from artillery and heavy mortars landed in the middle of the company's billeting area, though no casualties were suffered. At this moment, Grimm received an urgent message: "The Russians are attacking. Get your company ready for infantry operations immediately!" The alert could not have come at a worse time because he had split his company into platoons and scattered them in all directions to lay mines in different sectors. In some cases they had already started digging them in. Grimm got on the phone, alerted the infantry units to which the mine-laying teams were attached, and then guides from the infantry escorted the pioneers to designated assembly points. While running around, organising things, Grimm had another close call:

> When I was coming back in the motorcycle-sidecar to my company in the gully, two shells landed about ten metres behind and in front of us. After seeking cover, the bikes were remounted and then we dashed into the sand and detonation clouds and were able to crawl straight into our foxholes. Fortunately there were no casualties.

Gefreiter Vorherr was one of the drivers:

> We were barely a hundred metres away from [our gully], because that is the spot on the road where you cannot get through the sand, so you climb out and have to push. We had two motorcycles, a comrade and I, with my motorcycle behind his, in which the commander [Grimm] sat as we had to push. It was already dark because the Russians chased us so closely with six artillery shells that hearing and vision vanished in a split second. Luckily, there was so much sand that the things did not have much of an effect. We stayed under cover for a minute, then searched for our bikes in the dust and smoke and quickly got out of there. We were lucky once again!

A few hours was required to reassemble the entire company. Towards 2200 hours Grimm was told to report to II./Regiment 578 for infantry operations. He ordered his men take their bulky overcoats with them. He led his troops on a strenuous four-hour march on deep, sandy tracks and occupied reserve positions behind Major Rettenmaier's infantry battalion. "We had indescribable luck," Grimm wrote.

> We found foxholes and MG positions and therefore had cover, it only took a couple of minutes. Then, the Russians opened up with a heavy machine-gun that would have destroyed the company a few minutes earlier... We did not have a single casualty.

NCOs swiftly deployed their men for defence. Machine-guns were set up on both wings. The Soviets kept firing but did not attack. The overcoats, so burdensome on the march, were put to good use because the wee hours were chilly and every man was drenched in sweat. The expected Russian dawn attack failed to eventuate and the company was pulled out at 0600 hours. After reaching their billets, the company rested. Everyone caught up on sleep. Grimm had a few hours off for the first time in weeks. His assignment for the coming night was the construction and installation of seventy-five Spanish riders, a task easily accomplished with just one platoon and some Hiwis. The rest of the company had the night off. It was Beigel's turn to be rushed to the front because of an impending Soviet assault: "In the evening I had to march to the front with my company... The night passed quietly, as we had assumed, and there was no action." His company remained just behind the front-line as a reserve but he was missing the comforts of his home amongst the pines:

> I hope the Russians don't madly attack, so that we can get our rest and I can return with my company to our nice peaceful forest. My tent is there, my horses (about sixty of them) are there, and my tireless "Karl" [Obergefreiter Schaufler]– who cares for me like a mother does her child – is there. I've never experienced anything like it. He reads my wishes in my eyes. If something were to happen to me, then he would take care of my private matters down to the smallest detail.

Twitchy enemy forces and an expanding workload would end up keeping Beigel and his men away from their woodland sanctuary for a full week. For they moment, they were billeted in a gully filled with so many flies that they had to be continually fished out of their food. Grimm's men were further east in the same gully. They cut out all wood from their old shelters, hauled it to the new billet area and procured more lumber from the battalion's depot, but in spite of the abundance of timber, the new dug-outs were finished in a makeshift fashion because another transfer was imminent. "We're sitting in foxholes tonight," wrote Vorherr, "and we've stretched a groundsheet over our one. If we have to stay in such holes in the ground over winter, then we'll need to go deeper into the earth." The troops rued the decision to skimp on proper shelters as night frosts set in and they shivered through until sunrise. Needless to say, the dug-outs were upgraded the following day. Beigel slept in a proper dug-out, yet even he was troubled by the temperature-drop: "During the night it is so damned cold that you can barely sleep [and] every morning my water canister has frozen over." The spell of fine, sunny weather that set in and continued until the end of the month also came with bone-chilling night-time temperatures.

The division's sector was expanded further on 23 September. Compared with the frail state of its neighbour (76th Infantry Division), General Oppenländer's division was positively bursting with good health: of its nine infantry battalions, four were rated as strong, two medium and three average. Beismann's pioneer battalion was deemed to be strong with 90% mobility. The division was rated as suitable for any offensive task. The truth about its actual strength was shocking: on 27 September, the division reported a shortfall of 4,377 men – a quarter of its original strength – and this was exacerbated by the fact that no further reinforcements were expected. Regardless, it was considered one of Paulus' strongest infantry divisions, which is why it was ordered to relieve another battalion from 76th Infantry Division and stretch its right wing eastward to the Kotluban Gully. Major Beismann was directed to employ every available man to strengthen the new sector with mines and barbed wire.

Oppenländer also decided to straighten his own line: as it served no useful purpose, the Kulturstation bulge would be abandoned, but it was to be held until a shorter chord position was constructed. Haste was needed for the creation of this fallback position; Grimm was summoned to Battalion on 23 September to discuss his company's role in it. The enemy was to be denied anything of use. Work began at once. Once darkness descended, his men cleared two minefields totalling 64 mines, including three left in a volatile state after being damaged by shellfire. Intact mines would be used elsewhere. Although the area was barren and uninhabited, a "scorched earth" policy was implemented. Grimm reconnoitred his sector and listed all objects that required demolition: six derelict tanks, a kolkhoz, a threshing machine and an enclosed shepherd's cart. In the flat terrain, all of these could be used as observation posts, hence their proposed destruction.

Due to Regiment 578 being shuffled eastward, Oberst Winzer was forced to find a new command post, and Grimm was given the job of scouting a location and building it. A platoon began construction on 24 September and completed it the following day. Another platoon acquired Russian and German explosives, as well as primers and igniters, and carried out further preparations for demolition of the targets. The life of a pioneer was never sedentary. Shifted from job to job, working day and night, sometimes in construction, other times handling explosives. Oberst Winzer was grateful for Grimm's semi-permanent attachment because requests for assistance flowed in from all his battalions, forcing Grimm to dispatch platoons to different sectors. On the night of 25-26 September, his attention was focused on a new battalion sector. Two barriers of anti-personnel mines, each with a width of sixty metres, were staggered one behind the other, and laid behind those was a minefield with two dozen anti-tank mines. In addition, 108 Spanish riders were set up in the most forward line.

Throughout the night of 26-27 September, Soviet planes constantly bombed the division's sector while enemy patrols prodded the left wing. Towards 0400 hours a Soviet assembly was smashed by artillery fire. This was the background chorus playing while Grimm's pioneers removed a hundred Spanish riders from old positions and cleared three minefields (38 T-mines). The most crucial task was the attachment of detonation cords to explosive charges deposited in the demolition targets the previous night. Despite thick gunfire whizzing over their heads, the errand was carried out without loss. The charges were set off the next night (27-28 September). All soft targets were obliterated, as was a burnt-out T-34. The remaining five tank carcasses, four of which were KV-1s, were rigged for demolition. Several crates of explosives were packed into each tank to ensure complete disintegration: when the area was relinquished, there must be zero chance that any part of them could be used as an observation post. Crack! Crack! Crack! Crack! Crack! The ground shuddered as all five tanks exploded into their component parts. At midnight the infantry pulled back to the new chord position. Observers keenly watched the vacated area but Soviet soldiers did not feel their way forward until a full day later, when they were spotted digging in near the Kulturstation. Job completed, Grimm's company was shifted into a gully level with Winzer's command post and dug itself in.

The final days of the month passed with a similar routine: daytime briefings about new mine barriers, night-time operations to implement them. Grimm's final task for September was rigging up twenty tripwire T-mines with a 200 metre blocking width as well as laying a minefield with fifty anti-tank mines, all while under enemy fire.

– – –

It appeared to all men of the Bodensee Division that they were settling down into a comfortable routine. They knew they would not remain in their current sector because 113th Infantry Division began relieving 76th Division, as planned, and once that formation had been rejuvenated, it was the turn of Oppenländer's men to be withdrawn from the line and prepared for one final attack to obtain a defensible line north of Stalingrad. Then, it was a matter of fortifying that position and hunkering down for the long winter ahead. The prerequisite for this operation was the fall of Stalingrad. Expectations were high at the beginning of September that the city could be taken once enough forces were assembled for a proper assault, and this was accomplished by extracting divisions from quieter sectors and sending them east to General von Seydlitz, the corps commander responsible for capturing Stalingrad. General Kempf's panzer corps pressed into the city from the south-west. Both corps had made deep inroads by mid-September and even reached the Volga at two points, but compression of Soviet forces against the river only increased the ferocity of their resistance. The central and southern parts of the city were eventually taken. The hardest assignment was still to come: reduction of the Orlovka salient and conquest of the industrial north. After a redistribution of forces and insertion of another division (100th Jäger Division), the assault was relaunched on 27 September. An armoured spearhead sliced deep into Soviet lines while the salient was first ligated and then crushed. The month ended with German forces gaining the hills west of the city limits. Small-scale attacks continued to nibble away at Soviet defences but it was clear that von Seydlitz lacked sufficient strength to finish the job. The factories and their satellite suburbs remained unconquered.

None of this concerned Beismann's pioneers. All they knew was that Stalingrad was taking longer to fall than first expected which, admittedly, was affecting them because it delayed their final operation, but nobody complained because static defence was far less costly than attacking into built-up areas. Plus, lack of combat activity created the possibility of furlough. No other topic stirred the soldiers more than the thought of going home. Some positive news reached Beigel on 26 September and he could not wait to inform his girl:

I can give you a small sign of hope for a reunion in our beautiful homeland. Furlough will shortly be given to our division. One man from the company should be able to head off every five or six days. In comparison to previous quotas, that's a lot. Naturally, the first men to go will be those who've not had furlough for a year or more. I'm reckoning on furlough for myself in November. It's possible that I won't be allowed to head off until January or February, but it'll most likely be in the November-December period. That would be great! I'm sure you know how happy this makes me! It will be unbelievably nice. In my thoughts I'm already on the train from Kharkov to Berlin and then on to Munich. Should I surprise you at home or ask you to pick me up in Munich? Which would be nicer? It would be great for you if I was suddenly standing in front of you one day early in the morning. I want to have a few hours completely alone with you. You know, once at home, there will be so many questions from the first moment that we won't be able to have any time alone at all. Oh well, we'll see, hopefully it happens.

Three days later he had even better news:

In my previous letter I gave you hope about receiving furlough. I can now strengthen this hope. If nothing happens in the meantime, I'll be standing in front of you one day or night in mid-November or December and then everything will be fine. How that will make me happy! You know, I've already got my uniform sorted out. A new jacket, new riding breeches (you haven't seen them), a great pair of boots and a new overcoat. You'll be satisfied with me. I'll brush off all the dirt of the Russian steppe and the delousing station will make sure that I don't bring any "enemy tanks" home with me... Tomorrow, the first four men of the company head off. About ten more of my men will leave for the homeland in October! I've got two pioneers who haven't been home in eighteen months. If only I had a leave pass in my pocket!

To the division's men, increasing allocation of furlough was a sure sign that their participation in the summer campaign was winding down. How wrong they were.

Business As Usual (1-5 October 1942)

After four months of solid campaigning, the troops were jaded and homesick. Stalingrad was yet to fall, and until it did, they were in limbo. Most were vaguely aware of their mission but it was never placed within the greater scheme of things. As they huddled in dug-outs and huts on the night of 30 September, Hitler's unmistakable voice crackled from the radio in a nation-wide address that spread to the furthest limits of German-held territory. "Our Führer spoke yesterday," wrote Grimm to his wife, "and I heard part of his speech. Our enemies may take notice of the achievements. But we are conscious of our role." Normally, Hitler's long-winded monologues rarely reached the front-line, and if they did, their effect was minimal. But his speech at Berlin's Sportpalast fell upon receptive ears, primarily because he mentioned the troops fighting near Stalingrad. First, Hitler rattled off a list of achievements attained in the previous few months: pushing along the Don, reaching the Volga, attacking Stalingrad ("and we shall take it, too, you can depend on that" he declared) and entering the Caucasus. After mocking the Allies, he eventually focused upon a point that directly affected Grimm and his men. "The occupation of Stalingrad," he avowed, "which will also be carried through, will deepen this gigantic victory and strengthen it, and you can be sure that no human being will drive us out of this place later on." For soldiers fighting and dying on far-flung battlefields, affirmation of their sacrifice and triumphs by their head-of-state lifted morale and reinvigorated flagging confidence in victory. As loathsome as Hitler is now universally regarded – an irredeemable pariah, a synonym for evil – it was a huge honour for German troops to be addressed by their leader, akin to Tommies being spoken to by Churchill, or GIs by Roosevelt. The Führer's words electrified the troops, including officers like Beigel: "Our Führer has given us responsibilities and we must fulfil them. It is always a massive psychological boost when our Führer speaks to all Germans." Simply knowing that their leader was watching hardened the resolve of most soldiers to finish off the business at Stalingrad once and for all.

Life wasn't too bad. The sun shone from a clear sky, the Russians were quiet and no heavy-duty action was impending. A drop in temperature kept hands in pockets but the pioneers barely noticed the cold. While artillerymen idled near their pieces and infantrymen held the line, Beismann's men sweated and griped their way through a laundry list of tasks. Lay mines, install obstacles, string barbed wire, build command posts and dug-outs, and so it went on, day in, day out. "Everything is now completely different than during the advance," wrote Beigel. "Then, it was fighting, marching, thirst, heat, hunger and even "lavish" meals. Then, we only had three or four hours sleep. Days passed in a flash. Now, we have time to think." No-one was complaining too loudly. Better to perspire than bleed. Beigel was satisfied with a job well done:

> My men can feel proud that our infantry can disappear into the earth and spend
> the winter there... I'd like to see the Russians try and throw us out of our holes.
> You'd never believe how highly regarded the pioneers are within the division. When
> our men are deployed, then they can get whatever they want.

Decent accommodation for themselves was not neglected. Beigel had designed a comfortable lair:

> I'm having a beautiful bunker built for myself. Inside it I will hang two drawings of
> you that our artist has drawn from a photo. He didn't quite capture you but the
> drawings are very nice... When we have moved into our bunkers, then he'll create a
> large picture of you. I'll draw you a quick sketch of how I'm having my quarters built.
> Size is three by five metres, and two metres deep. The space will be divided in two. I
> live in section one. A bed will go in, then a bedside table with radio, a corner seat, a
> table and two stools. The room will be divided by a curtain. My orderly will live in
> the other room. Two windows will be built into it. I removed a battery from a Russian
> tank so that I'll have lighting. The bunker will only be deep enough in the earth so
> that the windows can still be looked out of. I've drawn it for you so that you'll know
> how I'll be living in four weeks time. I'll remain here until all of my men are in the
> ground. In addition to mine, I have to have another twenty-two such bunkers built.

Construction work was hard on the body but left the mind free to wander: Beigel was one of the worst affected. His mind kept turning to home and he could think of little else:

> Today, my commander asked me when I wanted to go on leave. As soon as possible,
> of course, I said, but he replied, only when the snow is gone. When I answered that
> that was too late, he laughed. I believe I'll be allowed to head home earlier. It is
> dreadful that furlough has now been blocked. All my thoughts are suddenly focused
> on furlough... I keep thinking about it. The months seem long again.

Priority for leave was given to fathers, husbands and those who had not been home in a long time. Beigel would have to wait. As a married man with a young child, Grimm was ahead of Beigel, but even he told Lotte to "not get her hopes up about furlough."

The shift from offence to defence was reflected in the battalion's unofficial ration strength. During the advance, prisoners were shunted off to the rear to avoid weighing down the unit with useless mouths. Now, however, manpower of any kind was an asset because all tasks involved manual labour. A steady influx of deserters swelled the ranks. Beigel's 3rd had more than doubled in size. "My company stands in front of me, 160 men and 250 Russians," he wrote on 1 October. Grimm's 2nd contained around 120 Germans on the first day of the month, including ten or so who were ill but receiving medical support in the hinterland. Infectious jaundice was rife. The company's tally of dead and wounded to date was frightful: 27 killed and 52 wounded – a third of its original ration strength, and almost half its combat strength.

An unfortunate consequence of the hectic workload is an end to Grimm's private reports: "There are no more combat reports from 1 October 1942 onwards. Either I did not reach the clerk at a command post or operations demanded so much that I did not have the time or strength to make a meticulous report." Constant deployment was dulling Grimm's desire to record his company's daily activities. He had already sent a large batch of reports to Lotte for safekeeping. "From them, you can see the type of hard days that are now behind us," he explained to her.

– – –

The Die is Cast: Off to Stalingrad (6-12 October 1942)

The OKH liaison officer attached to 6th Army reported back to his masters in Berlin on 4 October that "infantry strength of the divisions deployed in the city fighting must be regarded as nearly exhausted. A swift victory without the supply of fresh forces can no longer be expected. Without this influx, the attack can only be pushed forward slowly by constantly regrouping and focusing locally on taking one house at a time." The chiefs-of-staff of Army Group B and 6th Army – General von Sodenstern and General Schmidt respectively – discussed ways of getting more jackboots on the ground, and one of those involved the Bodensee Division. Schmidt stated that "we want 76th Infantry Division to replace 305th Infantry Division, and then use this to either relieve 60th Motorised Division or deploy 305th Infantry Division itself in Stalingrad." By 5 October, German attacks in the city were temporarily suspended. After discussing various options, army staff decided that "305th Infantry Division will be deployed in Stalingrad within the next week." The OKH liaison officer succinctly summed up the plan:

*In order to achieve a greater capitalisation within the grander scheme, the Army
has ordered the relief of the reasonably combat-worthy 305th Infantry Division by
the battle-weary 76th Infantry Division. The 305th Infantry Division can be ready
to attack the city on approximately 13 October.*

Just like that, with barely a day's deliberation, General Oppenländer's men were to
be plucked from their cosy winter positions and thrown into the Stalingrad mincer.

Army Group B laid down the future course of operations for 6th Army in a lengthy
message that could be summed up in two words: take Stalingrad: "The situation at
Stalingrad, whose complete occupation has again been designated as the Army Group's
most important task, demands the concentration of all available forces. It is imperative
that all other issues be pushed aside."

General Oppenländer received word to prepare his division for deployment elsewhere.
At 2030 hours on 6 October, the division issued an advanced order for its relief by 76th
Infantry Division over three nights beginning at nightfall on 8 October. "Our first-class
preparations for the construction of winter quarters were suddenly stopped," wrote
Beigel. Aware of the natural inclination of soldiers to look after their own, the division
sternly warned its units to hand over dug-outs and shelters without modification as well
as forbidding them from removing furniture and windows from houses and dismantling
bunkers and defensive obstacles. Incoming troops should find ready-made positions.
"Lucky swine," groaned the pioneers, "they're moving into fully furnished homes."

– – –

Paulus visited General von Seydlitz to brief him on the army's plans. Generaloberst
von Weichs had directed Paulus to relaunch the attack against Stalingrad by 14 October
at the latest, so this deadline was immediately forced upon Seydlitz. He therefore had less
than a week to incorporate two new divisions and prepare a large offensive. The biggest
headache for Seydlitz was supplies, particularly ammunition. The order to wear down
the Soviet defenders with continual artillery fire conflicted with austerity measures
designed to conserve ammunition and lay in stocks in preparation for a predicted spell
of bad weather and the forthcoming assault. Exacerbating this logistical nightmare were
petrol shortages, which affected the trucks hauling howitzer rounds from dumps west of
the Don to the artillery batteries in the hills around Stalingrad. The situation was so bad
that Paulus warned Army Group B that the offensive was being jeopardised.

The German leadership saw all its reserves being sucked into the Volga crucible but
could do nothing about it: Hitler wanted the city. Operation "Herbstzeitlose" (Autumn
Crocus), the sweeping attack to gain favourable winter positions on the Don-Volga

isthmus, was cancelled, so 6th Army's current position between the two rivers was to be upgraded to a "winter position". Soviet resistance in Stalingrad was compelling the Germans to cancel one operation after another. Weeks earlier, a buffet of potential operations lay before the Germans; now, they were mere sideshows, a waste of resources while Stalingrad remained unconquered. So much had changed within a month.

The relief of 305th Infantry Division began during the night of 7-8 October. Three more nights were required to completely replace the division. The rank-and-file were clueless about their ultimate destination. "We march off again tomorrow," Gefreiter Füssinger told his parents, "but it's not known where."

– – –

"Today is a very beautiful autumn morning," Grimm wrote on 8 October, "slightly hazy, but the sun will soon bring some warmth. It started raining last night and we thought that the bad weather had begun. Water in a bucket already has a nice layer of ice. It's pretty cosy in our hole in the ground." Everyone thought it a shame to relinquish their comfy dug-outs, but orders were orders. Grimm was faced with a mountain of administrative work in regards to the move but was pleasantly surprised to receive some assistance: "Yesterday, I received an officer in my company, Oberleutnant Staiger. He was wounded early in July and came back as a convalescent. So I have a deputy now."

Major Beismann was glad to see one of his young guns return. Staiger had been with the battalion from the beginning and was the first wounded officer to return, although he was not yet back to full health after being deafened months earlier:

I remained hard of hearing and should have found another assignment, but I requested that I be returned to my unit. This was authorised at the end of September. When I arrived at Pitomnik airfield, our unit was positioned outside Stalingrad.

On 4 October Beismann sent Staiger back to his previous company, the 3rd, but he felt like a foreigner: Hauptmann Häntzka was dead, most of his old platoon were absent from death or wounds, and the only officers present were the recent arrivals from 2nd Company, Beigel and Hepp. Allocation of officers amongst the three companies was now uneven: two in 1st Company, one in the 2nd and three in the 3rd. Beigel was asked to relinquish either Hepp or Staiger to Grimm. The choice was easy. Beigel did not want to upset his current order of battle, plus Hepp was a friend, so after barely one day with his old comrades, Staiger was ordered to leave. He remained on battalion staff for a few days, then joined 2nd Company on 7 October.

Conditions in the city were expected to be unsanitary and a threat to health, so Grimm, Staiger and the rest of the company received a cholera inoculation. The other companies also went under the doctor's syringe. On 10 October, leaving behind a small security detachment to guard the division's depot of pioneer materials, Beismann's entire battalion marched off to its assembly area. Radio silence was ordered. Beigel describes the move:

> *A march order arrived, which meant me and my company were moved seventy kilometres east of my old position. As usual, things happen quickly with us. While you at home go about your usual business on Saturday, climb into the bath in the afternoon and sit down to eat in the evening, your Wiggerl drove east through endless clouds of dust on the worst possible roads for several hours. Now we are in a gully that is filled in some parts by spruce forests. Up on the rim of the gully is steppe, like everywhere else. Not far away is a city. Seen there are large white buildings, apparently factories. Sixty tanks a day drove out of those buildings a few months ago.*

"I prepared my men mentally and training-wise for the battle," recalled Beigel "because Stalingrad would demand a lot from both body and mind." It is generally believed that the Germans were hopeless at urban combat and simply blundered into the city in serried ranks. Nothing could be further from the truth. The Wehrmacht was adept at fighting in built-up areas and any recognised shortcomings were swiftly remedied. The men of the Bodensee Division benefited from the experience of other divisions. As they entered the outer belt of Stalingrad, each company commander received a "pamphlet for urban combat" and was told to make it the subject of instruction. This leaflet reveals the German method of fighting in built-up areas and its principles, even today, are sound:

> *Success depends mostly on the independent actions of junior leaders. It is therefore important to put them in the picture, in detail, about their own limited tasks within the context of the planned military operations; this is the prerequisite not only for independent action, but in terms of overall operations, also for appropriate action.*

What is clear from this pamphlet is the role played by the lower ranks, a tacit admission that while senior officers could devise a plan, it was junior officers and NCOs who would bring success:

> *The fighting will usually be at close range; it breaks down naturally into a number of consecutive assault troop operations. The number of necessary assault troops and their composition is a matter for lower commands; they regulate the timing on the basis of detailed reconnaissance and planning.*

Cardinal rules were hammered into officers. Accurate recon and a detailed briefing of the battle plan was essential for success. The combat mission and nature of the objective would determine the composition and fighting style of the combat groups. Rigid tactics and repetition of failed attacks would be avoided. If an attack was stalled, switch the focus, attack from a different direction, try anything to catch the enemy unaware. Another component to successful city fighting was heavy weapons:

Commitment of panzers, assault guns, heavy infantry weapons and individual guns <u>in the forward line</u>, in addition to pioneer assault detachments with flamethrowers, is the norm. Guns are to be moved up manually with manpower (muzzle facing forward, a round in the barrel).

Then followed a principle still taught to soldiers today:

<u>*Fire from heavy weapons assures that the objective is reached by the shock troops;*</u> <u>*the shock troops enable the heavy weapons to advance into the conquered target*</u> <u>*area and setup to provide covering fire for a continued push by the infantry*</u> <u>*(pioneer) shock troops.*</u>

It is the nature of city fighting that pockets of enemy resistance will be bypassed and end up behind the advancing troops, as the Germans learned to their detriment in the previous weeks. Front-loaded offensives in an urban environment were destined to fail: a small number of assault troops, supported by adequate reserves to secure flanks and cleanse captured areas, was a more effective use of manpower:

<u>As a matter of principle</u>, basements and rooms on ground floors are to be eliminated with hand-grenades. By feeding in reserves with a pre-specified combat mission, the flanks of the seized area will be screened. Special mopping-up detachments are to follow behind each combat group; their task is to destroy bypassed pockets of resistance and immediately subdue any resistance that comes to life behind the battle front. Brutality is acceptable.

<u>Rooftop snipers</u> are annoying but not dangerous; their area of effect is limited.

At the conclusion of each local engagement, as well as after the attainment of the daily objective, the conquered area will be thoroughly mopped up by reserves set aside for this purpose. Surprises must be expected at all times. Locally knowledgeable enemy troops will use the sewers, underground passages and roofs to suddenly appear in the rear of our troops.

The men of the Bodensee Division were not being led to the slaughter. Soviet fighting methods were known, as were solutions. Battering-ram tactics would fail. Coordination of all arms, combined with the initiative of well-informed junior leaders, would carry the day. So much for the theory. The pioneers would soon put it into practise.

Responsibility for defence of the divisional sector was officially handed over to 76th Infantry Division at 1800 hours on 10 October. Oppenländer's men came under the control of 51st Army Corps at that same moment. The bulk of the division marched into the Gorodishche area the next day.

– – –

During a conference at corps headquarters at Gumrak attended by army commander Paulus, corps commander Seydlitz and several other division commanders, Oppenländer was not surprised to hear that his relatively fresh division was to play the main role in conquering Stalingrad's industrial district. In a nutshell, Gruppe Jaenecke (14th Panzer Division reinforced by Regiment 577, 305th Infantry Division minus one regiment, and 389th Infantry Division) would advance east, occupy the Tractor Factory, and then turn south to roll up the Soviet positions along the Volga all the way to central Stalingrad. As each sector was taken, the previous German garrison – assisting Gruppe Jaenecke with fire support as it swept southward in front of them – would shift forward and set up defences along the Volga. Elegant on paper, difficult to coordinate in reality.

No-one doubted that fighting through to the Tractor Factory would be costly, but once achieved, it was believed that the relentless application of force against the open flank would enable the entire Soviet defensive system to be rolled up in one fell swoop.

Storm's Eve (13 October 1942)

In light drizzle, Grimm's company was driven by lorry into the Gorodishche Gully during the night of 12-13 October, "our starting point for the battle" recalled Grimm. In 1978 he began typing up his Stalingrad experiences from memory, assisted by various notes, photos and correspondence collected during the war. If completed, it would be one of the most detailed accounts available, but every word of it was agony for Grimm. He would later write that "it was almost torture to report about operations in Stalingrad because dreadful details were remembered and my thoughts got stuck on them." He began his report in the third person but quickly switches to first person:

At around midday [on 13 October] at the battalion command post, our battalion commander, Major Beismann, issued an order for the assault on 14 October to his company commanders: Oberleutnant Schaate of 1st Company, Oberleutnant

Grimm of 2nd Company and Oberleutnant Beigel of 3rd Company.

Second company would attack at the spearhead of the battalion and the division via the stadium. Oberleutnant Grimm reconnoitred the approach routes through the Gorodishche Gully and made contact with a regimental staff – I no longer remember which one – for the purpose of being briefed about the assault positions. To some extent it screened entry into the northern part of the city, there was no other way in my opinion.

Before a big attack, radio silence was maintained as much as possible and so-called orderly officers were employed to transmit orders. Those called upon were veterinarians, paymasters, bandmasters and junior officers who had not been allotted for the attack. At irregular intervals the enemy took these approach routes under fire, whereby rocket-launchers (known as Stalin organs) became the most feared. The way to the briefing was lined by countless orderlies, either dead or wounded. That they were messengers carrying orders was apparent by their tidy uniforms – not threadbare at all – and they did not look emaciated like those of us who had hundreds of kilometres of marching behind us. Besides that, almost every officer in the forward line covered his shoulder straps and most wore enlisted men's marching boots, not officer's riding boots. Officers lay there wounded and powerless, moving their mouths as if they wanted to say something, but no medics could be seen far and wide and no-one else helped them, we just moved past them because an assignment had to be fulfilled. In a 'balka', which was a narrow ravine – this one was about two-hundred metres long, twenty metres wide at the top and looked to me like a tear in the earth with a skinny path at the bottom – I spotted the tactical sign on a pole. I crawled into the hole lying behind it. Inside it was a low cave, apparently scratched out of the sandy loess soil, which only allowed room to sit, not stand.

The purpose of Grimm's visit was to liaise with the unit currently holding the position and be briefed about the attack area. He was in for a shock:

By the light of a candle I recognised an Oberst and reported my mission to him. "What? You're supposed to attack there with your company? Utter madness! We attacked there with two fully replenished regiments and were wiped out, what's left of them you can still see out there" (he meant out in the balka). With tired words, totally exhausted, he gave directions and I headed back out into the iron-filled air. When I reached the supply train, the company fell in, specially equipped with close combat weapons. I permitted them to stand down, wrote to Lotte that my most difficult assault was just ahead of me and gave my photo to someone who was heading home.

Grimm felt like he was on a precipice. Beigel relished the opportunity to prove himself a great company commander in combat. In an earlier letter, he stated that "when I go into action with my new men, I'm convinced that everything will work out fine, like in my old platoon." His beliefs would soon be put to the test. Shortly before moving out, he dashed off a quick card to his girlfriend: "There is not much time to write. I'm always thinking of you. You'll always be my talisman." Grimm was able to compose a much lengthier letter, his farewell note. As always, he did not want to worry Lotte, so he refrained from expressing his pessimistic outlook – which is clearly visible between the lines anyway – but he did set his affairs in order:

My dear and good wife, my dear child,

When you receive these lines, another tough mission will be behind me and the swell of fighting will have faded away. Our Führer himself is supposed to have ordered our attack, so you will appreciate its importance. I still have to make the final arrangements with my company. Unfortunately, the weather has turned.

We are all aware of the harshness of the upcoming battle. Our struggle will be crowned with success because we know the Führer is behind us. Stalingrad will fall!

For operations in the east, the Assault Badge has joined my Iron Cross Second Class. The certificates are enclosed. The certificates for my Westwall Medal and War Merit Cross Second Class with Swords are in the folder.

I've recorded the entire Eastern operation in two notebooks. There's a small report for each day. After the war I'll receive many enquiries, which I can now answer based on these reports. These notebooks are therefore particularly valuable and have cost me a lot of effort. They are composed by me and are my property.

On Gisela's first birthday [16 October] the fighting will probably still be raging. But as it is raining heavily, the attack may be postponed. On that day I will remember you. Mail will be sparse in the next fourteen days. I will not be able to write.

Please greet your parents for me.

He then handed his camera to Unteroffizier Theo Moller to quickly take a photo of him, a keepsake for his baby daughter if he was killed. Grimm was fully kitted out for battle: overcoat, binoculars, combat harness with map-case and magazine pouches, and in his hand an unloaded MP-40. Oberleutnant Staiger, oblivious to the gravity of the moment, strolled into shot, a wry smile on his face as he made a light-hearted quip. He called over Gefreiter Nuoffer and they posed for a group photo. Despite Staiger's frivolity, nothing could expunge the deep-seated dread from Grimm's stomach:

Oberfeldwebel Platzer reported to me that the company had fallen in. I announced the Battalion Order for the attack, as well as the results of my briefing, and stated bluntly that we were about to face our most difficult attack, and much else. I also ordered the company to march in dispersed formation through the Gorodishche Gully into the assembly areas for the attack and told them where the closest first-aid posts were. I stipulated details for supply of ammunition and rations and for that purpose only took the two-wheeled infantry carts with ponies into consideration. The companies were informed upon arrival that the allocated assault divisions would attain the objectives in three days of attack and must adjust to it and then they would be relieved. Now, I do not know whether our commander volunteered us for this or whether he wanted to give himself courage.

I could not conceal my dejection. There was no room in my head for confidence. The thought would not go away. Thus something happened that the company had never experienced under my command. I went first to Oberfeldwebel Platzer and shook his hand, then went along the line, to every NCO and soldier. Platzer walked over to me because he must have detected that something unusual was going on and said quietly, so that the men could not hear: "You're very sad today, Herr Oberleutnant. You know that at least a few of them will survive." "Indeed, Platzer," I said, trying to answer in Bavarian, "but I don't know if we'll be amongst them." The company moved off.

About half way I was ordered to go to the battalion command post and was picked up by a motorcycle. Major Beismann, Oberleutnant Schaate (1st Company) and Oberleutnant Beigel (3rd Company) were waiting for me to arrive and I heard in passing that the attack order had been changed. Major Beismann read out the order from Army that had arrived shortly before, then the attack order from Division and then I heard it: 2./Pi.Btl.305 will support the attack of Regiment

576... My outlook immediately brightened and what my brain processed in these
seconds remains unforgettable to me... I believe God smiled down upon me. For
the attack, I was removed from the command authority of my pioneer commander
and subordinated to Regiment 576.

Grimm felt that his commander had given him a suicide mission. Beismann liked
Beigel and Schaate but never warmed to Grimm, and while it is hard to determine if
Grimm had a mild persecution complex, there is little doubt that both men disliked
each other. The change in plan forced upon Beismann by his superiors was like a death
sentence being commuted for Grimm. Instead of spearheading the attack against the
main Soviet defences, Grimm's company was allocated a subsidiary role:

A motorcycle took me back to the company which was still marching cautiously
and in a widely spaced manner. Due to enemy observation from the air, the
company was not allowed to stop because the approach route lay within range of
heavy enemy artillery and, about five hundred metres from the first row of houses,
within range of Russian rocket-launchers.

I was able to show Oberfeldwebel Platzer the street map of the northern part of
Stalingrad which I had received at the briefing and on which I drew the attack
sectors of the division. To the left was 94th Infantry Division, north of them was
16th Panzer Division and to the right was 14th Panzer Division. As far as I can
remember, the assault frontage of the division on the first day of the attack was
only about one kilometre and we were deployed with Regiment 576 – subordinated
to 389th Infantry Division on the left – and not with our pioneer commander,
Major Beismann. When we were subordinated to the infantry, we always received
the acknowledgment that we'd earned. Then there was also three aerial photos on
which every house, whether destroyed or not, was clearly visible. I kept one and
gave the others to Oberfeldwebel Platzer and Feldwebel Pauli. I brought the street
map home with me, the aerial photo was lost.

The overcast skies brightened in the afternoon and the sun began to peek through.
At 1700 hours, after a five-hour march, Grimm's company reached their assembly area
in a plantation just west of Stalingrad-North. Even though the orchard was blocked from
enemy view by a shallow ridge, the NCOs ordered their men to spread out and dig in.
Random artillery rounds occasionally screeched in. Looking around, the men noticed all
the signs of an imminent major assault: fresh units continually arrived; heavy infantry

weapons were emplaced. Grimm's men were also heavily armed and well supplied:

The company was equipped for the attack as follows: light machine-guns, submachine-guns, five-shot Mauser rifles, various pistols and every man with a rifle and plenty of ammunition. My equipment with submachine-gun, 08 pistol and so on, can be seen in the photo. The flamethrower section, with a strength of one officer and four men, had two flamethrowers and would follow behind. The section leader, armed with a submachine-gun, led the section into action behind Flamethrower I (Müller) with escort and Flamethrower II (Schwarzer) with escort.

Being at the focal point of the attack, the company was outfitted with close combat weaponry, so that even the NCOs often had six kilos of hand grenades and satchel charges hanging off them. The magnetic charge used to combat tanks was rare... there might not even have been six of them.

I cannot provide exact figures from memory as to the strength reported to me by Oberfeldwebel Platzer prior to our departure. Oberleutnant Staiger – whom I knew well from France – had been sent to me as a reinforcement by Battalion. However, I did not want to reorganise my combat unit that had been forged in battle and this was probably yet another "little piece of help" by Major Beismann. I solved this problem by ordering Staiger to supervise the flamethrower section, which would dig in and remain on call about 400-500 metres behind the company. Oberfeldwebel Platzer and Feldwebel Pauli would not have expected anything else from me. – The statement that the company came to Russia by train with a total strength of 210 pioneers, including officers, is dependable. How many replacements the company received for its heavy losses in the east cannot be determined because included amongst them were many convalescents (lightly wounded and sick men who were always gratefully accepted). Countless were wounded or killed before they even reached the front-line, that is, the most forward trenches. I believe that I'm able to remember that on 13 October, a figure of between 90 and 95 men was reported to me (without Oberleutnant Staiger). The company still had seasoned NCOs, above all my pillars of strength, Oberfeldwebel Platzer and Feldwebel Pauli. I risked leaving the supply train strong, thus not taking along grooms and drivers, as well as several convalescents, for the attack on 14 October.

Soviet counterfire began thinning out the division's ranks as they moved closer to the city. Hardest hit in the pioneer battalion was Beigel's 3rd. Gefreiter Erich Lang was struck in the chest by a shell fragment and, according to the company's Spieß, Hauptfeldwebel Hagen, his "life came to a quick, painless end." One of Lang's comrades, Gefreiter Eugen Bammert, was also struck down in the barrage. While Beigel's men went into action the following day, Lang and Bammert were being laid to rest in a cemetery at Razgulyayevka.

To help his commanders cope with the strain of battle and keep going for days on end, Beismann asked the battalion doctor, Oberarzt Dr. Wirtgen, for pharmaceutical assistance. Pervitin, touted as a "multi-vitamin" by its maker but in actuality pure methamphetamine, was doled out to the company leaders. Its effects were adrenaline-like, triggering a heightened state of alert, increased self-confidence, concentration and the willingness to take risks, while simultaneously diminishing sensitivity to pain, hunger and the need for sleep. Pervitin was viewed as a drug to be used when soldiers were likely to be subjected to extreme stress, and Stalingrad certainly fit that criterion. This was the first time it had been issued to the battalion, as Grimm recalls:

> The battalion doctor gave a small tube to us company commanders. We were to take a few tablets from it. Pilots received these before strenuous night flights. I had never heard of them before, also nothing about them since. As far as I can remember, I took six of them and put the rest in my pocket. The tablets kept me awake for almost three days and I had the feeling that I was freed from all pain and suffering. The duty and responsibility for correctly leading a unit was so enormous that one cannot think like that under any circumstances. I heard nothing about Pervitin during the entire war and I still don't know today – because it does not interest me – whether it was good or bad.

Final preparations for the attack were completed by 51st Army Corps. Assault troops formed up on the outskirts of the city; the ramshackle suburbs began a few kilometres ahead. The sea of houses was dominated by factories, smokestacks and multi-storey buildings. "The Führer himself ordered the attack," Grimm recalled, "which should cause the city to fall. There was to be fire magic such as I had not experienced before."

Maelstrom (14 October 1942)

After a cold, uncomfortable halt in the orchard, the pioneers shouldered their weapons at 2300 hours and continued their night march. Strobing artillery fire interrupted the blackness as both sides hammered away. Grimm's company paused in a gully at midnight while swarms of Soviet planes blindly bombed the surrounding area. Once the danger passed, they moved on, steered by a local guide. Their final position was reached at 0200 hours. Grimm continues his account:

In accordance with the orders of the battalion commander of Inf.Rgt.576, the company moved into its jumping-off position for the assault under cover of darkness without any noteworthy casualties. I first had to report in at the regimental command post, then to a battalion commander, whose name I've forgotten. The company would not attack as a homogeneous unit; the sector was too narrow for that. The infantry were to be supported during the attack, enemy dug-outs stormed and blown up, trenches rolled up, and above all, fighting from house to house.

To receive further orders for the attack, I was directed to stay close to a headquarters that had a land line. This was in a gully roughly 150 metres long, about twenty-five metres wide at the top, steeply sloping walls about fifteen metres deep and a narrow beaten path along the bottom. Alongside me were three messengers who would maintain contact with the platoons of Oberfeldwebel Platzer and Feldwebel Pauli. The enemy had previously used this gully for shelter and halfway up its sides foxholes had been scratched out or driven into the loamy earth, probably with spades. Since the slope was steep, the spoil served as steps so you could slip in head first. We believed we had found good cover here until called. Barely half an hour passed before I heard my name called. I forced myself to my feet as quickly as possible, out of the foxhole, slid down the slope on the excavated spoil and heard the drumming of heavy gun fire, immediately followed by the hiss of approaching Stalin organ projectiles. I ran the short distance to the headquarters, shells came down and I dived headlong into the hole, dragging some wires along with me. A man leapt in after me and bowled me over. He was as good as dead. It is simply indescribable what happened here in half an hour. The headquarters played no part in the battle because all the wires were ripped to shreds. After my later experiences in battle, I almost came to expect a headquarters to also have a radio post. The Russians always spotted the wires and disrupted them quick smart by dropping bombs from airplanes during the night or with heavy artillery.

And that is as far as Grimm got with this report. Unfortunately, it remains unfinished. He appended the following in May 1989:

I have carried on this manuscript from 1978 in my mind over hundreds of nights but cannot summon the strength to continue it. I cannot put my inner conflict down on paper: should I continue the report or is it better not to? I cannot fathom what is forcing me to continue the report. For me, the war continued for years and when I have a fever, when I'm ill, my thoughts – which are morbid, I have to say – always return to that four-week period in Stalingrad. Perhaps that will also happen during my final hours.

This paragraph speaks volumes. Grimm was a fastidious man and a conscientious writer who typed up hundreds of pages documenting his war-time experiences and penned thousands of letters during and after the war. Despite his penchant for writing, he couldn't bring himself to complete his report. Grimm faced a Catch-22: his failure to accurately record his Stalingrad experiences tormented him, but when he tried, traumatic memories were re-animated and they haunted him whenever he closed his eyes. In February 1985 he confided that "this drama will burden me until the day I die." Perhaps the completion of this report would have given him some closure? It will never be known because cancer ended his life in May 1990. Fortunately, details of Grimm's time in Stalingrad are scattered throughout his correspondence, and whilst they are not as detailed as his report would have been, they still provide many crucial details.

– – –

Flares hissed skyward at random intervals. Bombs and shells cracked throughout the night. The intention was to make it seem like a continuation of the previous week's attritional fire. Artillery, mortars, rocket-launchers and Mörsers steadily pounded the defences from 0400 hours. Soviet artillery responded. As kick-off time approached, the intensity of German fire increased. Thirty minutes prior to X-hour, bombers began dropping the heaviest bombs on the settlement west of the Tractor Factory. Stukas joined in ten minutes later, their staggered runs ensuring a Catherine wheel of dive-bombers over the target area. The Luftwaffe would carry out 1,250 sorties on this day.

A massive fireworks display erupted on schedule. The pioneers had never experienced such a fierce preparatory bombardment. The ground bucked under the merciless pounding. Recognition panels clearly marked the front-line for the Luftwaffe but they were more a precaution than anything else because targets were deep within the Soviet defences. Artillery and mortars worked over the buffer zone between Luftwaffe bombs and the German forward positions. Some men smoked furtively. Others glanced upward to watch the incredible spectacle. The sun glinted off canopies as Stukas tipped over. The sky was filled by diving planes, puffs of black smoke and whizzing shells. A light westerly breeze pushed the cloying clouds of dust and smoke towards the Volga, over the Soviet rear, thus leaving clear air above the front-line and enabling artillery observers to keep hitting their targets.

After the fierce preparation, Gruppe Jaenecke moved off at 0730 hours. It seemed impossible that anyone could have survived the bombardment but as soon as the panzers and accompanying footsoldiers moved forward, muzzle flashes flickered from every ruin, every gully. First to move in the sector of the Bodensee Division was Hauptmann Denz's I./Regiment 576, temporarily subordinated to Winzer's regiment for the first stage of the attack. Denz's task was to push north and eliminate any possible flanking effect on Winzer's troops. The main assault would not proceed until this was achieved. They swiftly blasted their way through the chimney graveyard, hurling grenades and spraying opposition with gunfire. Though stunned by the vicious preparatory barrage, the surviving Guardsmen of 2/109th Guards Regiment (from General Zholudev's tough 37th Guards Rifle Division) began to offer fierce resistance and took down several of Denz's soldiers. Nevertheless, the assault moved efficiently and relentlessly northward within a few minutes. As the Guardsmen were busy fighting for their lives, all flanking fire from this area was eliminated, so Oberst Winzer and Oberstleutnant Gunkel[1] were given the go-ahead to launch the main assault.

Both regiments had weighted their forces along their common boundary. First objective was the stadium because it would split the flow of the main attack if left in Soviet hands. A new barrage commenced and the troops set off. The combined attack aimed precisely at the boundary of 109th and 114th Guards Regiments. Every five minutes German artillery fire shifted fifty metres ahead, and in those five minutes, the fifty metres of ground had to be taken. Artillery observers monitored the attack; each infantry battalion had a minimum of two artillery batteries on call.

Gunkel's and Winzer's regiments worked together to approach the stadium, cautiously blasting their way down the slope towards the objective. A fresh relay of Stukas paved the way, followed by a furious artillery barrage. Binoculars to eyes, squad leaders surveyed the stadium: the playing surface was pitted by a few mortar craters, although the white goal posts still stood, and the wooden bleachers were mostly intact, as was the wire boundary fence. A large building, gully and spur railway in front looked to be greater obstacles. It was up to Regiment 578 to capture the stadium itself but Regiment 576 needed to clear the huts on the slope that overlooked it. As the ground

---

1   Oberstleutnant Werner Gunkel was the senior battalion commander of Infanterie-Regiment 576 and as such was temporarily leading the regiment while its commander, Oberst Krüder, was on furlough.

troops and assault guns methodically worked their way down the slope, they became more exposed to fire from the multi-storey buildings ahead, and when forward motion was stalled by opposition from a bunker, the pioneers were called upon to knock them out. Schaate's 1st was attached to Hauptmann Althenn's II./Regiment 578 as a shock company. During the advance upon a block of houses, the entire battalion was held up by a Russian bunker. Feldwebel Grosskreutz, leader of the 3rd platoon, was ordered to deal with it, and he in turn delegated the job to Unteroffizier Josef Schiele and his 8th Squad. Gefreiter Hermann Heeb, a 33-year-old Hessian from Wiesbaden, volunteered for the assault detachment. Heeb had distinguished himself throughout the summer with exceptional bravery and willingness to fight, so few were surprised that he put up his hand for this dangerous task. Despite frenetic rifle and machine-gun fire whipping about their heads, the assault detachment worked its way up to the bunker. Heeb heaved in a satchel charge that destroyed the bunker and its crew, but Schiele's squad paid dearly for this success: Gefreiter Josef Zeller and Pionier Hermann Beisswenger were both killed when struck in the head by mortar fragments. Because of Heeb's deed, Althenn and his battalion maintained momentum and continued the attack. Oberleutnant Schaate recommended Heeb for the Iron Cross but as with many deserving men in the battalion, it would be denied to him through a series of unfortunate circumstances.

At the tip of the attack, Beigel and his company advanced into a hailstorm of fire:

After a powerful fire preparation, we attacked into the occupied city covered by Russian as well as German drum-fire. The fighting was fierce and mad because house-to-house combat of this type is scarcely surpassed in cruelty. With my company I was subordinated to the spearhead battalions of the infantry. A thousand metres from the Volga. We attacked the city's houses from west to east. Fate took me from the battle. It is extremely fortunate when one emerges relatively intact from such a hellish place. The bullet went through the front of my lower right leg, piercing my shinbone. It was not that bad, however, and later on, nothing at all could be seen of it.

Leutnant Hepp assumed command of the company.

Grimm's 2nd Company reinforced Gunkel's Regiment 576 during its attack north of the stadium. While Denz's battalion remained behind to clear out the upper settlement, the other two battalions (Major Braun's II. Bataillon and Major Emendörfer's III. Bataillon) continued to battle their way down towards the stadium. Oberfeldwebel Platzer's platoon supported one infantry battalion, Feldwebel Pauli's the other. Assault guns were interspersed. The mixed combat groups slowly advanced from ruin to ruin, losing men all the way. The first soldier of Grimm's company to die was Gefreiter Josef Reisinger, a young Bavarian. Grimm would later console the father with these words:

Your son Josef fell during an attack while fulfilling his military duty, true to his oath for the Fatherland... May the knowledge that your son gave his life for the greatness and the survival of our people, Führer and nation be consolation in the heavy suffering that has befallen you.

The creeping artillery barrage quickly outpaced the attack, and as time wore on, opposition stiffened, despite assault guns and demolition teams blotting out one position after another. Gefreiter Hermann Moser was instrumental in helping the attack reach the stadium by destroying several resistance nests in close combat and opening the way for the infantry.

The assault prised the Soviet defensive line wide open. The 109th Guards Regiment resisted for several hours but was crushed in the middle and then carved up into small pockets. This irresistible phalanx of armour and pioneer-backed infantry surged eastward, smashed the left wing of 524th Rifle Regiment, seized a church and approached the stadium. The German attack then diverged: Regiment 578 and supporting armour moved along Kultarmeiskaya and Ivanov Streets toward the Tractor Factory, while the two battalions of Regiment 576 headed along Kooperativnaya Street towards Mokraya Mechetka Street. After crushing 524th Rifle Regiment, Gunkel's Regiment 576 lunged deep into the rear of 416th and 385th Rifle Regiments.

Gunkel's infantrymen stormed the massive apartment buildings that dominated all of Stalingrad-North. Fire from rockets, mortars and artillery reaped a steady toll on the attackers, but bullets came in flurries when a Soviet strongpoint was approached. Caught by one of these bursts was Grimm's stalwart deputy, Oberfeldwebel Platzer. In his last letter home, he had told his family "you don't need to worry about me, not every bullet finds it mark." Platzer was quickly hauled into cover. Eyes gazed fearfully at his blood-drenched abdomen. Everyone knew he was a goner. He writhed in pain, screaming over and over: "Commander, my pistol!" He wanted Grimm to give him a gun so he could put himself out of his misery, but his company commander was not there. "For the attack," Grimm recalled, "we pioneers were split up amongst the battalions, so I was not near him. Readers of these lines should ask themselves this question: what would they have done in my position?" Despite Platzer's heart-rending plea, a pistol was not placed in his spasming hand; instead, he was sedated. Pionier Gustav Schönberger helped evacuate his wounded platoon leader. Before the big attack, Schönberger had been promised the Iron Cross for his work during many nocturnal mine-laying operations, but had not received it. As Platzer lay there, eyes glazed, slipping in and out of consciousness, he recognised Schönberger during a moment of clarity and reaffirmed that he would get the medal. An assault gun was flagged down and Platzer loaded on the engine deck. He made it to a dressing station, and then back to the hospital set up in the Gorodishche church, but stomach wounds are feared for good reason. Platzer died in agony the following morning and was buried in the adjacent graveyard. The Catholic priest who presided over the burial wrote to Platzer's father:

I was by your son's side in the final hours, when he could no longer speak. He received the Last Rites and passed over quietly and calmly into the eternal home.

– – –

Gunkel captured his assigned sector by mid-afternoon, so he reformed his units and advanced westward to clear out more high-rise complexes. A death blow had been dealt to the Soviet units defending the area.

Obergefreiter Rinck scribbled some brief notes in his diary about the day: "Tough enemy resistance. Daily objective reached." The bulk of his entry was a long list of comrades killed or wounded during the attack. Killed: Gefreiter Josef Reisinger. Wounded: Oberfeldwebel Wilhelm Platzer, Obergefreiter Hepp, Obergefreiter Hans Pforte, Obergefreiter Fritz Prottengeier, Gefreiter Schingnitz, Gefreiter Anton Vogelbacher and Pionier Ferdinand Dammert. Obergefreiter Pforte, a squad leader, had been felled by a piece of shrapnel that pierced the right side of his abdomen and damaged the large intestine and pelvis. Another fragment struck him in the right eye and caused an almost complete loss of visual acuity. His injuries were so severe that he was still in hospital sixteen months later. Soon after, he was deemed physically unfit for active service, discharged from the Wehrmacht and classified as a level 3 disability. One consolation was a long-delayed promotion to Unteroffizier and the consequent increase in benefits.

The wounding of Obergefreiter Fritz Prottengeier underlines the tragic randomness of war. Some men received such gruesome wounds that it seemed their death was certain, yet they survived. Prottengeier was the opposite: struck in the neck by a tiny fragment, he was bandaged by a medic and walked from the field of battle unassisted. None of his comrades thought the injury life-threatening. As chaos reigned, it was a long time before his company discovered that he died ten days later. The head doctor informed his parents that he was admitted on 23 October but died of his "severe wounds" at 0630 hours the following morning "despite all medical help." In June 1943 Prottengeier's fianceé, Marianne, contacted Grimm to find out when, where and how he was wounded, even though it "is very hard for his parents and myself." Grimm was perplexed. Over a week had passed between Prottengeier's wounding and death:

> At the time, none of us thought that your fiancé would succumb to his wounds because he was relatively alert and well when, without any assistance, he left the battlefield [on the first day of the assault in Stalingrad].

This revelation floored Marianne because the quick death of her betrothed had now morphed into a lingering one. The distraught girl handed over the matter to Fritz's sister whose main concern was to prevent the truth from reaching her parents:

> Understandably, as my brother's former company leader, you must inform my parents, but please write very carefully, don't mention the date when Fritz was wounded. My momma lives with the belief that her Fritz was wounded on 23 October and died on 24 October (that is what the hospital wrote)... Under no circumstances should you tell my parents the complete truth, [they] could not bear it. Fritz meant the world to us.

Nothing is known about the intervening time between wounding and death, but a clue may lie in the place where he died: Leichtkranken Kriegslazarett 3/541 in Oblivskaya, west of the Don, was a military hospital that treated minor cases and was not the place to send a nasty neck wound. It is likely that Prottengeier arrived there soon after being wounded, and in an effort to ease the suffering of his family, the doctor lied about the date of his arrival. Thirty-one-year old Prottengeier was buried with military honours behind the village church in Oblivskaya.

— — —

The assault continued to make good headway on some sectors. With its left flank covered by Regiment 576 and its right moving alongside the surging armour of General Heim's 14th Panzer Division, Oberst Winzer's regiment was able to proceed with confidence, advancing along Komsomolsk and Kultarmeiskaya Streets, seizing one huge building after another, ably assisted by the pioneers of 1st and 3rd Companies. Unteroffizier Richard Hartmann, a hardened squad leader in Schaate's 1st platoon, was killed during this merciless fighting. Soon, Winzer's group was lodged deep in the Soviet line. The heaviest resistance now came from a massive H-shaped building that sat just north of the boundary with 14th Panzer Division. As Winzer's objective was the capture of sectors further north, this troublesome edifice was simply cordoned off. Little did the Germans know that several Soviet units had their headquarters in this heavily damaged brick edifice, known to its defenders as the House of Professors.

Oberstleutnant Gunkel suspended his attack into a district containing two dozen massive apartment blocks and a hotly contested school building. A couple of the multi-storey complexes were in German hands, so Gunkel ordered his men to temporarily take up a defensive posture. Oberst Winzer's regiment achieved its objectives. Progress was even better in 14th Panzer Division's sector: utilising a breach made by its panzers, a panzergrenadier regiment broke through a bunker line after nightfall, thrust through to the Volga and hedgehogged there. Regiment 577, temporarily subordinated to the panzer division, was also able to cross over the railway near the Brickworks. The Soviet defences in northern Stalingrad had been cut to ribbons. Casualties in the Bodensee Division were 3 officers and 84 men killed, 7 officers and 276 men wounded, and 15 men missing; ghastly for the men involved, yet numerically low in comparison to the stunning blow inflicted upon 62nd Army, which had suffered almost 10,000 casualties.

At 2300 hours, General Oppenländer issued fresh orders to his division: "Despite the strongest application of our weapons, the enemy has fought back stubbornly; he will also offer the toughest resistance tomorrow." The main objective was simple: "The division will continue the assault on 15 October and gain the Volga bank in its allotted sector." This task fell squarely upon the shoulders of Oberst Winzer. While Gunkel's Regiment continued to clear the workers settlement, Regiment 578 was ordered to sweep through the Tractor Factory and reach the river.

The German leadership shuffled around its units to prepare for a renewed assault. The pioneers were called upon to help laboriously clear the mammoth apartment complexes. The potential for resistance was terrifying. Each multi-storey building was fifty metres long and filled with stairwells, corridors and rooms, while beneath was a labyrinthine basement with high-set windows staring out across fire-blackened garden beds. The following account appeared in the German national newspaper "Völkischer Beobachter" and perfectly captures the nightmarish fighting that took place during the night of 14-15 October and the vital role performed by the pioneers:

A single flamethrower brought about the fall of the building complex. The flamethrower operator was from our third platoon, and four or five men gave him cover. The rest of us kneeled behind chimneys, which loomed in the blood-red sky like a finger raised in warning, behind ruins of walls and in hastily dug fox-holes, keeping the windows under fire and allowing our comrades to cross the street unhindered. Enemy riflemen in the windows were suppressed for a few seconds. The flamethrower team used the break in the fire to reach the block's main entrance. It was barricaded, blocked from the inside.

The team remained in the dead angle of fire from the house. One of the pioneers ran back to fetch two concentrated charges. The Soviets no longer showed themselves in the windows but hand grenades flew out onto the street, wobbling between the fragments of shattered window panes. In the glare of the detonation we saw the assault troop scramble into the shelter of a cellar hatch. – It seemed like ages until the comrade ran back over the street with the charges. Unconcerned about the dance of exploding eggs, he packed the explosives against the door. The assault troop quickly warned us. "Fire in the hole! Take cover", somebody called out, as if on peace-time exercises on a training ground. And in the bursting of the charge, which ripped huge chunks of masonry from the ground floor together with the door, a breach was blown open. A burst of submachine-gun fire hissed into the corridor, the assault troop jumped up and stormed into the interior of the apartment block.

Anxious minutes of idle waiting followed for those of us outside. There was no more firing in the house. The Soviets were silent. Nothing was heard of German weapons. – Then a blaze suddenly flashed behind the window openings on the top – the fifth or sixth – floor. Black smoke billowed out in spherical plumes. The flamethrower! Minutes later the same spectacle occurred one floor down. Before the flaming death flickered from behind the windows on the third floor, a brief shoot-out reverberated from the stairwell. The final shot came from a German submachine-gun. Hurrah!

A pair of shadows crept out onto a balcony on the fourth floor. They didn't last long because the fire also caught them there. As our assault troop smoked out floor after floor and finally came bounding out of the entrance, the building complex was well and truly ablaze. The surrounding area flickered in the light of the huge blaze. Outside, to the left and the right, bursts of fire rattled from German machine-guns. Attempts by the Soviets to intervene were quickly frustrated in the rubble of neighbouring streets now illuminated as bright as day.

From up on the balcony came an animal-like cry of fear. One of them swung over the railing in order to dash himself to pieces on the street. The others could not find the strength to hasten their demise. As walls began to collapse and iron girders in the central storeys – already glowing red – bent and groaned, a throng of defenders

burst out of the main entrance and the ground floor windows. A demoralised,
shattered horde who only remained upright with great effort. They emerged
without weapons, faces soaked in sweat, blackened with soot and distorted in
horror. They raised their hands, staggered, reeled, and fell down the steps into the
open. Forty survivors from three hundred men. For a quarter of an hour the sound
of whimpering and insane shrieks came from those trapped in the fire, buried
under collapsed walls, wounded by our bullets. The fire consumed them, without
anyone being able to help.

Despite lip service about holding out and fighting to the death, many Guardsmen relinquished their strongholds during the night, not just the rank-and-file, but also the commanders and their staffs. The command groups encircled in the House of Professors made a break for it. This exfiltration process continued throughout the night.

Resistance Crumbles (15 October 1942)

The sleep-deprived pioneers were greeted by welcome news as the new day dawned: during the night and in the wee morning hours, panzergrenadiers of 14th Panzer Division had broken through to the Volga east of the Tractor Factory and Major Brandt's Regiment 577 had pushed over the railway line near the Brickworks and then captured it. As a result, the southern half of the Tractor Factory was in German hands even before the scheduled operation began.

The attack against the rest of the factory commenced at 0700 hours. The Luftwaffe had been bombing it since 0630 hours, concentrating on the northern half and the vast settlement of multi-storey buildings to the east. When the infantry moved out, the artillery opened up a rolling barrage, shifting forward fifty metres every two minutes. Winzer's orders were dead simple: "Storm the northern part of the factory with the main effort placed on the boundary with 14th Panzer Division and push through to the Volga." In support were two companies of pioneers (Schaate's 1st and Hepp's 3rd),[2] assault guns, a company of towed 5cm anti-tank guns, forward artillery observers and a battalion of rocket-launchers. The pioneers were unnerved as they approached the factory. Why was it so quiet? Where had the Russians gone? Even their assembly areas were amongst buildings that he been hotly contested barely twelve hours earlier. Corpses were strewn everywhere but were living Red Army soldiers still about, lurking somewhere below ground? Only when the preparatory barrage shifted forward and

---

2   Leutnant Hepp assumed command of 3rd Company after the wounding of Oberleutnant Beigel, though
    only temporarily because Oberleutnant Staiger took over command in the coming days.

little resistance was encountered did the men begin to gain hope that the enemy had absconded. The wall of fire preceded the advance, first across the factory's forecourt, then into the massive workhalls, flinging corrugated sheets, ironwork and concrete chunks high into the air. Natural killing grounds were crossed with little difficulty. The contrast with the day's previous intensity could not have been greater. The lack of resistance was due to a cool-headed decision made by the opposing Russian army commander, General Chuikov: relinquish the tractor factory in favour of setting up a new defensive line further south. As a result, Winzer's assault pushed into a void.

Schaate's and Hepp's pioneers escorted the infantry into the empty factory grounds. Entering the Dzerzhinsky Tractor Factory was a fearful experience. Ahead and on the left were vast workhalls offering countless hidey-holes for snipers and machine-gun teams. To the right was an immense open yard filled with tank carcasses and turrets. The men felt they were walking into a trap, yet, they kept advancing, encountering only spotty resistance that was swiftly obliterated. Two hours was all it took for Winzer to capture the northern half of the factory. Momentum led to the easy capture of the settlement and installations east of the factory and the Volga was reached mid-morning.

In the meantime, Grimm's pioneers, in support of Regiment 576, continued the clearance of the workers settlement. An hour before the operation began, a shell landed in Grimm's assembly area: squad leader Obergefreiter Josef Ostermeier was wounded while one of his men, Pionier Wilhelm Käpernik, was killed instantly by a large fragment to the chest. Ostermeier recounted the death:

Pionier Käpernik came to our company, into my squad, on about 10 September. He was deployed with the company in the blocking position between the Don and Volga. He participated in the attack on Stalingrad on 14 October. On 15 October, at 0600 hours, during preparations for the attack, he was struck directly in the heart by a shell fragment. He was killed outright. I removed his valuables, which I handed over to Gefreiter Josef Maier. Because I was wounded by the same shell, I cannot state when and where Käpernik was buried. From comrades in the squad, nobody, except for me, is with the [reformed company in August 1943].

Ostermeier headed to the rear for treatment. His wounds were not severe enough for evacuation to the homeland but enough to keep him from front-line service. He would remain in hospital in Konstantinovka for the next six weeks. As recognition for his performance throughout the summer, but especially during the first day in Stalingrad, Grimm would award Ostermaier the Iron Cross First Class on 25 October.

Preceded by a ground-rocking preparation by Stukas, artillery and rocket-launchers, the assault squads set off at 0700 hours, attacking westward. Block after block was cleaned out with assault gun support. Resistance was light and the daily objective reached at 1400 hours. Covering parties were then put out facing the Mokraya Mechetka. The deep ravines were still off-limits: they were a known refuge for desperate Russian fugitives. Grimm's casualties until this point were tolerable from a statistical perspective but not for the men whose lives changed forever. One of them was Pionier Robert Ramsperger, a replacement who had arrived a month earlier. A sledgehammer blow to the head took him out of action and he was still in hospital a year later:

I was wounded on the second day in Stalingrad by a shot through the face which took my left eye. A chunk of my nose-bridge was also ripped off, and several teeth smashed in. I'm in hospital now, where I'll be for a long time before I'm released. It's a funny feeling, having to lie around when every man is needed outside.

Another man wounded during this mopping-up operation was 29-year-old Pionier Heinrich Wildum. According to Grimm, he had "repeatedly distinguished himself through his guts and courage during hand-to-hand combat in the northern part of Stalingrad as part of demolition squads and covering detachments. He considerably contributed to the destruction of enemy resistance nests." Shell fragments to his right leg took him from the battle.

During a mid-morning meeting at 51st Army Corps' command post, Paulus laid out his plan for the continuation of the assault. Oppenländer's infantry and Heim's panzers would attack south from the Tractor Factory along both sides of the railway, while elements of 24th and 14th Panzer Divisions would tie up Soviet forces by attacking the flank from the west. Initial attack objective: Barrikady Gun Factory and the Bread Factory. Paulus stressed that every success be exploited in order to roll up the entire Volga front from north to south, in one go if possible.

First, however, all small pockets of resistance needed to be eliminated. After a quick regrouping and swinging the attack front to the north, Winzer's regiment started cleaning out riverside sectors. Schaate's company supported one infantry battalion during its task to capture a cluster of buildings north-east of the Tractor Factory and mop up the point between the Volga and the Orlovka Brook. Proof that resistance was tougher is shown by the fact that 1st Company gained a Sturmtag for this day ("Attack on Block 23") and at least three of Schaate's men were killed: Obergefreiter Bernhard Lumpp (2nd Squad) and another two men from the already-weak 8th Squad (Gefreiter Ernst Hensle and Pionier Emil Zielke). Many more were wounded, including the leader of Schaate's HQ section, Obergefreiter Ernst Fahrion, struck in the right shoulder by a chunk of metal from a Russian hand grenade, but he remained on duty.

Leutnant Hepp's 3rd Company was on the other side of the Tractor Factory. Towards midday it helped clear a large block of houses, which cracked open the Soviet defence. The remaining buildings were rapidly seized and soon the entire southern bank of the Mokraya Mechetka was in German hands. Leutnant Hepp was wounded twice, first by a shell fragment, then by a bullet that lodged in his left upper arm. Once admitted for treatment, he fell seriously ill with dysentery. Oberleutnant Staiger assumed command.

The Dzerzhinsky Tractor Factory and its satellite settlements were soon wholly under Paulus' control. There was no time for the Bodensee Division to rest on its laurels: it was to be redeployed immediately. Regiment 578 was relieved from its positions along the Volga during the night and ordered to assemble amongst the high-rises west of the factory. Gunkel's Regiment 576 was ordered to remain in position along the high ground overlooking the Mokraya Mechetka. Grimm's pioneers helped to defend this line against Soviet counterattacks surging up out of the broad river valley. Grimm's casualties mounted. During the repulsion of one counterattack, Pionier Otto Sellmer, who contributed significantly to holding the embattled area while under fierce bombardment from rocket-launchers, was wounded by shrapnel in the right upper arm and shoulder. Pionier Schönberger, who had helped hoist Platzer on to the assault gun, was wounded in the evening. He asked about his grievously platoon leader upon arrival at the hospital at Gorodishche and was devastated to learn that he had died.

The toll for 2nd Company ended up quite high: three dead (Gefreiter Alois Kuchler, Pionier Wilhelm Käpernik and Pionier Franz Oberhardt) and at least eight wounded: Obergefreiter Josef Ostermeier, Gefreiter Otto Sellmer, Ellinger, Pionier Franz Ihle, Pionier Benedikt Klaiber (died of his wounds 9 November), Pionier Robert Ramsperger, Pionier Gustav Schönberger and Pionier Heinrich Wildum. Total daily casualties for the entire division were 53 men killed, 7 officers and 122 men wounded, one man missing.

Still stimulated by the handful of Pervitin tables ingested 48 hours earlier, Grimm worked non-stop through the night, arranging supplies, checking on his men in front-line positions, liaising with infantry leaders. At one point he received word that Major Beismann wanted to see him, but such requests were required to go through the current chain-of-command (from Beismann via Regiment 576 to Grimm). "Without permission, he summoned me to his command post during the night of 15-16 October, almost certainly just to make sure that I maintained contact with him." Fighting the enemy was hard enough; battling your own commander was another burden Grimm didn't want to take on, but he had no choice, so he told Feldwebel Pauli to look after the company during his absence and dashed off. The short trip to and from Beismann's dug-out almost cost Grimm his life. Soviet artillery impacts burst all over the place and night-bombers circled slowly overhead, dropping flares and then bombs when targets presented themselves. Grimm dived into cover every dozen metres or so and crawled through ruined buildings until finally reaching the comparative safety of a ravine. He was quite a sight when he presented himself to Beismann. As Grimm presumed, it was simply Beismann's way of keeping him on a short leash. He demanded more frequent reports, quickly updated Grimm on the actions of the other companies, and then sent him on his way. The return trip was nightmarish. The night-bombers rained down small incendiary bomblets that splashed unquenchable white phosphorous fragments all over the place. Grimm ran for his life, dodging and weaving until he escaped the beaten zone: "I had a lot of trouble getting the burning phosphorous off my boots."

Fortuitously, Beismann's unscheduled summons may have saved Grimm's life, or at least spared him a painful wound, because when he returned to the gully where he had been sharing a foxhole with his HQ section leader, Unteroffizier Theo Moller, he found his overcoat – left behind during his visit to battalion HQ – riddled with holes and flecked with Moller's blood. And it was not only Moller who was missing: Grimm's three messengers were also gone. Not until the next morning did Grimm learn that Moller was alive and well in hospital, carried there by one of the runners. Moller was not upset about his injury: "I now see very clearly that I actually had good luck with my light wound at the time because it got me out of the hell of Stalingrad." Moller was an enthusiastic photographer and had been recording the company's Eastern Front adventures: his Leica was "wounded" at the same time, but it had prevented another fragment from slicing into his body.

Into the Gun Factory (16 October 1942)

Throughout the morning the remaining units of Regiments 578 and 576 were relieved by 94th and 389th Infantry Divisions and assembled west of the Tractor Factory, out of sight of Soviet artillery but still within range. In the meantime, at 0800 hours, 14th Panzer Division relaunched the offensive southward. Major Brandt's Regiment 577 moved off from the Brickworks area to take the Barrikady Gun Factory and push through to a line south of the Bread Factory. Supported by assault guns, the regiment was deployed between the railway and the Volga, pushing forward along the main north-south railroad with its right wing. Brandt's men pressed into the northern part of the Gun Factory at 0900 hours. Just a few hundred metres to the east, a Russian counterattack from the factory's lower settlement stalled Brandt's left wing, but his right wing continued to plunge deep into the factory grounds, capturing the railyards and workshops along the western boundary. Some of the larger workhalls also fell into German hands. The panzers and grenadiers of 14th Panzer Division west of the main railway line were having a harder time after they encountered dug-in Soviet tanks and then took frightful losses from well-placed rocket barrages while stalled. The morning's promise of a swift victory dissolved. The attack needed bolstering.

The bulk of 305th Infantry Division was held in reserve, ready to follow up 14th Panzer Division. It was called upon sooner than expected. Brandt's Regiment 577 was returned to Oppenländer's control at about 1000 hours and the division tasked with leading the attack between 14th Panzer Division and the Volga. Gunkel's Regiment 576 received the order mid-morning and immediately set off. Supported by Grimm's pioneers, an anti-tank gun company and seven assault guns, the regiment was inserted into the line east of Regiment 577 with the goal of traversing the open ground between the Brickworks and the Volga and pushing through the settlement east of the Gun Factory all the way to the ramshackle housing quarter east of the Bread Factory. Gunkel launched his attack at midday. Artillery, rocket-launchers and Stukas blasted the route but resistance was stubborn from the outset. The turnaround in Soviet courage baffled the Germans. The pioneers cooperated closely with the infantry and assault guns, slowly moving forward along a railway embankment, step by step, and then slewing left at an electricity substation. Brutal fire pelted the attackers but their blood was up. Grimm's men worked in perfect unison, cordoning off individual bunkers and then smoking them out with hand grenades and demolition charges. Most of the Red Army soldiers died in their dug-outs, though 2nd Company took nine prisoners. Obergefreiter Rinck was lightly wounded in the right hand by shrapnel but had it bandaged and remained with his comrades. Gradually, the Soviet defenders were levered off the open terrain, but they still fought obstinately from trenches dug along the Volga cliff.

Resistance flickered to life in the Brickworks between Regiments 577 and 576. If it took hold, Gunkel's regiment would be in a precarious position. Immediate action was taken. Oberst Winzer's regiment was not yet ready for deployment, so Major Beismann's pioneers (minus Grimm's 2nd) were called upon to mop up the Brickworks. Assault guns were sent to help. Pionier Rudolf Mayer from 3rd Company would never forget this day. In the courtyard of the Brickworks, Mayer and his comrades climbed up on the assault guns – they were all headed in the same direction – and Major Beismann came with them as he wanted to check out the situation personally. The armoured vehicles attracted the attention of a Soviet observer who called in a barrage; minutes later, guns and mortars on Zaitsevskiy Island laid down a dense barrage. The pioneers leapt off the guns and sought cover. Mayer was just metres away when an artillery round landed right next to Major Beismann and ripped him apart. The intensive bombardment forced the assault guns to turn back. Beismann's death shocked his men. Most adored him, the few on his wrong side despised him, but everyone respected his professionalism and undeniable skill at converting a tender garrison unit, fit only for duty in occupied France, into a first-class battalion of combat engineers. The battalion's highest ranking officer, Hauptmann Traub, currently holding down the fort at battalion HQ near Gorodishche, was notified of Beismann's death and ordered to get to the front and take over.

Traub had not played a significant role during the summer campaign. After victory at Kharkov in May, he was granted furlough and did not return until late June. His main functions were fulfilment of special projects, assuming the role of adjutant while Leutnant Fritz was on leave and generally acting as Beismann's deputy. Even when company commanders were killed or wounded, Beismann kept Traub at battalion staff, instead opting to fill the vacant postings with junior officers. The workings of the battalion were intimately familiar to him and all ranks showed him the greatest respect. Enlisted men particularly liked him because he lacked the haughtiness and condescension of many senior leaders; a decade and a half of instructing young naval cadets had imbued him with the approachability of a patient teacher. No officer was more suited to leading the orphaned battalion at this moment than Traub.

Until Traub arrived to assume command, Oberleutnant Schaate took over the job of clearing out the Brickworks. Unteroffizier Wilhelm Emendörfer, leader of 3rd Squad, was killed during this mopping-up operation, while platoon leader Leutnant Hingst was wounded. After receiving medical treatment in the rear, Hingst would quickly return to the battalion and take up a position on the staff. A few hours was needed to rid the small industrial complex of Soviet stragglers, and as a result, Schaate's 1st Company gained another Sturmtag: "Attack on Brickworks in Stalingrad, south of the Tractor Factory."

By 1600 hours, Regiment 577 had gained the central street in the Gun Factory but Regiment 576 lagged far behind, still delayed by a few tenacious enemy groups holding out in the riverside ravines. Resistance in front of the regiment was broken an hour later and a rapid assault launched towards a fuel installation. Escalating hopes were quickly dashed by headlong Soviet counterattacks that sapped forward momentum and eventually halted Gunkel's men at the factory's north-eastern corner. As darkness fell, both regiments were busy fending off constant Soviet forays, one in the chaotic jumble of the factory itself, the other on the exposed ground atop the Volga cliff.

Who Holds the Barrikady? (17 October 1942)

Regiment 578 was brought forward during the early hours and inserted between its sister regiments. Oberst Winzer decided he could control his units more effectively by shifting his command post forward to the Brickworks. Two fresh Soviet rifle regiments crossed the river during the night: one of them managed to occupy its designated line before dawn, while the German attack that began at 0800 hours crashed upon the other as it was still preparing its positions. Boundaries between Oppenländer's three regiments ran parallel to the Volga. Artillery observers accompanied the infantry and Stukas circled overhead, ready to be vectored onto their targets. Each regiment was supported by assault guns and a pioneer company: Staiger's 3rd Company had been sent to the northern corner of the factory to support Regiment 577, while Schaate's 1st Company was attached to Regiment 578. Grimm's 2nd remained with Regiment 576.

The initial plan was for Regiments 577 and 578 to mutually support each other and push through the southern half of the factory. This miscarried right from the beginning because Brandt's units were already deeply lodged in the machine shops and warehouses, while Winzer's men started from a position outside the northern boundary fence. Regiment 576 was permitted to begin operations earlier than the official start time because it lagged behind and needed to take bunker positions along the Volga bank. After repelling a Soviet counterattack, Gunkel's men set about eliminating the stubborn defenders dug into the cliff. While doing so, the main attack began. "To the right is a barrage from planes, rockets and artillery," Rinck recorded in his diary. "Position improved. Attack carried further forward on the right." Grimm's men were fortunate to be outside the factory grounds because the fighting inside reached a level of bitterness never envisaged. Assault guns churned through the pitted rail yards to fire into Soviet strongpoints from barely a dozen metres away; snipers drilled exposed heads from all directions; deaths of friends led to spiteful retaliation. The previous days of fighting had been terrible but both sides rarely closed to within bayonet range: this all changed on 17 October in the Gun Factory. The combat was acrimonious, cruel and up close. German professionalism clashed heads with Russian determination. The grenade was the weapon of choice.

The ebb and flow of attack and counterattack generated a massive quantity of human flotsam. Bodies, and pieces of them, dotted the battlefield. Wounded men groaned but few could be helped until the fighting died down. Gefreiter Mayer from Staiger's 3rd saw lots of dead and wounded men even before entering the Barrikady. His unit was fired upon in the factory's congested stockyard. Unteroffizier Kurt Lanzenberger, his squad leader, was directly in front of him when hit by fragments from an explosive round. Lanzenberger recoiled and collapsed, but said that he had only been hit in the arm. Mayer and a Hiwi carried him to an aid station a kilometre away, handed him over to the medics and returned to the front. Mayer only learned the next day that Lanzenberger had died the same evening: a small splinter had also struck him near the heart, which caused his death. The twenty-two year old Stuttgarter joined the battalion's burgeoning graveyard.

The slogging contest caused another great tragedy to befall the division: a Soviet artillery shell caught Winzer and several members of his staff. Oberst Willy Winzer, husband and father of three, was killed instantly, his adjutant and orderly officer were both wounded. Grimm was staggered when he heard that Winzer had been killed. The regiment's senior officer, Hauptmann Wilhelm Püttmann, took command.

By 1730 hours, against strong enemy resistance, Regiment 577 reached the south-west edge of the factory, Regiment 578 the south and south-east edge of Workhall 4 and Regiment 576 a line running from the factory to the Volga bank. An hour into nightfall, Regiment 578 also reached the south-western edge of the factory. At this moment, it appeared to the German leadership that the two regiments had crushed the fiercest enemy resistance in the Gun Factory and reached their objectives, but they were in for a nasty shock. A determined Soviet counterattack pushed into a vacuum behind the spearheads of Regiments 577 and 578 and retook large parts of the factory. The introduction of this rifle division – Colonel Lyudnikov's 138th – into the battle on 16-17 October had taken the wind out of German sails and removed any possibility of rolling up the Soviet position all the way to Mamayev Kurgan. Oppenländer was ordered to defend his positions on 18 October and prepare to relaunch the attack the following day. Instead of moving southward parallel to the Volga, the axis of attack was to be swung ninety degrees to the east; newly drawn boundaries between all units now ran perpendicular to the river.

Robust progress was made in the first four days of the offensive, albeit at a ghastly cost, and news of the successful land grab filtered through to the rear echelons, creating a false impression, as evidenced in a letter written by Gefreiter Füssinger from 1st Company:

We have been in the tough final battle at Stalingrad for a few days. Only ruins remain. Four hundred aircraft and a vast number of guns helped clear the way for our assault pioneers and infantry. Perhaps it will all come to an end here today. We are on the Volga. If this is done, then we have nothing more to achieve this year and will probably head back home. Let's hope for the best.

Setback (18 October 1942)

The division's simple plan of defending its positions and carrying out preparations for an attack were hampered by Soviet resistance that flared up in areas conquered over the previous two days. German artillery had tried to protect the ground troops by placing a wall of blocking fire between them and the enemy, but the problem was that Lyudnikov's riflemen were already inside German lines. The battleground was not a flat surface; this arena was three-dimensional. Soviet soldiers could filter through below ground in sewers and conduits, or even overhead on catwalks and gantries. With movements masked by darkness and pelting rain, they infiltrated the northern part of the factory and effortlessly reoccupied buildings behind the German line, their presence only becoming known when messengers and supply troops were fired on. Resistance truly came to life when dawn cast its dismal light across the sodden factory. German officers sent small combat teams to deal with the intruders but the volume of fire showed that the workhalls were occupied by more than a few desperados. It was soon realised that the northern half of the Gun Factory was in Soviet hands; some German units found themselves in danger of being cut off. One of them was Schaate's 1st Company. They were holding an advanced position, but succeeding in keeping the Soviets at bay – for now. The leader of the second platoon, Feldwebel Hans Wolf, inspired his men during critical moments, even though he had arrived as a replacement just a few days before the big offensive. Schaate was fortunate because Wolf was an experienced soldier, having enlisted as a pioneer in 1936 and participating in the Polish, French and Russian campaigns, during which he was wounded several times. He came from a child-rich family; he was the seventh of fifteen children. In all, the Wolf family had five sons in the field. Even though he was new to the unit, the battalion was like a small Swabian village and men knew each other through friends. Just as the pioneers were forming up on the morning of 14 October, Gefreiter Vorherr met Wolf, as he told his parents:

> As we moved off, I ran into Hans Wolf. He arrived eight days earlier from Karlsruhe as a replacement and was a Feldwebel in our first company. I saw him for the first and last time and spoke with him, only briefly. Short and sweet, I said: "So, lots of luck and all the best!"

The wounding of Leutnant Hingst on 16 October thrust Wolf into the role of platoon leader. Just two days later, while looking around a corner with binoculars, a bullet thwacked into his head and killed him immediately. Life was cheap in Stalingrad.

The Soviet attack looked set to cut off 1st Company. Casualties mounted, particularly the number of wounded men. Some Red Army men infiltrated the German position. Obergefreiter Otto Schmid, leader of 1st Squad, grappled with a Soviet soldier in a life-or-death struggle, but was ultimately overpowered and stabbed to death with a bayonet. The situation was desperate. If all of his men were to be spared death or injury, Schaate needed to make a quick decision. "Pull back!" One man was instrumental in preventing the annihilation of the company, as Schaate recalls:

> [Obergefreiter] Gustav Glaser distinguished himself during the fighting in Stalingrad through courage and daredevilry. After his squad leader was killed, he took command of it [3rd Squad] on 15 October and was fully proven as its leader. On 18 October the company had to give up an advanced strongpoint because of ammunition shortages and heavy casualties. Personally firing a machine-gun for which all remaining ammunition had been handed over, Glaser covered the withdrawal of the rest of the company and in so doing destroyed several attacking enemy submachine-gunners and a machine-gun. Through his fearless actions, it was possible to carry back all the wounded. I myself was severely wounded during this attack.

Despite Schaate's assertion, not all of his men were accounted for. Pionier Josef Metzler, a young man but still very boy-like, disappeared during this hasty retreat; he simply vanished, most likely cut down, unnoticed by his withdrawing comrades. Schaate was more fortunate. A projectile – either a bullet or shell fragment, Schaate did not know – sliced straight through his arm, but it was a blow to the face that caused him the greatest pain; one lens of his spectacles shattered and a tiny sliver of glass was flung into his eye, permanently blinding it. Clutching at his face, he was guided out of danger by his men. Later on, doctors could not remove the tiny shard because it was not magnetic,[3] so it remained inside him, eventually migrating around his body over the decades until causing his death forty-five years later. Stabsfeldwebel Kurt Buchholz, leader of 1st Platoon, assumed command of the company.

The grapevine brought news of the carnage to Grimm and his men. Their sector was awful, but they were thankful they were not in the Barrikady. To provide some relief for Regiments 577 and 578, Oberstleutnant Gunkel was ordered to launch an attack with his Regiment 576. Armour and artillery support was quickly arranged. Hauptmann Traub had been ordered to blast obstructions in the factory so that the assault guns could manoeuvre, but this task – currently shelved because of the Russian attack – was not necessary in the sector of Regiment 576. At 1000 hours, the attack moved off from the right sector of the regiment's line. The assault was beaten off several times and the

---

3   Shell fragments are removed from eyes with magnetised probes.

infantry suffered heavy casualties. Grimm's company recorded no deaths although blood was drawn from Obergefreiter Rinck, as shown by his fatalistic diary entry:

Lightly wounded at 1500 hours. Shell fragment in right side of chest. Not bad. Attack progressed very slowly, everything is full of Russians. Bunker with two Russians behind the infantry was silenced. Heavy artillery and mortar fire. Direct hit ten metres behind me. Can only laugh.

Early in the afternoon, seemingly oblivious to the true situation, or just completely unaware of it, divisional staff issued further orders for the proposed attack the following day. The regiment commanders must have shaken their heads in disbelief.

Gunkel's attack drew off some heat and granted other regiments time to implement countermeasures. Despite the hectic situation, the division managed to realign its sectors: Püttmann's Regiment 578 took over responsibility for the southern half of the factory while Brandt's Regiment 577 transferred all disposable elements to positions near Halls 5 and 5a. Brandt's men were under the greatest pressure because most of the workhalls in their sector were in Soviet hands and needed to be retaken. Clearance began in earnest shortly after midday and was still in progress at dusk.

The 305th was forced to acknowledge that "the northern half of the Gun Factory and the adjacent houses appear to be, for the most part, in enemy hands. The situation in this area is unclear because no connections from Division to the regiments are available." Each day the division was getting weaker: of its 8,687 German personnel, just over a third were combat soldiers. The combat strength of the pioneer battalion was just 3 officers, 13 NCOs and 106 enlisted men, a total of 122. Apart from Hauptmann Traub, the only other officers in the line were Oberleutnants Grimm and Staiger. Naturally, Traub mentioned none of this in a small card to his wife Ella: "I am well. I marvel about my capacity in managing the many burdens. However, I have become very thin." This was just three days since taking over the battalion… and there was no end in sight.

Crisis of Command (19 October 1942)

Lyudnikov's unit diary asserted that "as a result of night actions, units of 138th Rifle Division have completely cleared the enemy from the Barrikady Factory." German troops still positioned within the factory walls proved otherwise. The assessment of nocturnal events by 51st Army Corps could not have been more different: "The night passed quietly with harassment fire, several enemy reconnaissance troop operations, and weak operations by the enemy air force." The reason for such a jarring difference is simple: divisional staff was clueless about the true situation. Oppenländer had suffered a nervous collapse and lost control of his division, but rather than notifying his superiors, chief-of-staff Oberstleutnant Kodré took ghost command and continued to issue orders in Oppenländer's name. Worse still, instead of reporting the seriousness of the situation, Kodré attempted to rectify the problem himself. Only when these measures failed was corps informed, but just about military matters; Oppenländer's breakdown was kept under wraps. The joint assault by the Bodensee Division and 14th Panzer Division, scheduled to take place on this day, was cancelled. The division's main task was to clear up the awful situation in the factory. Only then could the attack be restarted.

Grimm's company was kept busy holding the position and repelling counterattacks. "Heavy infantry fire," noted Rinck, "Attack rolled forward on the right. Metre by metre." Not all Red Army soldiers fought ferociously: 2nd Company collected four deserters. Throughout the day the company lost two more men to wounds: Pionier Otto Loos and Gefreiter Hermann Neumeier, the latter displaying exceptional courage in house-to-house fighting and during attacks on enemy resistance nests while under fierce fire. Neumaier was wounded in the head and feet by shrapnel.

Inside the horrific factory, the counterattacks made little headway against intensifying Soviet resistance. Some units of Regiment 577 were even encircled in the south-west part of the factory. Hauptmann Traub drew an incisive parallel: "The fighting here – and everyone agrees about this – will be the high point of this present war and can certainly be compared with the battle of Verdun in the First World War." Miserable weather compounded the horrific conditions and made the parallel with Verdun even more accurate. Light snow flurries ominously whipped across the city and added their moisture to the mud under foot. As a consequence of the appalling weather, Luftwaffe operations were suspended and the offensive stagnated. The division could not completely destroy enemy forces still located behind their front.

The day ended with both sides convinced that they held the factory except for a few enemy remnants holding out in the rear. The objective of 305th Division for the following day was to clear up the situation in the Gun Factory. Corps was unaware of the magnitude of this task but was nonetheless cognisant that any form of mopping up would take some time:

> The clearing out will continue tomorrow and probably also the day after that. With that, the division will not be available for the assault on the Steel Factory for the time being, especially since it has also been considerably weakened by casualties in the last heavy days of combat (around 500 men).

Pincer Attack (20 October 1942)

To cleanse the factory, the division formed two assault groups: the first, positioned in the southern half of the factory under Major Brandt, contained Regiment 577, elements of Regiment 578, two pioneer companies (Stabsfeldwebel Buchholz's 1st and Staiger's 3rd), part of the panzerjäger battalion, and six assault guns; the second group, outside the northern boundary of the factory under Oberstleutnant Gunkel, was much smaller, with elements of Regiment 576, Grimm's company, Radfahrschwadron 305, some anti-tank guns and four assault guns. The plan was simple: both assault groups would advance towards each other, meet up near Workhall 5 and then turn to the south-east and strike towards the Volga.

The proposed noon start time was pushed back to 1500 hours by various difficulties. One of those was a shake-up at Division HQ. Due to "health reasons," Generalmajor Oppenländer was finally forced to relinquish command. His nerves were shot. The slaughter in the Gun Factory had exacerbated his precarious mental state and he hit the bottle even harder than usual (he was, after all, nicknamed "Schoppenländer"[4] because of his fondness for wine). Word of Oppenländer's breakdown reached his superiors and he was immediately relieved of command and ordered home to recuperate. Leadership of the division was temporarily assumed by Oberst Hermann Barnbeck, the highly respected and decorated commander of Infantry Regiment 211 (71st Infantry Division) deployed in the southern suburbs of Stalingrad. Oberstleutnant Kodré was also advised that he would be relieved of his position as soon as a replacement arrived. Kodré was an excellent chief-of-staff and a brave combat soldier who had earned the Knight's Cross as a battalion commander, but as an ersatz division commander, he was in way over his head. Divisional leadership was a team effort, a dual system of command that required an inspirational leader to call the shots and a calculating staff officer to supply strategic guidance and convert his leader's orders into concise instructions. Kodré had assumed both roles during the Gun Factory battle. As a result, his impeccable record as chief-of-staff suffered, as did his health. This whole affair threw his career off-kilter.

These high-level goings-on had little bearing on the soldiers. Barnbeck thought it wise not to alter existing orders, so as ordered, the assault groups launched the attack – planned autonomously by Kodré – at 1500 hours. Grimm's company had shifted further to the right, away from the Volga, but did not participate in the attack. Together with other units from Regiment 576, they held the eastern end of the line while the

---

4   A "schoppen" is a term used in Württemberg as a unit of measure for wine, equal to 500 millilitres, or half a litre.

western wing attacked. Oberstleutnant Gunkel's assault group battled its way into the factory grounds. The terrain was nightmarish: a confusing network of railway sidings and rolling stock, trashed buildings and thousands of scattered gun barrels. Looming over everything were the skeletal remains of the workhalls. Lyudnikov's riflemen fought bitterly. Stukas and artillery smashed identified positions, a few moments of silence followed, then gunfire erupted elsewhere. Several small warehouses were cleared and resistance nests north-west of the factory, far behind the front, were eliminated.

Further south, Major Brandt's group skilfully grenaded its way through the workshops. Assault guns and anti-tank guns set up fire bases and blasted everything that appeared in the open while the infantry, supported by pioneers, swept through the buildings' interiors. The Soviets fiercely counterattacked at every opportunity. By nightfall, Regiments 577 and 576 had taken Workhalls 5 and 5a and cleaned out all the warehouses west of the factory's railyard. The division believed it now held the southern part of the factory and most of the north, though Lyudnikov's maps continued to show his division in firm control of most of the Barrikady, especially the northern half and even large sectors outside the factory fence. What was beyond dispute is that the Soviets still held Halls 4 and 6a.

Divisional casualties were surprisingly light for such embittered fighting: a dozen men killed, 1 officer and 16 men wounded, and one man missing. Nevertheless, these capped off a catastrophic ten-day period for the division. According to a report filed by 6th Army's head doctor, in the period from 11 to 20 October, 305th Infantry Division suffered losses of 15 officers and 434 men killed, 46 officers and 1277 men wounded, 5 officers and 81 men missing, and 46 men ill, a grand total of 1,904.

Oberst Barnbeck set about his new assignment with gusto. Reinvigoration and a renewed strength of purpose were instantly apparent. Fresh orders were issued in the evening. Regiments 578 and 577 were directed to shift their command posts close to the north-western edge of the factory before the attack commenced at 0800 hours the next day. The division's signals battalion was ordered to establish wire and radio connections to the new command posts. Proper command and control required faultless communications, so Barnbeck immediately rectified this shortcoming.

Grimm's Company Enters the Gun Factory (21 October 1942)

The objective for Brandt's and Gunkel's assault groups was to reach the factory's eastern boundary and then defend that line in close conjunction with Regiment 578 to the south – already holding the eastern workhalls in the southern part of the factory – and the bulk of Regiment 576 to the north. To prevent any resurrection of enemy resistance, both groups were directed to thoroughly mop up the factory after reaching their objective. As they were soon to discover, this was easier said than done.

The attack began at 0800 hours. Barnbeck stressed that the "unity of artillery fire with the movements of the infantry would be guaranteed by the deployment of forward observers in the factory" and not just on the ridge to the west. Under this expertly guided umbrella of artillery cover, the assault groups moved out. Brandt's assault groups surged northward and after hours of savage combat captured Halls 3 and 6b. Gunkel's group eliminated a bulge in the line, established a firm connection with Brandt's forces and pressed eastward into Halls 4 and 6a. Momentum rapidly drained away in the workshops' jumbled interiors. The attack stalled.

Barnbeck recognised the need for a single officer to control the actions in the factory, and the choice between Gunkel and Brandt was a no-brainer. Forty-six year old Gunkel, a veteran of three years of trench warfare during World War One, had only taken command of a field unit in June 1942 and while capable, displayed what a commander later called a "propensity for laziness." Brandt, on the other hand, was a go-getter and achieved superior results wherever he was deployed. Even though outranked by Gunkel, Brandt was given control: Gruppe Gunkel was dissolved and all its units attached to Gruppe Brandt. Major Brandt promptly reinforced the main attack. Grimm was ordered into the factory grounds for the first time. His company filed into position amongst the gaunt and twisted structures. The company's two flamethrower teams were with them. "For the attack," Grimm recalled, "my company was subordinated to III./Inf.Rgt.576, which in turn was placed under Major Brandt's Regiment 577." The addition of the pioneers was a considerable boost in offensive power.

The attack was relaunched at 1400 hours and led to gruelling hand-to-hand brawling. Rinck's laconic diary entry barely conveys the suffering and death inflicted on both attacker and defender: "Three workhalls and one factory taken and cleaned out. Heavy infantry fire. In position in Hall 4." The sequence, timeline and identity of every building captured has escaped the historical record, but Grimm's men succeeded in taking Hall 6a, or as they called it, "White Hall 6". Three or four pioneers lost their lives for its possession. The assault then became lodged in Hall 4, the factory's main mechanical workshop and largest building. Casualties in Grimm's weak company were harrowing: six killed and three wounded, two of them severely. The dead were Obergefreiter Hermann Schönbucher (shot in the head), Gefreiter Richard Rohr (ditto), Gefreiter Konrad Kugler, Pionier Rudolf Fiedler (shot in the head and right lung), Pionier Wilhelm Haug (bullet in the chest) and Pionier Alfred Schaz (stomach wound). The three wounded men were Gefreiters Ignatz Burkard, Josef Nätscher and

Hans Treu. Nätscher recovered from his light wounds and returned to the company a few weeks later, but Burkard succumbed to his severe injures the same evening and raised the day's death toll to seven, while Treu died at 0810 hours on 26 October from his neck wound. This reveals the intensity of the fighting: only one man out of nine survived bodily harm. Also indicative of the turmoil and confusion is the precise date of Gefreiter Rohr's death. Rinck recorded it in his diary as 21 October. Rohr's wife received notification that it was 22 October and that date is recorded in official documentation. Even one of the men who was with Rohr during his final moments, Obergefreiter Johann Schmitt, swore it was a different day:

> On 23 October I received the order to go into position at the front with the remaining men of 3rd Platoon. Gefreiter Rohr, who knew the way, was guiding me. As we were crossing a gap between two factory workhalls that was under enemy observation, we took machine-gun fire. Gefreiter Rohr, who was moving on ahead, was struck in the chest and immediately killed. Because it was daytime and the spot was under constant fire, he could not be recovered. Whether he was later brought back, I cannot say, because I was wounded the next day during the attack on Hall 6.

Grimm thought it happened even later:

> It seems to me that the date of 22 October is not correct because Rohr was killed by a sniper only after the capture of Hall 6 as he was showing two comrades the way to the company. Rohr's body was only recovered a few days later on my special order.

Although this assertion was made in August 1943, less than a year after the event, Grimm was mistaken. Rohr was definitely killed on 21 or 22 October and his body left behind when Hall 6a fell to a Soviet counterattack. Days would pass before his comrades recovered him for burial. The difference between reality and what was told to next-of-kin is interesting because Grimm notified Frau Rohr that her "husband was recovered during the night and together with other comrades buried at the heroes cemetery in the northern part of Stalingrad."

On the grounds of heavy casualties and the violent Soviet reaction, the main attack was discontinued at around 1500 hours. This decision was viewed gratefully by Grimm: "I owe my life to Leutnant Andreas, orderly officer of Major Brandt, because through his intervention, an insane, or one of the many insane commands to attack, was rescinded or limited."

Despite cancellation of the attack, isolated clashes continued for the rest of the day. Initially, 6th Army believed that "several resistance nests in the Gun Factory were again fought into submission, so that possibly the entire factory is in our hands." Wrong. Soviet units counterattacked and fought desperately. Both sides were locked together in Workhall 4 and the northern part of Hall 6a. Smoke hung over the battle area, limiting visibility, and in spite of their most strenuous efforts, German troops could not dislodge the Soviet defenders. Grimm tried everything:

> On the order of Major Emendörfer, I launched an attack with Pioneer Platoon 576 and both my flamethrowers after I was unable to disengage my company from the enemy. This first attack was driven off by the enemy. Even my flamethrowers were not brought to bear because during the first rush, the projector tube of the leading flamethrower was shot up.

Despite the close call, neither of the flamethrower operators, Obergefreiter Karl Schwarzer and Gefreiter Franz Müller, were injured. The fearsome fire projectors were sent back into reserve.

It was as this moment that something peculiar and unexpected happened. Major Emendörfer, a respected infantry battalion commander, vanished. He just wandered away from his command post without explanation. Word of his disappearance reached Division at the same moment Grimm was at Regiment 576 headquarters:

> It was on [21 October][5] that a call from Division came to Gunkel's command post. Sharp words were uttered during the conversation... I was in the same room. Oberstleutnant Gunkel hastily grabbed his helmet: "Come with me." "Jawohl, Herr Oberstleutnant," and after about 400-500 metres we reached the command post of Major Emendörfer. Precise location: in the north-west corner of the Gun Factory, west of the railway tracks, in a heating duct that was reached by climbing down a two-metre ladder. In this corner were car ports clad in corrugated iron. This covering was shot off and corrugated iron lay on top of the command post as overhead cover. According to orders, the front-line had to be walked along, but one did not go from trench to trench (that is to say, foxholes), but rather only section by section. Having accomplished this, I took over the combat group at the aforementioned command post. We did not see Major Emendörfer. At this point in time, the trench strength amounted to fifty men at most, not including Feldwebel Pauli and the fourteen men of my company who were employed as infantry to

---

5   Grimm originally dated the incident as 18 October but all other evidence shows it took place on 21 October.

prevent an enemy break-in. My four-man flamethrower section was under cover several hundred metres to the rear.

Grimm was clueless about the situation that landed him in provisional control of an infantry battalion. In fact, the whole command structure of Regiment 576 was in flux. For unknown reasons, possibly illness, Gunkel was forced to relinquish leadership at this precise moment and hand command to Major Braun. Members of battalion staff confessed to Grimm that Emendörfer had lost it and simply wandered off, stupefied, muttering to himself. Confirmation came in the form of a messenger:

An order conveyed to me by a Leutnant stated that I had to bring back Emendörfer with an escort. Major Braun, whom I knew well because I had been subordinated to him many times in France, issued the order.

Grimm carried out the task and uncovered the story:

After the attack [in Hall 4], Emendörfer suffered a severe nervous breakdown. Only at the last moment did I succeed in bringing him – he had completely broken down – away from the enemy. On the basis of my report, Major Braun ordered Emendörfer brought back [from the front] and that I take over command of the battalion until the arrival of Hauptmann Kempter.

Grimm would later write: "I don't know exactly what led to Major Emendörfer's breakdown but I did hear rumours." Unfortunately, Grimm does not elaborate, but it is not difficult to imagine how the remorseless killing affected unit leaders. In a post-war account, Emendörfer himself reported that in the first three days of combat in Stalingrad, six of the eight officers in his battalion had fallen through death, wounding and accidents in the frightful factory terrain. The divisional history tactfully states that "47-year old Emendörfer became sick as a consequence of the strain of the campaign." In a 1943 letter to General Oppenländer, Grimm provided some more details of Emendörfer's mental collapse:

The onset of darkness saved us both from snipers' bullets. During my search, I spotted Major Emendörfer, confused, out in front of our main line: "I'm a coward, I'll let myself get shot." I took him by the sleeve – which he unexpectedly allowed – and led him out of the immediate strip of death that had been contaminated by enemy anti-personnel mines and which had inflicted so many casualties on us.

In early November, after recuperating at the divisional train for just over a week, Emendörfer was evacuated from Stalingrad for treatment in the homeland. The

German Cross in Gold was awarded to him on 25 January 1943. Emendörfer's post-war memoirs are well-written and show excellent recall, but Grimm experienced a different side when he tried to talk to the man he rescued from a minefield inside the chaotic Gun Factory:

> *At a divisional meeting after the war in Geislingen [in May 1953], Emendörfer did*
> *not recognise me. I had so valued him as a battalion commander during the*
> *advance that I preferred not to burden him. I sensed that he was trying desperately*
> *to remember. So I turned back to my other comrades.*

– – –

Minor clearing actions, driven by a grass roots desire to improve local positions, continued until nightfall and destroyed several resistance nests in Workhalls 4 and 6a, but Soviet die-hards stubbornly refused to yield the eastern sides of both halls. When full darkness came, the Soviet riflemen initiated a series of skilful counterblows and forced the Germans to abandon their hard-won gains in both halls. Rinck recorded this with characteristic understatement: "Left wing pulled back ten metres during a Russian night attack. Strong attacks. Artillery and mortar fire." The fact that ten metres was considered a retreat is significant as it shows every metre in the fiercely contested Barrikady was equal to hundreds on a regular battlefield.

Surprisingly, despite all the talk of severe casualties, the division's losses were fewer than expected: 8 men killed and 32 men wounded. Five of the dead came from Grimm's 2nd Company, thus constituting a majority of the division's deaths. Dreadful mêlées raged throughout the night.

Missing in Action (22 October 1942)

During the night, while German artillery took potshots at targets east of the factory, signallers laid wires from Major Brandt's command post to the combat groups deep in the factory. Reliable landlines were absent over the previous week and radios were fickle because the massive steel frameworks of the workhalls adversely affected transmission. These erratic communications had a direct effect on Grimm; isolated and dependent upon unreliable radio signals and vulnerable messengers, he disappeared from the view of his superiors and was even thought to be missing in action:

> *I had no connection whatsoever to the outside world and only by means of the*
> *portable radio set could I receive radio signals. The batteries were conserved for*
> *that purpose and the radio operators were forbidden from listening to the radio*

under threat of punishment. The radio operators moaned to me that they did not
have and could not get a supply of batteries.

Worse still, the pioneer battalion's paymaster, Oberzahlmeister Max Keppler, whose family lived in the same town as Grimm, reported to his wife that Grimm had been reported missing, so Frau Keppler had to conceal her condolence and put on a smile when she saw Lotte Grimm as she was not allowed to say anything.

Daybreak brought lively Soviet artillery and mortar fire, interspersed with strong barrages. The bloodletting recommenced at 0420 hours when Major Brandt renewed the attack to win back the Soviet-occupied areas of the factory. The fighting followed the same pattern as previous days with German gains being nullified by Soviet counterattacks. Grimm's infantry battalion did not participate in the assault. After relentless attacks the previous night had forced them out of Hall 4 and back across the road to Hall 5, they held position throughout the day and night despite fierce attacks. Most assaults were beaten back with hand grenades. The range was simply too short to use other weapons effectively. It was nerve-racking manning defensive positions amongst the shattered remains of Hall 5. On the other side of the road loomed the menacing façade of Hall 4 replete with hundreds of potential embrasures and sniper hides. In between were bare birch trees, mounds of iron parts and half-finished gun barrels, nondescript industrial machinery and ranks of small-scale rolling stock that once trundled around the factory's internal railway. In short, no-man's-land was a cluttered place that offered hidden approaches. The goal was to see but not be seen. Sentries gingerly chipped away at the thick brick walls to create observation slits. Any movement out front was pelted with grenades. The task of holding the positions became even more thankless when light rainfall began. "Very cold," Rinck wrote in his diary. "Made a fire in a corner." He also recorded a grim event; a commissar who fell into their hands was summarily shot. Grimm's company incurred much lighter casualties on this day. Four men were wounded: Gefreiter Kurt Mayer, Pionier Kohler, Gefreiter Franz Winkelmann and Gefreiter Brunner, the last two suffering only light injuries and remaining at the front.

Although orders from Battalion and other paperwork did not reach Grimm, a handful of Iron Crosses – sent to each unit by Division – did. The intention was to boost morale and directly correlate actions with reward. Grimm recalls distributing them without corresponding award certificates:

On 22 and 23 October 1942, at the command post of III./Inf.Rgt.576 in the Gun
Factory, I was informed that all men who had fought in the front-line in Stalingrad

since the first day of the attack should be submitted for the Iron Cross Second Class. After that, I submitted Iron Cross Second Class recommendations to Battalion on slips of paper from a message pad. As far as I can remember, there were still six men. In the meantime, a number of Iron Cross Second Class and two First Class were allocated to the company without prior recommendations. The recommendations were handed in later. The first priority was to decorate the same men who had been wounded during daring surprise attacks on bunkers and resistance nests.

One of the men Grimm submitted for the Iron Cross Second Class was Gefreiter Winkelmann, still with fresh white bandages on his wounds. Soviet propaganda belittled this medal by saying that "every second German soldier has an Iron Cross, but every second Soviet soldier has a mortar," yet what a man had to go through to earn this award, especially at Stalingrad, almost defies belief. His recommendation read as follows:

Winkelmann distinguished himself through exceptional guts and courage in the period from 14 to 21 October during the storming of the stadium, the house-to-house fighting north of the stadium and particularly in the fighting in Hall 4 of the Gun Factory as leader of a demolition squad during the suppression of resistance nests.

Surviving one week of hell was enough to earn this medal, yet according to Grimm, only six men managed this feat, all of them front-line soldiers who fought hard, not just on one day, but repeatedly. Ordinary footsoldiers realised that the Knight's Cross was beyond their reach; it was the preserve of generals, panzer leaders, men who changed the course of a battle or seized vital objectives. None of that applied here. The Gun Factory was an apocalyptic realm totally devoid of potential glory. The men just kept performing their duty and hoped to emerge unscathed. Nobody thought about chasing tinsel for their chests, but when a medal was awarded, it had huge significance. Those who wore the tri-colour ribbon in their buttonhole knew the Iron Cross was not showered upon the troops as suggested by Soviet propaganda. It represented weeks of death defiance, courage and the blood of fallen comrades. Enlisted men who wore the Iron Cross First Class possessed a certain aura: they had done something extraordinary to earn such a prestigious award. When two of the glittering crosses were allocated to Grimm, he bestowed them upon Obergefreiter Josef Ostermeier – wounded 15 October – and Gefreiter Johann Wäschenfelder, still at the front. Three more were awarded to soldiers from the other companies: Oberleutnant Schaate and Gefreiter Hans Siegel of 1st Company, and Gefreiter Walter Glatt of 3rd Company.

– – –

Despite repeated attempts to completely clear the factory, Brandt's groups made little headway in the face of stiff resistance. In the evening, 305th Division reported: "The enemy defended their sections of the Gun Factory so obstinately that our attack could not penetrate, despite every sacrifice."

Raising Combat Strength (23 October 1942)

Major Braun did not allow Grimm, a pioneer officer, to control an infantry battalion for long. He selected a replacement from his staff: Hauptmann Hans Kempter, the commander of his HQ company. Kempter described what he found at the command post of III./Regiment 576 in the Gun Factory when he took control from Grimm:

> *I was led to the command post by a messenger. There, I immediately oriented myself about the mission and deployment of the battalion, about the enemy situation, gained an insight into the terrain and took the first necessary steps. Then I reported to Regiment that I'd taken over the battalion. The command post was in a covered hole in the ground and was so close to the enemy that the regiment commander could hear a Russian "hurrah" during our phone conversation.*

The Soviet shouts audible through the phone line emanated from an early morning raid that captured a few houses and kept pushing southward in the face of numerous German counterattacks. At 0400 hours Kempter's combat group, including Grimm's pioneers, laid down small-arms and machine-gun fire from northern Hall 5 and threw grenades at the attacking Soviet detachments. Just over an hour later, two other Soviet groups went over to the offensive with the ambitious goal of advancing along the factory railway towards the central gate. They picked a bad day to launch an attack: the sun was incandescent, the air clear as vodka and the Luftwaffe was out in full force to support a new offensive aimed at the Krasny Oktyabr Steel Factory. Twenty-six battalions of German artillery initiated a three-hour long preparation at 0515 hours; the Luftwaffe chimed in ninety minutes later.

Providing the main weight of the attack was Generalmajor von Schwerin's 79th Infantry Division, a new arrival in the city. Its orders were to capture the steel factory. Excellently supported by pinpoint-accurate Stukas, the infantryman set off at 0815 hours. The Bodensee Division's right neighbour, 14th Panzer Division, began its attack on the Bread Factory at the same time. Once captured, a combat group commanded by Hauptmann Püttmann – two infantry battalions, a pioneer company and an anti-tank gun company – would join in for the push through to the Volga. As all assault guns were escorting von Schwerin's troops, a panzer company was sent to support Püttmann.

During the night, mine-clearance troops allocated to the panzers had created a crossing over the railway tracks and blasted obstacles in the factory grounds to clear a path. Püttmann's group stood ready at the Gun Factory's eastern fence from 1030 hours onwards and awaited 14th Panzer Division to appear on their right. The rest of 305th Infantry Division held position.

The morning's shimmering blue skies disappeared behind banks of low-lying cloud and Luftwaffe operations were suspended from 1030 to 1300 hours. Kampfgruppe Püttmann was still patiently waiting to commence its attack just as soon as the left flank of 14th Panzer Division moved level, but progress in the Bread Factory was slower and more costly than expected. In the meantime, Major Brandt launched several small-scale assaults to keep the Soviets occupied. Grimm's weary pioneers spearheaded one raid across the jumbled roadway into Halls 4 and 6a. Obergefreiter Rinck summed up the results: "Attack two times at midday repulsed. Back in old position." Other German groups pushed much deeper into the Soviet defences. The ebb and flow of senseless fighting, attack and counterattack, continued all afternoon, and the company gained another Sturmtag: "Counterattack in the Gun Factory near Hall 6."

News from 14th Panzer Division was not good; obstinate resistance in the Bread Factory was retarding the advance and several more hours was needed to reach the southern part. Until the panzer division reached the designated line, Kampfgruppe Püttmann's attack could not be triggered. The main attack by 79th Infantry Division, however, achieved an objective that appeared increasingly out of the reach of the Bodensee Division: it reached the Volga. The exciting news flashed through the trenches. During the night, however, those that reached the river could not hold their vulnerable positions and were compelled to pull back lest they be cut off.

Back in the Gun Factory, Kempter's appearance did not thrill Grimm:

When Hauptmann Kempter arrived, he wanted to lay down the law on the question of pioneer operations. He pointed out that the supply train should be combed through and indicated how non-compliance would be punished. He got wise to the fact that I exchanged a few men with the supply train, who came back later. Then, when he learned that the flamethrowers weren't in the front-line, an argument broke out. He listened as I dictated to a messenger a written order for them to come to the battalion command post.

Measures to raise combat strength did not originate with Kempter; he simply disseminated an order from Division to extract manpower from supply trains. "After the entire supply train and MuMa had been combed out," Grimm wrote, "I had a combat strength of fourteen men on 23 October." Some of the men caught by this sweep were Unteroffizier José Karpenkiel and Gefreiter Xaver Sperl. The latter was a dependable man who had already distinguished himself time and again "through exceptional guts and courage" during September's defensive fighting. Sperl was one of the men Grimm had been holding back at the supply train and his return was warmly greeted. Karpenkiel, on the other hand, was an enigma, an unknown quantity. A soldier of some experience, with the Polish and French campaigns under his belt, as well as a stint in the navy, he arrived as a replacement in early September. His rank and awards, including a wound badge, marked him as an old hand. As company commander, Grimm was privy to personnel records and Karpenkiel's was not a glowing one:

> Among other words in his assessments were "boaster and a show off." As a result of
> a shortage of junior leaders, Karpenkiel was entrusted with leadership of a squad,
> which revealed his lack of pioneering-technical knowledge. At the end of September
> or beginning of October, Karpenkiel was detailed to a training course for pioneer
> squad leaders. The medical officer of this training course sent him back to the
> company with the indication that Karpenkiel was not up to the physical demands.
> He rejoined the company when it was employed in the Gun Factory and only had
> a combat strength of fourteen to sixteen men.

The complete lack of NCOs forced Grimm to put Karpenkiel in charge of an assault detachment. Time would soon tell if this was a wise decision.

Karpenkiel was not the only troublemaker with whom Grimm had been saddled. The other was "a difficult fellow" by the name of Franz Müller, a 35-year-old father of seven. Paternal fecundity was one of the few extenuating circumstances that excused a draft-age male from front-line service, yet Müller found himself in the Eastern Front's hottest crucible, and only two men in the company knew why:

> Gefreiter Müller was supplied to 2nd Company as a replacement "on probation",
> as far as I can remember, only after the Don crossing. The papers were with the
> unit train and I was verbally informed about it by my Spieß (Hauptfeldwebel). As
> the father of seven children, he must have committed numerous criminal offences
> to be sentenced to "probation at the front." To ensure that nothing affected him, the
> Spieß and I had to keep silent about his probation. I continually received complaints
> about his uncomradely behaviour towards other men, so I put him into another

squad. The NCOs and platoon leaders complained about my "problem child." Only
in my presence did he show himself to be a courageous man... When I took Müller
to task, he said "they accuse me of things, which they are not allowed to mention
at all," and with an eloquence that could, in my opinion, only have been learned
through many court appearances. I therefore kept him behind the front-line in a
flamethrower team.

Müller had already shouldered the cumbersome flamethrower on 21 October and
been committed to action, but never had the chance to torch anything. Packed off to the
rear again with the rest of the flame team, he was called forward again on 23 October
as a result of Kempter's badgering.

Stalingrad's rear areas were short on suitable accommodation, but almost anything
was better than the medieval conditions in the derelict industrial complex. Hauptmann
Traub gave his wife a vague idea of the conditions, but did not go into detail:

Our soldiers bunk down in the cellars, the many factories are nothing but large
fields of rubble, and it is very hard to get through this mess. I wonder if we will still
be here during the winter or be reassigned to another sector along the front... The
sun usually shines during the day but the weather is freezing at night. We'll all be
happy when this shit is over.

Gefreiter Vorherr, an inveterate grouser, held nothing back from his parents:

Tonight, it's already our tenth night in the battle of Stalingrad. A layman has no
idea what that means! I've been lucky to get through it OK, and I hope that God
stays with me in the coming days... I've already lost many comrades in the last ten
days; it's horrifying what's gone on here in the Gun Factory. For ten days no sleep,
no washing, and how much longer will it go on? Because of that, lice are now
appearing. That would not be so bad but now it is cold at night. Not one house here
is intact, often no longer one brick on top of another, everything has been smashed
by our Stukas or, where we are, the Russians have clobbered everything. I'm now
sitting in a machinery room below the factory while shells whistle overhead.

Vorherr then told his parents that he had received a card from a friend recently
wounded in the factory: "Everyone here would dearly prefer such a wound rather than
a few more days here. You are glad every day when you've survived it." With that,
Vorherr was emotionally spent. "It's difficult for me to write about such events..." Just

a few hours later, he narrowly escaped death in Hall 5: "I was one centimetre away from being shot in the head by a ricochet that bounced off a girder in a workhall and smacked into my helmet. The mark is still clearly visible on it." This close call shocked the life out of Vorherr. "I went back and reported in sick." Likewise headed rearward was Hauptmann Traub. After a solid week in the front-lines since taking over the battalion on 16 October, Traub was allowed to head back to the rear for a brief spell:

I have returned to headquarters to keep the shop rolling because the adjutant only returns from leave in early November. Two company commanders and some other officers are wounded, so the total strength among the battalion's officers at the moment is just five.[6] The days in Stalingrad have been really exhausting. For example, during this lengthy period, I have been unable to wash or shave and look like a villain. I have survived the exertions well and feel cheerful in my current quarters, a Russian hut some eight kilometres outside Stalingrad.

Traub was able to clean up, have a good meal with some wine, and fell into a deep sleep despite the windowpanes being rattled by Soviet planes scattering bombs nearby.

Day of the Flamethrowers (24 October 1942)

Heim's panzer division relaunched its attack to conquer the eastern half of the Bread Factory at 0630 hours. Just under an hour later, Regiment 578 was ordered to employ a battalion to mop up enemy fragments between the Gun and Bread Factories, occupy some houses north-east of the gully and clear the gully itself. Throughout the day, 14th Panzer Division won several buildings in the Bread Factory in challenging close quarters combat. The main focus for the Bodensee Division was the stubborn Soviet strongholds of Halls 4 and 6a. Hauptmann Kempter sets the scene:

Written in the Armed Forces Report of 18 October, amongst other things, was that assault troops had taken all of the "Red Barricades" Gun Factory by storm. It turned out, however, that one workhall, namely Hall 6a, was still in enemy hands. On 24 October I received the order to take this workhall, together with the administration building and the surrounding area, clear out the enemy and then hold them. Hall 6a, a factory building with countless machines and subterranean rooms for heating equipment, etc., was fortified and firmly held by enemy forces. The outside grounds

---

6   The five officers were Traub, Grimm, Staiger, Homburger and an unknown fifth, possibly Hingst. Traub did not count the battalion's non-combat officers: the doctor (Oberarzt Dr. Wirtgen), veterinarian (Stabsveterinär Dr. Pompe), paymaster (Oberzahlmeister Keppler) and technical inspector (Inspektor Zeller).

were mined. It was therefore clear that this would be a difficult task, would require careful preparation and necessitate the deployment of all available means.

The means available to Kempter – in terms of manpower – were woefully inadequate, as shown by Grimm's inventory of forces assembled: "Deployed for this attack was Leutnant Gillardon with the remnants of III. Bataillon (a total of about 30 men), as well as myself with my company (16 men) with the subordinated Pioneer Platoon 576 (about 7 or 8 men)."

The assault troops were powerfully armed and weighed down by satchel charges and blocks of explosive, while every man carried as many grenades as possible, tucked into boot-tops, belts and retaining straps on entrenching tools, or in purpose-built musette bags. The biggest trump card was the four-man flame team. The two men shouldering the incredible heft of the flamethrowers – Grimm's troublemaking parolee, Gefreiter Müller, and Obergefreiter Schwarzer – had little comprehension of the havoc their fire-spewing weapons were about to wreak. Flamethrower operators were not brutish sociopaths eager to inflict agonising deaths on other humans. Müller admittedly had criminal tendencies, but he did not volunteer to man the weapon; Grimm assigned him as a form of punishment and to sequester him from the other men. No sane man wanted to piggyback a flamethrower into battle. The other operator, 29-year-old Austrian Obergefreiter Karl Schwarzer, was an ordinary guy, a member of the company since France, but it is not known how he ended up in this role. Younger, naive men often found themselves crewing a Flammenwerfer because wiser heads knew how to avoid being assigned to it: some feigned ineptitude during training, others kept their heads down at the right moment, but most stated their unavailability due to their positions of responsibility. In any case, officers were loathe to appoint experienced combat soldiers to the weapon, so it fell to newcomers and those who were non-essential to the fundamental integrity of the unit. Contrary to popular belief, operators would not be engulfed by flames if a bullet or shell fragment punctured the pressurised tank, thanks to the ignition system of the Model 42 flamethrowers. Whereas the Model 41 used a hydrogen jet to light the fuel, the newer model used a magazine loaded with ten blank 9mm pistol cartridges that loaded, fired and ejected in automatic succession with each pull of the trigger. Since the same trigger operated the fuel ejection and firing mechanism, the result was "hot firing", the jet of Flammöl – a viscous black oil that smelt like creosote – being ignited the instant it left the nozzle (the US M2 flamethrower was able to "wet fire", that is, squirt fuel without igniting it, due to separate triggers for ejection and ignition). As a result, the Model 42 had no naked flame, even when the trigger was pulled.

Stukas and artillery worked over the entire Soviet defensive area, then Major Brandt's assault groups commenced the assault at 0800 hours by swiftly penetrating Hall 4 from the west and south. This vast assembly hall was broken up by internal walls and further subdivided by massive concrete pylons, conveyor belts, work stations and narrow-gauge railways. The acres of tin sheeting on the roof had been shot to pieces over the previous week, allowing sunlight to cascade in, while here and there large sections of roof trusses had collapsed. The tangled interior was a death-trap to the meek and careless, but Brandt's assault troops advanced with intrepidity and purpose, pressuring the defenders from two directions. By dominating the open areas outside the building with observed machine-gun fire, the Germans forced the Soviet riflemen to remain inside to try and halt the two-pronged assault. The Soviets were outclassed. Defenders repelling an attack from one direction were taken by surprise from the flank. Cleverly sited machine-gun teams laid down deadly fire along passageways to isolate small groups of defenders, and in several locations, created a vicious cross-fire. The man-high piles of debris and girders enabled infantry to close and clobber the harried defenders with grenades. Stay put or flee. Both choices led to death. Very few Red Army men escaped. It was all over in a couple of hours: Brandt seized the workhall.

The second phase now began, and it fell to Hauptmann Kempter and Oberleutnant Grimm. Their first objective was to subdue Soviet soldiers in what Kempter called the "administration building". This two-storey component at the southern head of Hall 6a contained three stairwells (one in the middle and another at each end). The upper floor was completely destroyed, as was the central stairwell. Experienced soldiers briefed Hauptmann Kempter on the building's layout and suspected enemy positions. His assault groups, led by Grimm and Leutnant Gillardon, formed up in Hall 4 for the attack:

After discussing the plan of attack with the company commanders from a spot where it was possible to survey the terrain, orders were given and, after a short artillery preparation around midday, the companies launched the attack. Soon, however, any movement congealed because resistance was offered from well-camouflaged and fortified positions which forced everyone into complete cover. I had to locate the main resistance nests and silence them with machine-gun and mortar fire. At first, the fighting moved around the ground floor of the administration building, where the penetration into the stairwell was rather difficult.

Weight of firepower applied in the right areas forced the defenders to beat a hasty retreat downstairs to the basement. Soviet soldiers were never more dangerous than when backed into a corner. "There was a fierce struggle in the cellars," recalled Kempter, "particularly around the large heating systems. We simply had to have these, so that our troops pushing into the large workhall would not be attacked from behind." Kempter ordered Grimm to clear out the cellar. Grimm called over the flamethrowers from Hall 4. Protected by other pioneers, Müller and Schwarzer worked their way along the outer walls of the building, spewing destructive fire into the cellar through ground-level windows. Demolition squads worked alongside them, pushing satchel charges and grenades through any opening. Gefreiter Müller may have been a troublemaker, but he "notably proved himself" during this attack. The brutal application of "blowtorch and corkscrew" tactics, as Americans would later coin them, gradually broke resistance, and the cries of the trapped yet violently resisting Red Army soldiers grew silent. The administration block could not be declared secure until someone went down there. Acrid smoke from Flammöl and burning bodies billowed up stairwells and out windows. Mixed teams of infantry and pioneers descended into the Stygian murk. The cellar was filled with heating equipment and ducts. Pillars of sickly sunlight splayed through the smog. A few dark objects, ablaze, generated more smoke than light. Muzzle flashes suddenly erupted and bullets splashed off the walls. More grenades. Flamethrowers shrieked. Bitter wrestling ensued in the underground passages. The Germans ultimately prevailed and Grimm was proud of his company's role in this success:

> *I was able to deploy my company and Pioneer Platoon 576 with success. Only with the assistance of the flamethrowers and my demolition squads was the hall taken... Infantrymen who took part in the attack confirmed the following about the employment of the flamethrowers: the Russians howled like lions.*

Kempter's recounting of this attack at post-war reunions caused some chagrin amongst the pioneers, as Grimm later explained to fellow veterans:

> *I cannot tell whether or not Hauptmann Kempter recognised me at the divisional reunion. He called together the survivors of his Kampfgruppe for a report about his attacks that took place in the halls of the Gun Factory. The deployment of the flamethrowers should have been described more accurately. Since I saw no other pioneers there, I walked away because I was unable to listen to him. Within my tight circle of comrades, I scoffed at this and said that it could be reckoned that some people wanted to earn another medal after the war!*

The job of capturing Hall 6a was only half-done. With their rear now secure, the assault groups reformed and pressed into the workhall itself. Hauptmann Kempter picks up the story:

After [the administration block] had been taken out of the battle, the attack into
the large workhall continued. It was a battle from machine to machine with rifles,
pistols and hand grenades. An inconceivable din filled the hall which had been
screened outside so that the Russians there could not offer assistance to those
inside. The hall measured seventy by one hundred metres. We subsequently
succeeded in clearing out the enemy and then occupied the hall and surrounding
area. Despite the toughness of the fighting, our casualties – if I remember correctly
– were kept to within reasonable limits. Congratulations for this success came from
all sides.

Grimm's pioneers had worked like a well-oiled machine. Obergefreiter Rinck jotted
in his diary: "Attack through remaining workhalls with flamethrowers. Objective
reached at 1700 hours." The conquest of Hall 6a had taken five hours, three more than
required for Hall 4. A Sturmtag was granted to 2nd Company ("Attack and seizure of
Hall 6 in the Gun Factory"). Contrary to expectations, Grimm's "problem children" had
performed well. Unteroffizier Karpenkiel fought bravely as a squad leader while
Gefreiter Müller managed to suppress his rebellious individualism and worked perfectly
as part of a team. Despite screeching columns of flame advertising their presence to
Soviet snipers, both flamethrower operators survived the day without a scratch. Such
barbarous fighting could not pass without some sort of toll being paid. From Grimm's
shrunken company, Pionier Merle was killed and five men wounded: Gefreiter Josef
Beck, Gefreiter Johann Wäschenfelder – recent recipient of the Iron Cross First Class
– Gefreiter Franz Winkelmann, Pionier Alfons Flotzinger and Pionier Josef Geiger. All
valuable men. Winkelmann had been wounded two days earlier and stayed on, but now
fresh injuries necessitated his evacuation. Pionier Flotzinger was wounded in the
stomach and chest by shell fragments. Gefreiter Beck, a riflemen in 6th Squad and one
of its original members, was so severely wounded that everyone thought he would not
make it. In June 1943, he reported to Grimm that he was still alive:

When I came to the replacement company and met some members of the old
company, all of them said that they thought I was dead, but that's not the case. For
six months I lay in hospital in Berlin and could not get in contact with anybody.

The Bodensee Division's losses were remarkably light: three men killed, one officer
and 58 men wounded. Lyudnikov's division, on the other hand, was gutted: 74 dead and
289 wounded. Although Soviet records remain silent on the subject, many of the men
recorded as dead were in fact alive and in German custody as POWs.

A "New" Officer (25 October 1942)

Nights were never quiet. German artillery harassed river traffic, landing stages and suspected assembly areas along the riverbank, while Soviet planes hummed around in the darkened skies and artillery fire plastered German-occupied factories. Soviet storm groups also attempted to press into the Gun Factory but were repulsed in close combat.

According to Rinck's diary, the company changed position behind Hall 6a, a move that happened at the right time because they soon witnessed a "heavy Russian attack to the right." The situation was far more serious than Rinck realised; to support an attack, Soviet sappers had undermined a wall in Hall 6a at 1030 hours. German defenders were thrown into confusion when part of the outer wall erupted in a massive explosion and then collapsed into a sink-hole, but they quickly rallied and repelled the surprise attack. It was then the turn of Gruppe Brandt to continue its assault to cleanse the factory grounds in a series of small targeted raids. Unteroffizier Karpenkiel claimed a Sturmtag on this day – "Reconnaissance patrol for Inf.Rgt.576" – for a small operation in the factory's north-eastern corner. The terrain was terribly exposed, criss-crossed by railway lines and hemmed in on two sides by a concrete boundary fence atop a small embankment held by Soviet riflemen. This minor foray generated serious losses in Grimm's company. Obergefreiter Karl Schwarzer, the flamethrower operator, was killed, and four men were wounded: Obergefreiter Johann Schmitt, Gefreiter Kurt Böhmert, Pionier Karl Kaltenbach and Pionier Konrad Spiewok. Another man forced to leave the company was Gefreiter Karl Seel. A week earlier he had fallen ill because of the terrible sanitary conditions, but on his own wish stayed with the company when they moved into the factory on 21 October. Although drained of strength, he continued to fight and distinguished himself during the brutal hand-to-hand clashes in the workhalls. His illness worsened so much that on 25 October a medical officer ordered his admission to hospital. Reluctantly, he joined the other wounded soldiers and was carried to the rear.

In the meantime, other assault groups mopped up behind the line. Small teams carefully probed each building and set off explosives in cellars, underground ducts and sewers. The intention was dual-purpose: kill any defenders still hiding there, and block those routes. During one of these operations, Feldwebel Emil Grassl, a platoon leader in Staiger's 3rd Company, was killed when hit in the head by shrapnel. Grassl was originally a member of Grimm's company and acknowledged as one of the best NCOs. A wound in June did not keep him out of action very long and he was amongst the first to be awarded the Iron Cross Second Class. As a proven deputy platoon leader, he was transferred to 3rd Company earlier in the summer to command his own platoon and now he was dead, another victim of the Barrikady.

The aim for the coming days was to continue the assault toward the Volga with 79th Infantry Division, 14th Panzer Division and the right wing of 305th Infantry Division. Gefreiter Füssinger from 1st Company was appalled by the ongoing slaughter:

We are still stuck in Stalingrad and in the final battle for the vast Gun Factory, which itself is as big as Ravensburg and Weingarten combined. Unfortunately, many comrades have met their deaths and my company only has about thirty men. Stalingrad is like Verdun from 1914-1918, that is, it's hell. I also have a lot of difficult tasks because bringing supplies into the forward line is not always easy. And one must keep having good luck. Our division has become so weak that it can be assumed that we can no longer be deployed this winter. Where are our winter quarters, in Russia or Germany, who knows? When will this killing end?

It was not only the men in the front-line being affected by Soviet weapons. In the afternoon, a devastating blow befell the division when an artillery shell landed squarely on a bunker at the divisional command post in Gorodishche, killing the unfortunate chief-of-staff – who had only taken over the role from Kodré two days earlier – and the divisional adjutant.[7] Reams of valuable paperwork, including many well-deserved recommendations for medals, were destroyed. Kodré was still in the 6th Army area, so he was called back to resume his old job until another replacement arrived.

The battalion received a welcome reinforcement in the form of an officer. Twenty-one year old Leutnant Hans Zorn arrived after a twelve-day journey, first by train via Warsaw and Kharkov to Stalino, and from there by Ju-52 to the Chir railhead, a major supply depot for Paulus' army, where he was able to hitch a ride to Gorodishche. Zorn was no stranger. He had been a member of the battalion since its formation and enjoyed the peaceful year of occupation duty in France. He showed a lot of promise during a training course in January 1942 and was soon accepted as an applicant for a reserve officer's commission, promoted to Unteroffizier in mid-April and sent to Germany for further training. After completing a three-month course at Pionierschule II in Dessau-Rosslau, he was promoted to Leutnant on 1 October and granted a brief furlough before departing for the Eastern Front on 12 October. Zorn reported in at the battalion command post and was quickly filled in. The state of the battalion, particularly his old 1st Company, shocked him because every officer from France was gone. Hauptmann Traub was a familiar face, but his physique had changed; months of campaigning and

---

7   The IIa (division adjutant) was responsible for all matters dealing with replacements, personnel matters of officers such as promotions, decorations, punishments, as well as the rosters, war rolls, and casualty lists.

meagre rations had thinned him out. "Zorn, get some rest, you'll take over 1st Company tomorrow." Gefreiter Zrenner recollects the return of Zorn:

> A *"new" officer reported in direct from the officer's school, but he was not unknown to me. He stood in front of me in a neat officer's uniform. To advance his career, he had to command a combat group. I told him the night before to take off his epaulettes or at least conceal them. Opposite us were a lot of snipers who observed us and were also very keen to capture officers alive, whom they recognised by their epaulettes.*

Proof of Life (26 October 1942)

Guided by a messenger, Zorn reached the factory before dawn. A cleanly-shaven officer in a fresh uniform juxtaposed incongruously with the mutilated workhall and its troglodytic denizens. Red-rimmed eyes stared from bristly faces at this outsider. Zorn did not recognise anyone. "Where is your leader?" Zorn asked. "Down there," one of them replied, pointing to an underground entrance. Zorn ducked into the narrow opening and introduced himself to Stabsfeldwebel Buchholz. "I'm here to take over the company." Buchholz briefed Zorn on the state of the unit and handed over a three-page document with the name, rank and position of every combat soldier in the company. It was dated 10 September but small notations were pencilled in beside most names: dead, wounded, sick, on leave. Back then, its combat strength was two officers, thirteen NCOs and 107 men, supplemented by a further NCO and twelve men who arrived as replacements prior to the big offensive on 14 October. Six weeks later, all that remained was three NCOs and 25 men. Word spread that a new commander had arrived and those from France – the few still alive – vouched for Zorn and the fact that he had risen from the ranks.

All of the battalion's companies were now led by officers, but Grimm's responsibilities were broadened by circumstances outside his control. Although Kempter was in command of III./Regiment 576, practical control of the battalion's combat groups was held by Leutnant Gillardon, a burden soon to fall to Grimm: "When Gillardon dropped out due to illness, I led the entire Kampfgruppe in four attacks from Hall 6a in the direction of the Volga bank." Grimm's group fell back across the scrambled terrain into Hall 6a with Soviet riflemen right on their tails. Gunfire rippled along the battle-beaten façade, forcing the attackers down, but they inexorably shifted closer behind grenade broadsides and tight mortar salvos. The distance closed. Hand grenades flew back and forth. The Soviet riflemen stubbornly clawed closer, and soon,

Grimm's cache of hand grenades ran out. In desperation, his men lobbed Blendkörper,[8] frangible glass smoke grenades normally used to combat tanks. Incredibly, this barrage of "phosphorous grenades", as they were deemed in Soviet reports, finally broke the back of the three-hour Red Army assault. "The effect of these grenades was very painful," reported one of Lyudnikov's regiment commanders. "When phosphorus got on the body, it burned. Sand was not effective in quenching it. The fires were extinguished only by water, which could not always be found up front." After repulsion of this attack, Grimm riposted with one of his own; at 1420 hours, his assault groups debouched from Hall 6a, instigating two more hours of frenzied fighting before he pulled them back to the workhall. "As a consequence of flanking fire and high casualties," recalls Grimm, "the railway line east of the Gun Factory could not be held." Total casualties within the Bodensee Division were four men killed and 27 men wounded. Unteroffizier Karpenkiel was severely wounded while breaking into a trench system between Hall 6a and the boundary fence. Grimm helped him from the battlefield. He claimed this day as a Sturmtag ("assault detachment") and catalogued his wounds as follows:

> Shrapnel from an explosive shell got me in both legs and on the left arm and hand. Now I look really knocked about. On my left hand I've lost my index finger, and the middle finger is half stiff. This has put an end to my profession. Shrapnel is still stuck in both legs. Because of these wounds, I've now been designated unfit for military service.

Karpenkiel's spell at the front only lasted three days but he made a big impression on Grimm, so when requested, Grimm attested to his performance:

> As a result of his dashingly led attacks at the most difficult position of the combat sector, I submitted Karpenkiel for the Iron Cross Second Class... Also, I consider that the prerequisites for the bestowal of the Assault Badge have been met because Karpenkiel was in the attacks on 24, 25 and 26 October 1942 in the forward line under violent enemy influence with a weapon in his hand when he broke into enemy-occupied trenches and bunker positions.

---

8   A Blendkörper, shaped like a large electric light bulb and filled with 260 grams of amber-coloured titanium tetrachloride, was designed to be thrown against tanks so that the glass smashed, allowing the chemicals to create a spontaneous volatile reaction. The caustic smoking mixture would penetrate to the interior of the tank and either incapacitate the crew or force them to abandon their vehicle. The effect of the Blendkörper was potent in confined spaces.

Thanks to Grimm's input, both medals were awarded, and when Karpenkiel was discharged from service in late 1943, Grimm once again provided a sworn declaration in order to obtain a promotion and consequent increase in benefits:

Unteroffizier Karpenkiel only belonged to my company for a comparatively short time. All of the company's written records were lost through the fall of Stalingrad... In the barely two months that Unteroffizier Karpenkiel was in my company, I was unable to accurately assess a man like this... I can confirm that against all expectations Karpenkiel really proved himself as a squad leader during my company's toughest battle in Hall 6 of the Gun Factory. His daredevil acts on one hand and the loss of almost all NCOs on the other certainly prompted me at the end of 1942 to recommend Unteroffizier Karpenkiel for promotion to Feldwebel. In the time he was with my company, he didn't do anything wrong, his leadership was good. In his previous assessment there were words like "boaster and a show off!", which I can still recall quite clearly, but I must confirm that he did not talk so big. In view of the fact that Unteroffizier Karpenkiel has really proved himself in the face of the enemy and his leadership in my company was good, I request approval of his promotion to the next highest rank.

Also injured during the attack from Hall 6a was Pionier Thomas Lais. The fact he was even up front was a sign of the desperate need for manpower. Conscripted in February 1942, Lais joined the battalion in mid-August, but after moving into defensive positions north of Vertyachii in September, he constantly griped about a pain in his knee. Some thought he was feigning, but Grimm sent him to the medical officer anyway. Something was amiss, so he remained with the rear echelon to recover, but was caught in the dragnet and sent to the front on 24 October. Lais reported to Grimm in the factory, complete with a bandaged knee. For three days he fought, hobbling through Halls 4 and 6a, until he aggravated his knee problem. In the scheme of things, his injury was a trifle, but Lais kept groaning with pain, so Grimm ordered him back to the supply train for treatment. A medic decided to send Lais to a main dressing station, where he lay for a night before being transported rearward on 27 October and eventually ending up in a Dresden hospital specialising in tuberculosis cases. This disease, discovered by the company's medic, was the reason for his evacuation. In July 1943, Lais lodged a claim for the wound badge and Grimm was forced to write a diplomatic assessment as to whether "a non-battle injury is legitimate or not."

Nightfall suppressed the offensive desire of both sides, so for the first time in a fortnight, Grimm had the opportunity to write to Lotte:

Since the current mission has lasted longer than expected, I send you a sign that I'm alive... The struggle is hard and grim. We look like old fighters, unwashed for days and unshaven. Of course everyone is infested by vermin.

Don't be angry with me if I make you wait so long for news, I just wanted to write to you when the battle was over.

I'm writing from the command post, the Russians fire incessantly from the Volga. I would never have dreamed that I'd ever see this stream! Now I care so little about it.

Hopefully, with God's grace, I'll get through the fighting in good shape.

In marriage, a burden shared is a burden halved, but Grimm filtered his anguish and torment because he was determined not to upset his wife or cause her undue worry. Having a husband at the front was bad enough; knowing he was experiencing hell-on-earth might emotionally cripple her. Gefreiter Vorherr had no qualms about laying himself bare in letters home:

Have now been through eleven days and nights of battle without sleep, as a foot messenger. I was the only one of three drivers up front. I've got such a cold from the freezing nights, in which we must lie in the open, and am suffering from diarrhoea, so that everything went into my trousers. It was evil. Because of that, my calves are sore and I almost can't walk. I was like that for five days, but it was not going away, so I reported myself sick. On top of that I now have a cough and am choked with phlegm. The fighting around Stalingrad continues, I never thought that it would go on for so long. Who will be able to take responsibility for what this costs men? It will be a miracle for whoever comes out of this in one piece... When this thing at Stalingrad is over and done with, there won't be many of us left. There is still only a few hundred metres to the Volga, but it is fortified. Every metre of ground, every piece of machinery in the factory halls must be bitterly fought for. Our panzers cannot be targeted on anything. Then there's the continual bombardment from the other bank of the Volga with every available weapon. If it does not end soon, then I'll have to head back to the front, into this awful carnage... There's not many of us left, everyone is wounded or dead.

A Day Off (27 October 1942)

Grimm's Kampfgruppe was granted a day off. The majority of the fighting on this day would take place in the southern half of the factory. After several days of extraordinarily hard fighting in neighbouring sectors, conditions were right for 14th Panzer Division to reach the Volga. The objective for Kampfgruppe Püttmann was also to reach the river, form a defensive line from the cliff-tops back to the factory's rear gate and then, after realigning its forces to face north, roll up Soviet positions east of the factory. Based on resistance encountered over the previous week, this objective was ambitious, to say the least.

Kampfgruppe Püttmann and Heim's panzer division slowly pushed forward in brutal house-to-house combat. The pioneers of Zorn's 1st and Staiger's 3rd bolstered the infantry assault teams. Vicious fighting, mostly with grenades, raged throughout the morning, and only toward 1000 hours did the Soviet defence begin to crack. A breach was gradually widened in both directions and the fuel installation taken. This was considered a major success. Bitter fighting north of the massive cylindrical tanks continued well into the afternoon. Kampfgruppe Püttmann had achieved its first objective but was incapable of rolling up the Soviet position along the factory's south-east edge due to the pigheaded defence, though as planned it erected a defensive line from there back to the factory's rear gate. Infantrymen immediately began digging defensive positions along this line. The aim for the next day was to mop up the riverbank east of the factory.

After a week of service, Oberst Barnbeck handed the reins of the division to its newly appointed leader, Oberst i.G. Bernhard Steinmetz. As chief-of-staff of 8th Army Corps, Steinmetz had frequently dealt with the division and knew many of its officers.

Remorseless Fighting (28 October 1942)

Kampfgruppe Püttmann was ordered to mop up the riverbank during the night, particularly below the escarpment. The intention was to clear all enemy positions and block the riverbank to the north and south, but resistance was just too bitter, and was about to become even harder; their opponents launched an audacious counterthrust under the cover of night. German soldiers were scared by these nocturnal forays. Somehow, nightfall turned ordinary Red Army soldiers into terrifying phantoms. By exploiting this fear and raiding sensitive sectors, Lyudnikov's storm groups swiftly recovered the cliff-top positions and only stopped at 0100 hours when they stumbled into superior German forces. They transitioned to the defence and had dug themselves in by dawn.

Lyudnikov's deft nocturnal manoeuvre threw German plans into disarray. The assault to finally capture the Steel Factory and mop up the Soviet bridgehead east of the Gun Factory was delayed because of this penetration on the boundary between 14th Panzer Division and 305th Infantry Division. Püttmann dispatched some units to deal with this crisis while pressing ahead with his planned attack. The division's forces were formed into two groups: Angriffsgruppe Püttmann[9] in the south and Angriffsgruppe Brandt[10] in the north. Both massed their forces on their common boundary – the factory's main west-east road – and worked in unison to unhinge the Soviet defences, though the weakened attack collapsed after several attempts.

A two-hour lull fell across the sector while battle swelled further south. The Soviet penetration in the boundary was eliminated throughout the morning and then the attack relaunched. Combat groups slogged forward in formidable hand-to-hand combat and captured several blocks of houses. Grimm's pioneers likewise went back into action: "My combat strength on 28 October was eight men." Examination of available documents, however, shows at least 13 men of 2nd Company still in the line. Regardless of this discrepancy, the weakness of Grimm's company is beyond question, yet it did not prevent their redeployment. Every unit in the division was feeble. Men who had toughed out the brutal engagements in the workhalls now succumbed to relentless deployment. Pionier Paul Klöpfer, a 37-year-old from Pfullingen in the foothills of the Swabian Mountains, was one of the few to survive a fortnight's fighting unscathed. Grimm described his actions as "daredevilish" during close quarters scuffles at the stadium on 14 October, near the Tractor Factory on 15-16 October and during the week in the Barrikady. On this day, 28 October, he moved against a Soviet bunker in the factory's north-eastern railyards. Clutching a satchel charge, he crawled up to the bunker, yanked the pull igniter and hurled it into the embrasure. He hugged the ground to avoid the explosion. Seconds passed. Nothing. Something had gone wrong. The charge failed to detonate. Just then, a grenade bounced out of the bunker. Eyes wide, Klöpfer jumped up and began to run but the blast bowled him over. Comrades who were providing cover-fire dragged him into shelter while others finished off the bunker with more explosives. Another casualty during this action was Pionier Karl Fasser, a young Swabian who still lived with his parents. With his trusty MG-34 he had covered numerous demolition teams during risky raids, often laying down sustained fire when

---

9   Angriffsgruppe Püttmann was composed of Regiment 578, two battalions of Regiment 576, Gruppe Pressmar, Staiger's 3rd Company and a panzer company.
10 Angriffsgruppe Brandt was composed of Regiment 577, III./Regiment 576, Grimm's 2nd Company and Radfahrschwadron 305.

everyone else was ducking their heads during mortar bombardments. Like Klöpfer, he had endured two weeks of remorseless fighting, only to be taken down during the attack against the dense bunker belt east of the factory. Both Klöpfer and Fasser would recover from their wounds and rejoin the military in 1943.

Undeterred by pressure all along the line, Soviet units had held firm and did not pull back. This Soviet capacity for steadfast resistance did not seem to figure in planning because the ambitious aim for 29 October was the "destruction of the enemy east of the Gun Factory between the fuel installation and the Brickworks".

– – –

Cold weather set in. Compared to the dreadful winter of 41-42, the Wehrmacht was now far better prepared. Depots were already stockpiling warm clothing and boots and more rolled in over the coming weeks. Hauptmann Traub was the beneficiary of Major Beismann's forward planning:

Some days ago I missed out on a vest with cat's fur sleeves, which would have given good service during winter. Fortunately, our commander had requisitioned a thick fur coat. I have taken it. I think wearing it will help me to survive the winter. At the moment we still have good weather. Today there was lovely sunshine along with a stiff east wind. Maybe this winter will not be as bad as the last.

Try telling that to Gefreiter Vorherr. The tone of his letters was one of abject misery, deprivation and concern for the future:

I am still not in the best of health because of the cold. Yesterday I had to spend half the night riding around on the motorcycle until 1 o'clock this morning. It was bloody cold and my legs froze. My cloth pants are ripped open and I only have heavy cotton pants. My underpants are really thin and the wind whistles through, but all this driving around was simply nothing for me. Had I known that I'd have to drive around for so long, I would have pulled on my oversocks, they're warm... We're not too badly set up for winter, but we should already have been in winter positions and not first have to build them when it freezes, as we have to do now! We will live through this winter and won't be able to wash for many weeks: where will we be able to wash in the wind and the cold? In a brook? What have we done to deserve being in this country? If the fighting does not come to an end soon, then I reckon I'll be sent back to the front. Oh well, others have to do it.

Bloody Metres (29 October 1942)

Fighting continued. Lyudnikov's divisional journal states that fourteen German attacks were repulsed on this day. By mid-afternoon, barely any progress was made in the bitter house-to-house fighting. The records of 6th Army show that just "two houses in the row of buildings south-east of the Gun Factory were taken." Oberst Steinmetz was sickened by the slaughter for minuscule gains, so he informed his superiors that the division no longer possessed the strength to carry out further mopping-up operations. It was decided to discontinue the attacks on this sector. Casualties were ghastly: one officer and 37 men killed, two officers and 93 men wounded, and five men missing, a total of 138.

Unwarmed by the heat of combat, Gefreiter Vorherr was suffering in the rear. "As long as we have not moved into winter positions, it is a misery to have to live in this area." The weather was so chilly that he was forced to write letters in the cab of a lorry "because you can't write out in the open":

> It's miserable here, not a house or warm room in the cold wind! At the supply train, where I'll be perhaps until the day after tomorrow, we have a hole covered with boards and earth, where we sleep at night, but during the day no-one knows where to hang out. If I have to go forward again in a few days, as I certainly believe I will, then there will be no rest day or night. Prepared for the end (death) at any moment because there's only hand-to-hand fighting here. None of us can understand why we've been left here so long on our own and not been relieved. It looks as if we'll have to stay here until not one of us is left. We've still only got a few hundred metres to the Volga but we just can't do it with our meagre forces. Whoever is lightly wounded and gets out can be glad.

Vorherr's thoughts constantly dwelt upon an escape from Stalingrad. If not through wounds, just one alternative remained: "Every month nine men go on leave, as I have learned, so I'll receive furlough next year, although we drivers – according to the company commander – will only go after everyone else, as he has already said." Men leaving on furlough were eyed enviously by Vorherr: "Lucky are those who've received a leave pass and can spend a few months in the Reich over the winter while we'll probably have to spend the winter in our foxholes." Vorherr thought Grimm harsh for denying him leave but seats on homebound trains were strictly allotted, so commanders selected men on several criteria, including parental status and time served at the front. Fairness also dictated that soldiers who had stared death in the face for weeks on end be granted priority in the furlough queue. One of the men heading home was given an envelope by Grimm to post once back in Germany. This, his second letter home from Stalingrad, was simply proof he was still alive. He had neither the time nor desire to write more:

> *From my command post in Stalingrad I want to give you a small sign of life today.*
> *By the light of a Hindenburg lamp I'm now reading a newspaper again for the first*
> *time in a long while. The Russians are firing with artillery and so forth, so the lights*
> *are flickering. Otherwise the fighting is hard... Hopefully I'll get through it by the*
> *grace of God... Just now the Stalin organs are drumming.*

Respite (30-31 October 1942)

The division's sector was held by three groups, each formed along regimental lines: Gruppe Brandt (Regiment 577) in the north, Gruppe Püttmann (578) in the middle and Gruppe Braun (576) in the south. A minor reorganisation carried out during the night returned small detachments to their parent regiments and reforged unit integrity lost a fortnight earlier. The only battalion not under its own regiment was Hauptmann Kempter's III./576, still positioned in the factory's northern workhalls as part of Gruppe Brandt. Grimm's and Staiger's companies were also with Brandt while Zorn's supported Püttmann. In accordance with Steinmetz's order, the division temporarily discontinued the attack. For the next few days, defence was the main priority:

> *Positions won during the attack are to be held, their technical reinforcement is to*
> *be tackled immediately. "Positional improvement" by vacating tracts of land or*
> *buildings may not be made. Through the deployment of heavy weapons, enemy*
> *assemblies will be smashed, buildings and fortifications that could interfere with*
> *subsequent attacks suppressed by observed fire. Group leaders must create a*
> *reserve, no matter how small, to intervene in local break-ins.*

Combat actions along the front-line were non-existent as infantry on both sides kept their heads down, grateful for the respite. The pioneers were put to work strengthening the defences, mostly mines and wire entanglements, but these tasks put them in harm's way and led to the deaths of two men: Obergefreiter Josef Männle from 2nd Company and Gefreiter Franz Josef Schelb from the 1st.

During the night of 30-31 October 389th Infantry Division took over the left wing of Gruppe Brandt, including Hall 6a, and released Kempter's battalion from this charnel house. The contraction of the Bodensee Division's sector was in preparation for a planned attack against the Gun Factory bridgehead on 2 November.

Casualties for the division on this day were fifteen men killed, one officer and 36 men wounded, and three men missing, capping off a bloodsoaked ten-day period for 305th Infantry Division: from 21 to 31 October, it suffered losses of 6 officers and 201 men killed, 14 officers and 688 men wounded, 28 men missing, and 5 officers and 72 men ill, a total of 1,014. When combined with the previous ten-day period, the division suffered just under 3,000 casualties in twenty days. It had been bled white… and there was no end in sight.

Each pioneer company was now the strength of a squad. Grimm's original complement of combat soldiers at the beginning of the Stalingrad offensive – about one hundred men – were mostly dead or wounded. Only eleven of them, including Grimm, were still at the front. Even a dozen or so replacements harvested from the supply train were gone. The casualty rate for 2nd Company was around 100% in just seventeen days. It was the same for Zorn's 1st and Staiger's 3rd. Beismann was dead. Unity was lost. The battalion had been changed forever.

CHAPTER 8

Washing Hands in the Volga (1-2 November 1942)

Thirty-two year old Pionier Georg Klima from Myslowitz in the Silesian highlands felt like an outsider. His Polish hometown only came under Hitler's control during the opening weeks of the war, so his German was heavily accented, but this was compounded by a speech impediment and resultant mockery. The introverted Volksdeutsche[1] entered Beismann's Swabian battalion as a replacement in early September. Unexpectedly, he distinguished himself during the fighting in Stalingrad, first as a messenger and while bringing up close combat weapons during the many assaults, and then, when every able-bodied pioneer was culled from the rear echelon, as a combatant in the hand-to-hand combat in the Gun Factory. "He fulfilled every position in an exemplary manner in spite of fierce enemy activity," Grimm wrote in a recommendation for the Iron Cross. During the night of 1 November, Klima was on an errand, carefully picking his way through the shattered factory; turning on a torch was suicidal, so he relied upon the twitching light of parachute flares to scout his route. Sheets of corrugated iron and broken machinery lay everywhere. When a flare fizzled out, he dashed forward, but his feet suddenly stood upon nothingness and he tumbled down a deep shaft, landing with a splash at the bottom. Foul-smelling sludge sloshed onto his face and mouth, and he even swallowed a little while flapping about madly, scrabbling for a handhold. An evil stench overwhelmed him: he was in a sewerage pipe choked with effluent and putrescent corpses. After clambering up the rungs and scraping off the odious muck, he continued on, much warier, but soaking wet and reeking. The following day he fell terribly ill and was sent to the rear. It was not just shot and shell that could incapacitate a man in the Barrikady. Klima's condition worsened, so he was admitted to hospital and then shipped back to Germany. His Stalingrad nightmare was over.

– – –

Oberst Steinmetz arrived at the command post of 51st Army Corps for an hour-long conference with Paulus, von Seydlitz, two more generals and their respective entourages.

---

1   Volksdeutsche = a person whose language and culture had German origins but who did not hold German citizenship.

The subject of discussion was how to continue the assault. Steinmetz was ordered to completely capture the Volga bank in his sector in conjunction with 389th Infantry Division. The task was challenging: what once would have been considered a minor action was now designed as a two-phased operation. First, Regiments 576 and 578 would reach the river, thoroughly clean out the cliffs and establish a south-facing barrier. Then Regiment 578 would quickly prepare for a new thrust to the north-east in conjunction with Regiment 577. One overriding message was made clear to the assault leaders: "The Volga bank must be reached under all circumstances." The order filtered down to battalion commanders like Hauptmann Kempter:

> The ground was flat for a few hundred metres, without cover, and then fell steeply 40-50 metres and ended in a 25-metre wide strip along the Volga. The escarpment and riverside strip were unknown territory to us because they had not been looked over and offered surprise opportunities of all kinds. My battalion was reinforced for the attack. Heavy weapons could not participate because they were unable to observe the steep slope and the shoreline.

Although Grimm's men did not participate, the attack deserves to be described in full as it would prove to be a dress rehearsal for a later operation that heavily involved the pioneers. The attack was launched at 0600 hours, twenty minutes after the sun had risen. By 0930 hours, Regiment 576 had struck fierce resistance south of the fuel installation and was unable to reach the bank of the Volga. North of them, Regiment 578 was able to occupy a few blocks of houses. After a quick regrouping, the assault was relaunched and realised some initial triumph, as Hauptmann Kempter describes:

> After distributing orders to unit leaders, the attack began at the appointed time, supported by mortars and heavy machine-guns. The Russian garrison of the flat, mined terrain was overrun in the first attempt; at times, a hand-to-hand struggle ensued with cold steel. I moved forward with the first wave because I had to decide on how to proceed from atop the cliffs. After I saw just a few Russians in the steep slope, there was only one thing to do: slide down and advance across the riverbank to the Volga. The initial negligible resistance was quickly broken. I washed my dirty hands in the Volga. Soon, however, our lives were at risk when the Russians emerged from their secure dug-outs in the cliff-side and fought back, whereby we were down below on the Volga, the Russians around and above us on the steep slope. Pressure was also applied on us along the riverbank from the north and south, so gradually

we were facing a superior force. There remained no alternative but to vacate the shoreline and in the falling darkness – which began at 1500 hours – climb up the escarpment and prevent the Russians from following up. Luckily, the Russians on the eastern bank could not intervene because they would have shot their own people. Little by little we had to fall back to our starting position. The only thing achieved was gaining an insight into the occupation of the cliffs and the shoreline. Large depots and advanced staffs were located in the cliffs. Our casualties during this operation were relatively low and our wish, to be on the Volga, was over.

During this attack, a sniper shot Hauptmann Kempter through the earlobe, but he remained up front. Major Braun was not discouraged by the ultimately futile charge on the fuel installation. Lessons were learned. They just needed to be applied to future attacks. Püttmann's assault teams also tried again at the same time, renewing the attack at 0945 hours with support from ten assault guns and ample artillery and aircraft cover. One infantry company and four assault guns attacked the Apotheke, a four-storey L-shaped building anchoring the centre of Major Pechenyuk's 650th Regiment. After a two-hour struggle, the house was set on fire by incendiaries and Püttmann's men rushed into its southern wing. Lyudnikov ordered Pechenyuk to hold his positions on the right flank while launching small counterattacks on the left. Pechenyuk ultimately prevented a penetration to the Volga on his sector, but had a tough time on the right:

Enemy tanks cruised along the main defensive line and fired directly into buildings, setting fire to houses on Pribaltiskaya Street and Prospekt Lenina. By 1015 hours up to a company of infantry rushed into the house on the corner of Prospekt Lenina [House 71] but 650th Rifle Regiment liberated the house with a resolute counterattack. All further advances by the enemy were stopped.

The men of Brandt's Regiment 577 were poised in Halls 3 and 4, ready to launch their attack when Püttmann's combat groups attained their initial objectives, but as that never happened, Brandt assisted with some minor forays that seized a few houses. Eventually, the offensive stalled up and down the line. The day's effort was costly for the Bodensee Division: for total casualties of 44 men killed, 3 officers and 116 men wounded, and 1 man missing, it only gained a few buildings and some small patches of ground. Moreover, once darkness fell, counterattacks against Braun's men near the fuel tanks led to the loss of gains achieved at such cost during the morning. Püttmann's men also relinquished their hold on the Apotheke. The toll taken on burnt-out units was ghastly. The division reported its condition to 6th Army: of nine infantry battalions, three were weak and six battleweary. The pioneer battalion was also listed as battleweary. "The division is suitable for defence only," concluded the report.

The division was let off the hook. Only the southern wing of 389th Infantry Division was expected to continue the assault. Oberst Steinmetz ordered his units to "establish defences along the line reached today." An indication of how narrow the German strategic vision had become was the fact that the division now dispensed with its 1:20,000 gridded maps and began using a hand-drawn sketch at four times the scale. Each building was numbered, while major landmarks bore memorable names coined by the troops. Steinmetz's order accurately plotted the course of the front-line, from house to house, until it merged into the factory. Each regiment was encouraged to echelon its forces in depth, construct strongpoints in and behind the fighting line, and form small local reserves. "At night," he continued, "the main battle line must be more densely occupied to prevent infiltration at any price. It depends primarily on the fact that every nest and strongpoint, even if temporarily cut off, must be held with the utmost tenacity."

Settling In (3-7 November 1942)

The men worked tirelessly on the defences throughout the night. Soviet big guns pounded the factory area, as did the Red Air Force, while German howitzers targeted the eastern bank and the Volga, and infantry guns battered positions in the slender bridgehead. With this fracas drowning out noises along the battle line, the pioneers lugged rolls of wire forward and helped the infantry erect barriers. Mines were also laid. Tools were downed occasionally to repel sporadic Soviet raids north of the fuel tanks.

After being injured several times, one of Grimm's stalwarts, Obergefreiter Rinck, finally succumbed to a serious wound on 4 November. As fatalism set in, he stopped writing in his diary. Recording the names of dead comrades made him despondent. The shell fragment that punctured him also saved his life; it was enough to get him sent home for further treatment. Grimm submitted him for the Iron Cross Second Class:

Obergefreiter Rinck repeatedly distinguished himself through exceptional guts and courage as a squad leader during the attack near the Tractor Factory and particularly during the attack from the Gun Factory to the Volga bank under violent enemy influence with rocket-launchers and heavy mortars. He led his squad prudently and in an exemplary and brave manner.

As Rinck was carried rearward, his battalion commander headed forward after ten days in the rear. Bedevilled by a hacking cough and cold symptoms, Hauptmann Traub spent most of the time attending to a mountain of paperwork. "With some impatience

I am awaiting the return of our adjutant, who must be coming back from leave in the next few days. He will be shocked on returning to the battalion to see how many have dropped out in his absence." On the other hand, Traub enjoyed some of the perks of his new position. "You must be surprised to get another airmail letter so soon," he wrote to Ella, his wife, "one advantage a battalion commander enjoys. He can allow himself this pleasure." A more visible sign of this privilege was parked nearby:

A further example to illustrate this: Beismann had a so-called commander's van built. This means we found a small Russian omnibus on the road and over weeks of labour installed a new motor, benches, tables, lighting and other luxuries, creating a superb motor home. Beismann wanted to use it as his living and sleeping quarters but was unable as he fell in combat before it was finished. The van is parked outside my door. I could use it except it's too cold at present, but next year it will give me great service.

Traub knew full well that the war was far from over, but when his wife mentioned a neighbour's belief that it would go on for another decade, he retorted: "She asks if the war will last another ten years? The old toad should not talk this nonsense." Nevertheless, the bus was a constant reminder to Traub of the ongoing struggle, new responsibilities and his old commander:

It would be a delight for me, despite everything, if he (Beismann) was still here. In the meantime, I received a very sad letter from his wife, who got the bad news of his death surprisingly quickly. She wanted to know more explicit details about his demise. I wrote in detail to her before getting her letter. I had his grave photographed so I could send her the pictures later. We can be very sorry for the poor woman, she is alone with her two children, but one becomes quite inured to such a fate.

In some ways, Traub was glad to leave this sad correspondence behind and return to the front. He ventured into the northern part of the Gun Factory to visit Staiger and his men:

Early today I was at the forward posts, visiting the division's three regimental commanders as well as my third company, and returned here dead tired. I can cope with everything, but running between factory ruins, throwing myself down for the artillery, mortar and machine-gun fire does seriously affect an old man like me. I really believe I'm the oldest in the division. But my men were happy that I was with them, which, after all, is the important thing.

Traub was not wrong: at 46 years of age, he was in the running for the title of the division's senior combat soldier, but the aches and pains – resulting from old injuries, not age – troubled him little: "My knees and hand are fine. I don't notice anything with my hand. My knee hurts a little when I get up, but once I'm in my stride, all is well." Soldier's wives are understandably anxious, especially when their husbands mention gunfire, but Traub exuded confidence in his letters. He had survived the Great War and would make it through this one, too: "Anyway, don't worry about me, an old combat fox like me is not that easy to get."

— — —

On 4 November, Steinmetz issued a warning to his troops based on intelligence gathered from disillusioned Red Army soldiers who crossed over to German lines:

> According to statements by deserters, the enemy has the intention of retaking Stalingrad on the day of the Bolshevik Revolution, 7 November. Even if the value of such statements is questionable based on past experience, a resurgence of hostile offensive activity must be expected from 6 November onwards.

Preparations were made: officers ensured their sentries maintained the highest state of vigilance; reserves were kept ready behind the line; fire emplacements of heavy infantry guns and artillery pieces were converted into strongpoints able to repel an infantry attack; assault guns were put on call. The pioneers toiled hard to fortify positions deep behind the forward line. The main provision, however, was artillery. Any indication that the Soviets forming up, any sign at all, was to be blasted and a curtain of blocking fire dropped in front of the German line.

All of these precautionary measures proved to be a waste of time. Rather than attacking the Germans in the Barrikady, Lyudnikov's soldiers celebrated the day of the October Revolution with vodka, congratulatory decrees from higher headquarters and the longed-for issue of cold-weather clothing.

A replacement for chief-of-staff Kodré finally arrived on 5 November in the form of Oberstleutnant Paltzo, a bookish staff officer from army headquarters. Kodré packed his things, bade farewell to Steinmetz and other staff members, and was driven west to catch a train. He took with him a small memento: copies of divisional orders written during his tenure as chief-of-staff, an indelible record of one division's descent from a powerful formation to an impotent combat group drained of blood and vigour.[2]

---

2   These divisional orders, now filed at the Bundesarchiv-Militärarchiv in Freiburg, have proved invaluable during the research and writing of this book, for they contain details found nowhere else.

Hauptmann Traub also welcomed another arrival with open arms: "Fortunately, my adjutant [Leutnant Max Fritz] has returned, he relieves me of a lot of the daily routine details. We all hope we will soon be pulled out of the line so that we can recover our strength." Traub was greatly relieved by Fritz's return: "I can arrange my affairs more comfortably and don't have to do everything myself." Pre-occupying his mind was an elevation in rank, and he pestered his wife to provide the symbols of office: "Ella, what is happening with the major's epaulettes? Hopefully you have already sent them as I expect my promotion will soon be confirmed." Unbeknownst to Traub, his idle mind was about to be turned to the next big undertaking.

The Lesser Evil (8-9 November 1942)

Days earlier, five pioneer battalions arrived in Stalingrad for a final offensive.[3] The seed for deploying these engineering specialists sprang from the mind of Luftwaffe firebrand and ground warfare dilettante, General von Richthofen, and it found fertile soil in the form of Hitler. The initial plan was to commit these pioneers against the Lazur Factory bulge, the largest Soviet foothold in Stalingrad, but an alternative soon arose: clear the bridgeheads east of the Gun and Steel Factories first. Army leadership was not keen on either operation because even though the pioneers would form a potent strike force, sufficient infantry support was lacking. The pros and cons of each objective were tossed around for a few days. "Give us a decision and an order," remarked General Schmidt, Paulus' chief-of-staff, "and we'll carry it out!" Army Group B and 6th Army agreed about which bridgehead should be taken first:

> We regard several small bridgeheads to be less dangerous than the one large bridgehead around the Lazur chemical factory, particularly once the Volga has frozen over. This large bridgehead may then represent an operational threat.

The final call, however, was up to Hitler. The decision was phoned through to Paulus' HQ: "The Führer has ordered that the 'lesser evil' (the bridgeheads east of the Gun Factory and metallurgical works) be eliminated first." And with that, the men of the Bodensee Division were destined to be dragged into another bloody assault. Steinmetz formed his division into regimental combat groups. Oberstleutnant Paltzo trekked through the industrial wasteland to give the regiment commanders a heads-up on the forthcoming attack. No preliminary order was issued, Paltzo just casually said: "You might want to give some thought as to how we could get through to the Volga."

---

3   The history and deployment of these battalions in the forthcoming battle is covered in great detail in one of the author's previous titles. See Mark, Jason D., Island of Fire: The Battle for the Barrikady Gun Factory in Stalingrad, November 1942 – February 1943.

The concentration of so many pioneer battalions necessitated the appointment of an overall leader, and this role was assigned to Major Josef Linden, head of the army's pioneer school. Steinmetz and Paltzo thoroughly briefed Linden about the situation and informed him that five pioneer battalions were at his disposal.[4] In addition, two more with the northern neighbour[5] would cooperate. "With the exception of the 305th and 389th Pioneer Battalions," Linden recalled, "the battalions were completely combat-ready. The 305th Pioneer Battalion was at about one-third strength and the 389th Pioneer Battalion about half-strength. All battalions had Eastern Front experience from many battles and were fully qualified for their impending task."

On the frosty but dazzlingly radiant morning of 8 November, all of the pioneer battalion commanders were summoned to the Bodensee Division's forward HQ in the Schnellhefter Block. The newcomers were strangers to Hauptmann Traub, but he introduced himself and chatted. Linden began the meeting at 0930 hours. After a brief orientation, he issued terse orders: "Get pioneer assault groups ready! Commanders will be at my disposal in three hours at the advanced command post of 305th Infantry Division for a reconnaissance." The battalion commanders hurried back to their units and set their officers to work. Specialist tools-of-the-trade were readied: flamethrowers, explosives, elongated charges, detonators, smoke-pots and wire-cutters. Boxes of hand-grenades were stockpiled. They were left in no doubt that heavy-duty action was headed their way.

Linden and Steinmetz surveyed the factory from an observation post before venturing into the devastated complex to scout the attack area with the three regiment leaders. Sector commanders briefed them on local landmarks, enemy defences and potential pitfalls. Having gathered all the information needed, Linden made his proposal: each infantry regiment was to be supported by a pioneer battalion[6] while Regiment 578 in the middle – which Steinmetz had designated as the main spearhead – would be additionally supported by Traub's battalion, placed behind the 50th Panzerpioneer Battalion as a reserve. Linden also nailed down other details:

The assault would be launched on 11 November 1942. For the time of attack, I chose the first light of day after the appropriate preparatory fire. Strong pioneer assault groups (at least platoon strength) would push into starting positions under

---

4   These were 50th Panzerpioneer Battalion, 162nd Pioneer Battalion, 294th Pioneer Battalion, 305th Pioneer Battalion and 336th Pioneer Battalion.
5   The 45th Pioneer Battalion and 389th Pioneer Battalion.
6   The 294th, 50th and 336th Pioneer Battalions were with 305th Infantry Division while the 162nd and 389 Pioneer Battalions supported 389th Infantry Division.

the protection of this fire, then the artillery would be abruptly advanced: after this, the pioneers would advance as the first wave and penetrate to their objectives while overcoming the enemy in known positions. Once there, they would immediately set themselves up for defence. Infantry would follow as the second wave and clear the enemy from the intermediate ground and subsequently take over the defensive line.

My proposal was accepted and details of its implementation were discussed with the regiment commanders. Never before during the war had so many pioneer battalions been concentrated for a single attack in so small an area. The commanders of the pioneer battalions employed on the sector of 305th Infantry Division were precisely briefed on site by me about boundaries, objectives and known enemy positions... The commanders briefed their company leaders on their assignments. All care had to be taken while doing this so that the Russians did not notice movements in the front-line and discern our attack objectives.

Their Soviet opponents suspected nothing. To them, it seemed their foe was settling down for the winter. German soldiers were no longer seen dashing from house to house. Instead, as evidenced by smoke drifting up from every building and basement, they were huddled next to fires kept going around the clock. Gefreiter Füssinger spent every night carting supplies to his comrades in Zorn's company, so he was feeling the chill more than most:

It has now become quite cold here. It is rather unpleasant on the open plains! We should have received a rest long ago but it is not yet completely finished in Stalingrad. Only a small strip, along the Volga, is still occupied by the Russians. At present, men on leave from the east are at home... Today, one would be happy if a bed and a house was seen again. No more houses are standing here in Stalingrad and the city will not rise again for 20-30 years. Now we hope for the best for the future, in particular that the slaughter here stops.

This hope would soon be dashed as another bout of killing began. The freshly arrived pioneers knew that a big attack was imminent, but the exhausted soldiers in the front-line were not informed until late in the piece. Hauptmann Rettenmaier, leader of Regiment 578, remembers the moment he received official word:

The order for the attack to the Volga bank came on 9 November. It stated: "The bank of the Volga is to be taken securely into our hands and the encircled enemy destroyed. As reinforcement, Kampfgruppe 578 will receive the 50th Pioneer Battalion and is allocated the 305th Pioneer Battalion."

The addition of a full-strength pioneer unit alleviated some of his concerns about the forthcoming assault and the inclusion of Traub's battalion was a bonus, although Rettenmaier was not yet fully aware of its headcount: "All I knew," he later remarked, "was that the 305th Pioneer Battalion was very much weakened. There may still have been eighty to a hundred men in it." Despite Linden's plan to keep the battalion in reserve, circumstances dictated its deployment in individual companies: Grimm with Regiment 576, Zorn with 578 and Staiger with 577. Local knowledge possessed by the Bodensee pioneers was needed to provide guides and advise the newcomers. They would also bolster the infantry assault groups.

Paltzo's casual "advanced order" had prompted Rettenmaier to gather advice from those with first-hand experience of their renascent adversary, so he had a plan ready to go, though he modified it slightly due to the additional pioneer units. While an artillery barrage suppressed and screened targets with smoke, the Apotheke and the Commissar's House would be taken in surprise attacks. When Braun's Regiment 576 reached a predetermined point near the fuel tanks, Rettenmaier would then initiate the second phase: a thrust through "Index Finger Gully" towards the riverbank and House 79.

Leutnant Zorn's company was to play a leading role in Rettenmaier's plan despite Linden's initial intention of keeping Traub's battalion in reserve. Zorn's mission could not have been more daunting: take the Apotheke in a surprise attack. Infantry would move up with him. At the same time, the 50th Panzerpionier Battalion would take the Commissar's House, though its commander brusquely dispensed with any infantry support. Once these two defensive cornerstones were taken, the gate to the Volga would be open via the Index Finger Gully. For this second phase, one combat group, reinforced by Zorn's pioneers, would reach the water's edge and press further up the river, while a reinforced battalion attacked House 79.

Infantry units made room for the incoming pioneers. Hauptmann Püttmann – back in command of his old battalion after the more senior Rettenmaier took control of Regiment 578 – instructed his men to prepare for relief by the 50th Panzerpionier Battalion the following day. Leutnant Zorn was ordered to de-activate the touchy anti-personnel mines ahead of the battalion's sector. As twilight descended mid-afternoon, Leutnant Zorn and his pioneers left the safety of their building armed only with mine probes and entrenching tools. Zorn had created and updated the mine plan as they were laid over the previous weeks, so he knew where each lethal jack-in-the-box was located.

Complete darkness set in and they commenced their delicate task. A gunfight suddenly erupted around them and flares hissed into the night sky, illuminating the area and bathing everything in flickering light. Whilst hugging the ground, Zorn and his men deftly disengaged trip wires and removed detonators. All men returned unhurt and Zorn immediately notified Hauptmann Püttmann that the minefield was now inert.

A snapshot of Zorn's 1st Company on the eve of the new offensive reveals some telling details. A trickle of reinforcements had boosted trench strength to 1 officer, 4 NCOs and 24 men, while the supply train was almost as strong: 4 NCOs, 12-15 men, and another NCO and 5 men convalescing from light wounds. With just 29 combat soldiers, 1st Company possessed the strength of a platoon, though most were newcomers. When the company entered Russia in May 1942, it was composed of three platoons, each with three squads (strength of each squad was 1 NCO and 14 men). By November, the squad with the most original members was the 1st with four, followed by 6th Squad with three and 2nd Squad with two. All other squads only had one, while 9th Squad had been completely wiped out by death or wounds. Also noteworthy is the fact that the company had lost all of its original officers and combat NCOs: those present on 10 November did not originate from the company. Nevertheless, they were able soldiers and would lead their squads with skill.

Sick Leave Delayed (10 November 1942)

The past week had been very quiet for Grimm, and for that he was grateful. Chronic blood-streaked diarrhoea and stabbing intestinal pains drained his strength, leaving his face sallow and brow fevered. The summer campaign had whittled away his stocky build, but it was Stalingrad, which fused corporeal fatigue with mental exhaustion, that had taken the greatest toll. Cavalier notions of fighting for Germany's glittering future were engulfed by images of sickening deaths and gnawing guilt; as commander of the unit, he was like a father to his men, but could do nothing to prevent their demise. Willpower had kept him going a few more days but even that reserve was now tapped out. Crippling illness was the last straw. He managed to get a letter to his wife:

After twenty-one days of fighting in Stalingrad, I must leave the battlefield. I'm suffering badly from stomach and intestinal infections and my new battalion commander, who was my predecessor, has more sympathy for me [than Beismann]. I have been so miserable since then that I cannot write, let alone read. Today, however, I have been able to dictate. At the request of my new boss, the division has authorised sick leave for me. My train will leave on the 19th of this month and will presumably be on the rails for six to eight days. Until then, I hope to regain a bit of health. Perhaps I will be home before this card.

Barely was his letter sent when new instructions arrived:

I was not a little astounded to receive the order that I was subordinated to Major Braun for the attack on 11 November. We already knew each other from France. I was ordered to support the infantry after reaching the steep cliff of the Volga bank. I reported at his command post the day before. My route to him in the Gun Factory, in which I'd been for weeks, was about 150 metres. A guide led me to a massive annealing and tempering furnace which had large heating chambers, an overhead coal supply and an underfloor receptacle for ashes and slag. Next to it were gigantic presses and drop forges. It is simply unimaginable how impressed I was to read upon them the names of German companies in embossed, red-lacquered letters... The finished, pressed-out baseplates, probably for Russian heavy mortars, had been dragged by our infantrymen to the entrances of their dug-outs as overhead cover. It was suspiciously quiet and there was no immediate enemy action. By candlelight, under the furnace, where there was even a tiny trestle table set up with a seat for me, I received the following order: for the s.I.G. auf Selbstfahrlafette [heavy infantry gun on a self-propelled chassis], employed for the first time during the war, a crossing over the railway tracks was to be immediately readied so that they avoided damaging their tracks. Flamethrowers (two units), rolls of plain wire, explosives and several anti-tank mines would be brought up during the night and placed in readiness. A shelter was set up in heating ducts, under the collapsed ceiling joists of the workhall (about 15-20 metres high). This was carried out without casualties and I stayed with Major Braun in his command post under the furnace.

Grimm tried to soldier on but maintaining his composure was difficult:

During the night I once again considered the thought that this would be my last attack because I was so exhausted that climbing back up the steep riverbank did not bear thinking about. My men were to blow up the entrenchments to the left and right and erect a makeshift blockade with wire and mines down to the waterline.

With a new offensive about to begin, the division was able to review its losses over the previous ten days. Nine days of defence and one major attack ripped another four hundred men from its ranks. How many would the next ten days claim?

The "Final" Offensive (11 November 1942)

Pioneers filed into assembly positions, artillery observers climbed into elevated roosts and officers conducted last-minute checks. Oberst Steinmetz and chief-of-staff Paltzo moved into an artillery observation post with commanding views of the factory. At 0340 hours the western horizon flashed, accompanied moments later by a ground-shaking percussion. A wave of shells shrieked overhead and crashed upon Soviet territory. Rocket tails were visible for a split-second before they exploded, unseen, down behind the riverside escarpment. The pioneers gazed at the awesome spectacle. Adrenaline began to pump and butterflies stirred in stomachs. No sooner had the preparatory bombardment commenced than the Soviets swiftly replied with heavy counterfire. Grimm was still in Major Braun's command post beneath the furnace:

> *I suspected that the Russians had been expecting our attack because a destructive*
> *artillery barrage began. Large calibre shells slammed onto our furnace. The light*
> *went out and Major Braun grabbed me every now and then and shouted "Are you*
> *hit?" Our exits were buried by piles of brickwork.*

Sturmkompanie 576 moved out knowing its inspirational leader was entombed beneath a house-sized electrofurnace. Leading the way were the assault teams of the 294th Pioneer Battalion, while Grimm's men remained in Hall 6e, ready to rush forward with wire and mines to barricade the Volga beach. The cannonade cracked overheard as assault troops poured out of the factory grounds and through a junkyard of stacked ingots and gun-barrels. As they bobbed and weaved through this wasteland, helmets glinting in the bombardment's glare, they attracted the attention of the Soviet defenders. Desultory small-arms fire intensified to accurate volleys. Bullets pinged off scrap metal. A battery of assault guns joined up with the assault groups and the attackers pressed forward into vicious fire. Headway was sluggish. Momentum dissolved when two of the armoured escorts erupted in flames and another stopped in its tracks, reversed, and then sputtered back into the factory. All had fallen victim to a Red Army soldier and his anti-tank rifle. The attack ground to a temporary halt on this sector.

In the meantime, Braun and Grimm were rescued from their crypt. Braun got straight back to business. His adjutant filled him in and he issued new orders. Grimm, still dazed by his entombment, learned that his men were still waiting in readiness:

> *After hours of silence, I needed a lot of time to register that my pioneers had not*
> *been called into action and it also took quite a while to clear the two exits, with*
> *assistance from outside. We didn't know what had been happening outside.*

After regrouping, Major Braun personally led the new attack toward the fuel tanks at 0930 hours. This time flamethrowers were committed. Grimm's miscreant, Gefreiter Franz Müller, once again went into action with a pressurised fuel cylinder strapped to his back. As expected, Soviet resistance was savage, but worse still, terrain already captured became the scene of fierce clashes. The cause was soon identified: Red Army soldiers were using sewers and other subterranean passages as secure shelters and launching counterattacks from them. A large underground tunnel ran from the east face of Hall 6e along the boundary to the central street where it swung west into the heart of the factory. It had four access points on this straight stretch and needed to be cleared out. A dirty job like this required a weapon of terror, so Müller was sent in, as one of his comrades recalls:

> Müller was given an order to attack the sewerage system, that is, go down every manhole, the entire length of the street. The Russians were causing us a lot of trouble there, they had barricaded themselves in the cellars, and in the sewerage canals, and that's where we started… He climbed in; we gave him cover, good cover, with rifle fire, and he climbed in and cleaned up the entire length of the sewerage canal. They have to do that alone… Like a fireman using a hose, [he] worked his way along [the entire street] and cleaned out everything. After that, of course, we could resume the attack.

Müller briefly described his actions in a post-battle letter:

> During my operations in Stalingrad as a flamethrower-operator, I was in action six times with success, the last time on 11 November with the assault pioneers of 294 under Feldwebel [Oskar] Dickler.

After the way had been burned clear by the flamethrower teams, Braun continued his attack. The fuel tanks and river were within reach. Nasty fire and collapsing comrades were ignored as Braun dashed into the fuel installation at the head of his men, overpowered the defenders with grenades and spades, and quickly dug foxholes in the oil-soaked earth. Small groups pushed up to the edge of the scarp. Other troops also reached the riverside cliff further north. Groups of pioneers slid down the loamy slope but disturbed several nests in the cliff-side that would not surrender. The men of Grimm's company, led by Feldwebel Adam Pauli, had orders to erect wire obstacles down to the water and lay mines on what they called the Sandbank, but it was far too hazardous to be carried out. Pionier Ernst Kempter, a recent returnee after recovering from a stomach wound back in June, was killed.

Zorn's men were in action right from the beginning. Under cover of the barrage, they silently filed out of their assembly position in House 53, Rettenmaier's advanced command post. A kindergarten was on their right. Looming ahead was the Apotheke, their target. Careful observation revealed that its entrances were choked with rubble and man-made barricades, and were probably mined, too. Rettenmaier decided to avoid them altogether: "The most important thing was to create new openings into the house. We had the means for it. Enormous explosive charges would be positioned and detonated at the scheduled time." Zorn was allocated this crucial task that could literally open the door to victory; failure might doom the entire operation. Covered by the infantry, the pioneers stole up to the Apotheke, positioned their explosives, inserted detonators and then withdrew, trailing the detonation wires behind them. The Soviets inside the Apotheke had noticed nothing. At 0355 hours, massive detonations rocked the air, instantly shrouding the building in smoke and dust. Charging across open ground, the assault groups hurdled through the seething holes and caught the Soviet garrison unawares. Each floor was rapidly cleared and forty-five stunned prisoners taken. The surprise had come off splendidly. Below ground level, however, resistance still flickered: Soviet troops in the basement refused to give up and resisted every attempt to smoke them out. After casualties were suffered, Rettenmaier suspended further attempts to storm the basement.

The operation by the 50th Panzerpioneer Battalion to seize the Commissar's House did not run as smoothly. They too planned on gaining access with demolition charges, but would do so without infantry support, a decision made by Hauptmann Gast, their ambitious young commander. Rettenmaier watched their attack go in:

Armed with mines and other explosives, they plunged into the darkness to place them in and around the house. They tried to find an entrance or other opening into which they could place their charges. However, nothing could be found because everything was bricked up or so skilfully camouflaged that any search in the dark was in vain. For that reason, the pioneers waited for the dim light of dawn.

That was a fatal error. A blind man would have noticed the pioneers lurking outside the building. The Soviet garrison set about ruthlessly gunning down their besiegers and Rettenmaier was witness to the ensuing panic:

Many pioneers fell, lifeless, far more crept back, exhausted and bleeding, and formed up for a trip to the hospital. They had achieved nothing, except perhaps gaining experience that this way would not result in them taking their objective.

Failure to capture the Commissar's House threw a spanner in the works because it dominated the entire settlement. Its garrison could observe and fire in all directions. Even though Rettenmaier's men had all but captured the Apotheke, accurate sniping from the Commissar's House trapped them inside, together with their prisoners. "Daylight came," Rettenmaier recalls, "and our pioneers could no longer leave the house or let themselves be seen in any doorway because death lurked behind every opening. Who were the prisoners now?" Incredibly, Zorn's company did not lose a single man to death or wounds all day.

– – –

Staiger's 3rd Company watched the assault teams of the 336th Pioneer Battalion plunge into the murk raised by the rolling barrage and fight their way up to the houses on Pribaltiskaya Street. Following behind were the infantry combat groups of Regiment 577. The targeted two-storey houses had been battered into ruins, yet fierce small-arms fire lashed the pioneers. At one house, satchel charges were tossed into cellar windows and the pioneers charged down the stairs past dying Soviet soldiers. Try as they might, they could not reach the upper floors; four attempts were rebuffed. Reinforcements were required. Staiger's company was summoned to support the attack on House 66. He led his men through the curtain of artillery fire to assist their brothers-in-arms. While doing so, one of his NCOs was thrown to the ground by a chunk of shrapnel:

> A Feldwebel by the name of Heiduk was in my battalion. His name sticks in my memory because he suffered a grievous stomach wound from a shell, a wound from which he could not be saved. He knew that and said to me that if I made it home, could I locate his bride and bring her his final greetings.

Staiger and his men reached the building and found the men of the 336th Battalion "only on the lower floor. [We] could not capture more. The Russians were on the upper floor. The Russians, however, were only a bunch of stragglers." Hand grenades rolled down the staircase were caught and thrown back. A hurricane of fire raged around the building. Fighting as one, the pioneers from different battalions captured the rest of the house. The Soviet garrison resisted to the last man. Rather than be captured, a few threw themselves off the roof. No pause was to be given. As soon as this house was in German hands, another assault group struck what they supposed was an exposed flank. In reality, they crashed into the main defences of Lyudnikov's bridgehead.

– – –

After being rescued from the furnace, Grimm's weakened state was apparent to all. It was obvious that he was in no state to be on the front-line:

Whether by chance or by orders, I don't know, I was spoken to by a doctor... He suggested that I go back to the regimental command post and get through to divisional HQ by phone. Because of enteritis and blood in my stools, he wanted to have me admitted to the hospital in Kalach but didn't get away with it, and I also didn't want that because I would then not be returned to my unit. A solution someone suggested, which I had not even thought of, was to send me off for ambulatory treatment while on leave and return to the battalion when it was being reformed after the battle. But was this the solution? Of my few men, I still only had one NCO up front, by the name of Pauli, and we simply had to be withdrawn soon. Alone, in my condition, back to the supply train?

Hauptmann Traub gave Grimm permission to head to the rear:

I made my way back to the supply train in the Gorodishche Gully with Obergefreiter Schenkel. There was only one route possible during daylight: from the west side of the Gun Factory to its north-west corner – everyone knew where it was – then along the railway line (sitting on them were freight wagons, loaded with lathes, for as far as the eye could see) to the western side of the Tractor Factory and finally past the stadium to a gully that was almost unrecognisable because of the shelling. We were zeroed in on time and time again, but we overcame this unbelievable exertion without being wounded.

Grimm was in a sorry state when he arrived at the supply train, his mind so fever-addled that he only learned about his wretched condition after the war:

I might probably have suffered most from bugs in my company because the lice and bed bug bites had left my wrists so swollen that I could not take my jacket off. I had only been able to rid my shirt and pullover of vermin in a totally laborious and ineffective way: after a few hours, the collar of my jacket was white with lice eggs and I was able to scrape these off with the back of a knife. I ask myself, why doesn't one read about these plagues in war literature?

His orderly helped him out of his lousy uniform and prepared some hot water:

I was able to have a wash in a dug-out (due to the cold weather), put on some fresh clothes, cut off my beard and have a shave. I was given a rough haircut by an inexpert hand but was too worn out to have a photo taken. It was only now, by

means of the many reports that I had to sign, that I saw how many casualties the
unit had suffered. After a great battle, it was authorised to write the notifications
on a typewriter. It affected me so deeply, I would have much rather sunk into the
ground. The artillery strikes and bombing barely disturbed me that night.

The assault groups of the neighbouring division achieved much better results than the Bodensee Division. Major Krüger's 162nd Pioneer Battalion and the infantry assault groups of Regiment 546 reached the Volga and trapped the remnants of a Guards regiment and approached Lyudnikov's dug-out as dusk was falling. Every member of Lyudnikov's staff was called to arms to protect the headquarters. Despite some setbacks, 389th Infantry Division gained the Volga along a 500-metre front, taking the streets north-east of the Gun Factory and then cleaning out the area. The pocket containing the Guardsmen collapsed under incessant pressure and only eight men, all wounded, made it out. Lyudnikov's bridgehead had been reduced by half.

Rettenmaier was displeased with the day's outcome. Huge casualties had been suffered for one or two houses and a few hundred metres of shellpocked ground. "We had to realise that it was a different type of warfare," he said. Nevertheless, Lyudnikov's division was encircled and Oberst Steinmetz felt significant headway had been made. It was not a killer blow but it seemed fatal.

Major Braun's men cleared stubborn Soviet soldiers from the gullies and cliffs around the fuel tanks and blocked every approach route from the south with barricades and mines. These barriers were erected by the men of Grimm's company, led temporarily by Feldwebel Pauli. Leutnant Erwin Hingst, currently on battalion staff while convalescing from wounds, was ordered to take over the company.

Together with its three battalions of pioneer reinforcements, the Bodensee Division suffered heavy losses: two officers and 83 men killed, three officers and 271 men wounded, a total of 359 men. The five fresh pioneer battalions had a 25% casualty rate. The German leadership sought answers as to why the attack had failed to completely eradicate the Soviet bridgehead. Major Linden laid it out for his superiors:

To achieve quick results in this difficult terrain, with such a strong enemy, it is
essential to bring up an infantry regiment to reinforce 305th Infantry Division.
This difficult task can only succeed with the help of these fresh forces. I know that
the divisions along the Volga have all had heavy casualties and are tied down in
their sectors, but it should be possible to fly in a reinforced infantry regiment, just
as the pioneer battalions were flown to Stalingrad.

As had been pointed out before the attack began, infantrymen were needed to consolidate the gains, regardless of how many pioneer battalions were loaded into the attack force, but corps commander von Seydlitz's reply was terse: "There is no infantry available." Linden rejoined: "Herr General, the pioneer battalions employed here must be regarded as specialists. Under present circumstances, these battalions will bleed to death if the immediate supply of infantry forces is avoided." While sympathetic, General von Seydlitz said nothing could be done: "Our task now is to consolidate and hold the positions we have reached on the Volga. All the forces at our disposal must be employed to achieve this." Fresh orders went out in the evening: the attack would continue on 13 November. The main objective for the Bodensee Division was the Commissar's House, for 389th Infantry Division the Red House. "With good successful progress along the Volga," stated the corps order, "both assault groups will meet up and then destroy the remaining nests of resistance. The 162nd Pioneer Battalion is subordinated to 305th Infantry Division with immediate effect."

Cleansing Actions (12 November 1942)

The fighting on this day – cleansing actions in attained areas – was on a smaller scale but no less bitter. Some units suffered heavier losses than the previous day, like Zorn's company: it escaped casualties on 11 November despite its lead role in the attack on the Apotheke, but on this day, it lost six men during minor raids to mop up some small pockets. Two of Zorn's men – Gefreiter Valentin Warth and Pionier Ernst Kruse – were killed, while Gefreiter Paul Merk, Gefreiter Helber and Gefreiter Josef Bertsch were wounded. Leutnant Zorn was wounded too, hit in the upper left arm. Feldwebel Langendörfer assumed command of the company, a position he would only hold for a few days before wounds took him from the battleground. Leutnant Hingst's 2nd Company also suffered a few casualties, including Gefreiter Xaver Sperl, one of the men caught in a rear-area comb-out and sent into the line on 24 October. Since then, he had fought with distinction, particularly during the close quarters combat east of Hall 6a in late October. He was admitted to the main dressing station and then transported home, another lucky man with a "Heimatschuss". Grimm was also moving slowly westward:

In the morning, I was driven to Gorodishche in a sidecar by Umele, meeting only the adjutant Oberleutnant Fritz and a newly arrived doctor [Oberarzt Dr. Donatus Wörner]. He gave me medicine for my intense stomach pains. The leave pass had already been written out for 13 November, from Chir. I was able to solve the problem of getting there (about 120-150 km) by making a delivery in a motorcycle-sidecar.

Grimm was hoping to chat with Hauptmann Traub before his departure but Traub was still up front preparing for coming operations. There was little time for Grimm to rest. A mountain of paperwork awaited his consideration. The most draining task was writing letters to next-of-kin: "The correspondence, in particular my notifications to the families of the company's twenty-five dead, which I wrote in the time from 12-14 November, was handed over to the postal service for forwarding."

The Commissar's House Falls (13 November 1942)

Units were shifted around during the night in readiness for continued operations. The 162nd Battalion was moved into the sector of Regiment 578 to bolster the main effort and constrict the pocket. Braun's Regiment 576 (with 294th Battalion) would be left behind on the right wing to fend off Soviet relief attacks from the south, Regiment 577 (with 336th Battalion) would take some of the stubbornly defended houses along Pribaltiskaya Street and Prospekt Lenina, and Regiment 578 would renew the assault against the Commissar's House with the 50th Panzerpioneer Battalion and penetrate along the Volga, into the Soviet rear, with the 162nd Pioneer Battalion.

The attack began long before the sun crested the horizon. Darkness was essential if the Soviet defences along the cliff-tops were to be overwhelmed before the defenders of the Commissar's House intervened. The assault troops assembled in House 79 and the Apotheke. Heavy artillery, rocket and mortar fire howled in and the groups moved out. Small clusters of pioneers dashed through the gloom, subduing Soviet sentries, grenading dug-outs, pushing aggressively towards their objective. One assault group from the 162nd Pioneer Battalion, led by the audacious commander of its 3rd company, Oberleutnant Alfons Schinke, caught the Soviet defenders flatfooted and infiltrated right up to Lyudnikov's command post. Lyudnikov assembled every last man to foil Schinke's raid. They rose up into a counterattack and flung themselves at the surprised Germans. The resulting collision blunted the attack and then quickly imperilled it. The Soviet defences came to life all over the sector and Schinke found himself cut off. A desperate fighting withdrawal prevented the complete obliteration of his combat group, but losses were high. The bold advance up the Volga had failed.

Rettenmaier resumed the offensive with an equivalent of two battalions. His objective was the Commissar's House. No other civil building could have been more suitable as a strongpoint than this one. Its brick walls were up to a metre thick in some places and the ends of both wings resembled castellated keeps. The floors were steel-reinforced concrete and beneath the building was a claustrophobic maze of cellars and garages linked by narrow corridors. Also in the basement was the entrance to a tunnel that ran all the way into the factory. The failed assault two days earlier had been a bitter pill for Hauptmann Gast. For this renewed assault, his pioneer squads were allocated locally-knowledgeable guides from Regiment 578. The Commissar's House was battered by artillery and direct-fire weapons. Every window and breach in the walls was in German sights. Fires flared up throughout the building and emitted dense, suffocating smoke. Panzers and assault guns rolled forward to batter the building. Salvo after salvo hammered the second storey and its garrison. A risky frontal attack was then launched in an attempt to catch the Soviet defenders off-guard. Intrepidly bypassing the well-defended entrances at the end of each wing, the pioneers and infantrymen crept into the enclosed forecourt and blasted their way into the main entrance. Observation had shown that Soviet defences were concentrated on the ground floor with only support weapons based on the second floor, so the German plan was straightforward: outflank the Soviet defences from above. The pioneers cleared the Commissar's House from the top down. They then brought to bear everything at their disposal to eradicate the Soviets in the labyrinthine basement. Explosives blew holes in the concrete floor, into which went flamethrower nozzles, satchel charges, grenades, petrol-filled jerricans with explosives attached, smoke bombs and caustic Blendkörper. Only ten defenders, all wounded and burnt, escaped. The dread building was in German hands. With its neutralisation, the Germans gained a substantial amount of operational freedom.

Hingst's company was deployed near the fuel tanks, but a couple of its men participated in the struggle for the Commissar's House. Gefreiter Karl Küster had belonged to the company since March 1941 and entered the front-line at Stalingrad on 21 October, earning praise as an assistant machine-gunner. Keeping a machine-gun fed in battle was easier said than done. He lugged around several cans of ammunition and ensured the belt flowed smoothly. While fulfilling this role during the attack on the Commissar's House, he was severely wounded. Joining him on the long haul to the aid station was Gefreiter Anton Reinel. Grimm had recently submitted him for the Iron Cross Second Class. The Wound Badge would now be added to his collection. The fate of one more 2nd Company soldier revolved around this action: Pionier Georg Russ went missing and maintains that status to this day, but according to one of his comrades, Gefreiter Egid Kölmel, he was killed on 14 November.

Farewell to a Coward? (14 November 1942)

Just before he left, Grimm handed over death notifications for dispatch to next-of-kin, but forthcoming events would intercept those letters and prevent the families from receiving them. Glad to be rid of this sad burden, Grimm bade farewell and headed off:

On the morning of 14 November, I once again failed to meet up with battalion commander Traub. We drove to Chir in my Morris jeep after acquiring some "black market" petrol. The doctor received a motorcycle-sidecar from my company, and together, we took three wounded men with us as far as the main hospital in Kalach-on-Don. In Chir I got some warm noodle soup, filled my canteen with tea and obtained some march rations... The train only departed once night fell. In the middle of each wagon was a round iron stove already fired up, next to it was a container with coal. I climbed on board with wounded men and jaundice cases but there was only space for a few to lie on the bare floorboards, so I moved. Fleas began attacking me before the train moved off. Flea and bed bug bites caused me to come up in red welts. Lice bothered me little because I'd left them behind at the supply train. The delousing in Lozovaya was the only stop on the entire trip to Stuttgart and that only lasted six hours at most. Wagons were processed one at a time, papers examined and stamped. Enlisted men were separated from officers and officials bearing the rank of officers, everyone was fed and given rations for the trip. But in between came the most important thing. In batches of about 30-40 men, we went into a dressing room, all of our luggage was taken away, then we removed our clothes, showered together naked – drying room next – into the adjacent room to receive still-warm clothing, then we got our luggage back. In between was a strict cordon. Unthinkable that someone who had not been deloused or was not allowed to return home be permitted on the Germany-bound departure platform.

Grimm's train was underway for just over a week and arrived in Stuttgart on the morning of 22 November. Fully dressed and wearing an overcoat and pistol, he weighed just 96 pounds (43.5kg), was suffering from infectious enteritis and had feet so swollen by fluid that he had been unable to remove his boots since Lozovaya.

One man not sorry to see Grimm go was Gefreiter Vorherr. His desperation and helplessness compounded daily until it transformed into covert anger directed at Grimm. At the beginning of the month he wrote:

I'm still with the supply train because I won't voluntarily go forward. Others should be at the front for eleven days, like I was! Then they'll gain an insight into the matter. This was a trick at the time by the company commander because he did not like me, he said: "You, come with me," but if something happened to me, he would be responsible for it because Battalion expressly stated that it was forbidden for drivers to be deployed there. But ours had no qualms in this case, he is a big coward, he does not go forward with his men like a commander! In the eleven days, I didn't once see him at Battalion. Now he's holed himself up in the bunker of an infantry battalion and when he has to go out, he first asks whether there is still heavy fire! Because he sits on me – when I am well enough to go back to the front – then I must go forward, if he demands it... Such a commander makes everything more difficult because he's heading off on leave in fourteen days, so he has to see to it that he makes it through this period in good shape.

Grimm was never aware that his name had been besmirched in the worst possible way, but in his correspondence, he unintentionally provides counter-arguments to Vorherr's slur:

Instead of the originally planned three days, I had been on the attack or in the front-line for 29 days, never more than fifty metres from the enemy (in order to avoid the continuous impacts of heavy artillery, Stalin organs and mortars), twice had no contact for several days, had no rations and without the possibility of carrying our wounded men back. Twenty-nine days with no possibility of receiving water in which to wash, no razors, sleeping only on bare concrete or in a large pipe – without any bedding – despite the onset of cold weather, no supply of underwear or any other clothing, no mail. Days on end without any warm food.

Plaintive words of a coward trying to justify himself? It is not this author's place to determine if Grimm was a coward or not. Only those who fought alongside him know the truth and have the right to judge. Based on available records and impressions gained, however, the author believes a few points may extenuate Vorherr's accusation. First and foremost, simply remaining at the front, especially in Stalingrad, required courage. Hauptmann Traub once summed up the general feeling by belittling an acquaintance who had avoided front-line duty: "The chap must be happy that he hasn't been posted here because he would shit his pants with fear." Men at the front held shirkers in contempt, and it is on this point that Vorherr's hypocrisy is apparent: suffering from diarrhoea, exactly like Grimm, he called in ill and left the front, while Grimm was only removed from combat on doctor's orders. A comparison of their physical states also provides a good comparison of the severity of their illnesses: Grimm was severely incapacitated – doubled over in pain, with blood in his stools – and could only write letters through dictation; Vorherr was able to pen lengthy letters in his own hand. If time at the front is a measure of fortitude, then Grimm would win hands down: Vorherr was there for a total of eleven days, Grimm for twenty-nine.

Secondly, Vorherr did not possess all of the facts when he impugned Grimm's character. His statement that Grimm had "holed himself up in the bunker of an infantry battalion" was true but only because Grimm had been ordered to take command of the infantry unit, thus explaining his extended presence in the bunker. Naturally, an enlisted man like Vorherr could not have known the full details of Grimm's activities.

Thirdly, not every soldier loves his commanding officer. Grimm, a martinet whose foibles made him human, only ever presented his authoritarian side to his subordinates. He was different, however, when alone with a fellow officer, such as when he was visited by Oberleutnant Staiger at a low moment: "I once saw Richard Grimm crying bitterly about the miserable situation of his troops and from fear of what was coming. Grimm had truly broken down. It was enough to make you cry!"

The preceding trio of paragraphs seem to show the author siding with Grimm, but it is in fact a sober analysis of the facts. Vorherr was acid-tongued and renowned for telling it like it was, and in this case, he was voicing an opinion held by other men. Grimm never ran away from battle, but suspicions that he was "gutless" took hold amongst some 2nd Company members. Why? The reason is simple. Vorherr contended that Grimm did "not go forward with his men like a commander," and leading from the front was a virtue revered in the Wehrmacht. In Vorherr's mind, those that did not must, by definition, be cowards. Everyone in the battalion, from Beismann downwards, was able to compare Grimm's combat performance with Beigel's gung ho attitude. Beigel epitomised the German officer who led from the front and was a golden boy in the eyes of Beismann. Grimm was not. As a married man with a one-year old daughter, it is understandable that Grimm was more cautious. He was a competent officer: not great, but not bad either. It must be remembered that the battalion's officers were not highly trained general staff officers or military geniuses, yet they were dealing with situations beyond their control – sometimes their capabilities – and were often unable to make the perfect call when a problem arose.

Would Grimm be judged a chicken by modern standards? No, probably not, in fact, officers who charge forward at the head of their men might be regarded as reckless. Grimm was not a hot-head ready to plunge into enemy fire at the vanguard of his troops, although he did lead quite a few attacks, particularly in the final week of October. He directed the flamethrowers during the conquest of Hall 6a, controlled the assaults in the north-eastern corner of the factory and even helped carry wounded soldiers from the front-line. Of course, Vorherr was not at the front to witness these attacks, and he may actually have been the one to start the rumours of Grimm's alleged cowardice. He had a grudge and no qualms about spreading hurtful scuttlebutt. Leadership demands more than rank and bestowal of responsibility; enlisted men must trust their commanders. Stalingrad eroded that confidence. Up and down the line soldiers realised they were trapped in a deadly situation and their officers could do nothing about it. Victory could not be achieved, withdrawal was out of the question and the only chance of escape was furlough or injury. A third option was available to Vorherr: Grimm simply had to say the word and he could head back to the rear because he should never have been at the front anyway. Grimm needed a messenger, so he kept Vorherr close at hand. It is easy to see why Vorherr felt so hard done by.

Six months of front-line service besmirched by the gossiping of one disgruntled soldier. Fortunately for Grimm, these whispers never reached his ears.

– – –

The battle was but a dull rumble on the horizon for most men in 6th Army. As a supply soldier, Gefreiter Willi Füssinger lived in the rear and escaped the perils of front-line combat, but every evening he undertook a task just as likely to end in his death or maiming: hauling forward supplies to his company near the Apotheke:

Nobody can imagine the fighting in Stalingrad, it is the heaviest in this war, and whoever has experienced it and gets out will be able to proudly say that they had a guardian angel, and many have learned to pray here. Taking supplies to the front is always very difficult.

Combat soldiers in Stalingrad had the greatest respect for their supply troops. Night after night Grimm watched men from his faithful supply train traipse into the factory, carrying everything on their backs. Motorised vehicles were not used; instead, horses pulled small two-wheeled carts laden with ammunition and food to a small depot west of the factory, and from there it was hauled forward using nothing but manpower:

It is simply indescribable with what bravery and under what unbelievable casualties our supply train brought us ammunition and rations on the pony-drawn infantry carts – only at night – and transported our wounded back, and how they kept coming back despite knowing what lay ahead of them. Only once, at night, did I witness the arrival of ambulances at the Gun Factory. For us, the "rear" was the dressing station not far from my supply train.

Accompanying the supply troops back to the rear after the nightly delivery was Hauptmann Traub. Three days of gruelling fighting was behind him:

We have had heavy combat in the past days. We had to take each house from the Bolsheviks one by one in hand-to-hand combat. The Russians have turned the remaining houses into fortresses. Despite this, we have reached the Volga in some places and I believe the fighting in our sector will end within a week at the most. Then, the metallurgical works and chemical factory south of us must be captured, which will take somewhat longer, and then Stalingrad will be completely conquered. The fighting here – and everyone agrees about this – will be the high point of this present war and can certainly be compared with the battle of Verdun in the First World War. We still believe, however, that the Russians are at the end of their tether. Some days ago we took prisoners from a naval NCO school in Vladivostok. It is noteworthy that they have to put such people into action.

No-one begrudged Traub returning to his HQ in Gorodishche because 14 November was his birthday:

After four days in Stalingrad, I returned to our quarters, one good reason being to thoroughly wash myself and shave for my birthday. The other reason is that I wanted to read my incoming birthday mail and finally, tonight, as a celebration for the day, I want to get a good night's rest... After thoroughly completing my washing and shaving, I feel reborn. With the birthday mail, there was unfortunately nothing of value, neither from you nor anyone else. I now hope to get the expected mail tomorrow. I will have a good sleep tonight and not allow myself to be disturbed by the Russian pilots.

Before climbing into bed, however, he held a small birthday celebration attended by the battalion's adjutant (Oberleutnant Fritz), paymaster (Oberzahlmeister Keppler) and doctor (Oberarzt Dr. Donatus Wörner). They enjoyed a few bottles of wine and closed festivities by draining a bottle of Sekt. "The other three officers[7] – more I don't have in the battalion – were at their posts and naturally could not be with us," he explained to Ella. "However, I hope that some of the lightly wounded officers will soon return." Alcohol-enhanced exhaustion dragged Traub down into a deep slumber: "There were no enemy planes, I would not have heard them anyway, I was too tired." The following morning he awoke to a small pile of mail: "The many letters I have been ardently awaiting have arrived... In addition, my birthday parcel also arrived." Ella had sent him

---

7   The three officers were Leutnant Hingst (2nd Company), Oberleutnant Staiger (3rd Company) and Leutnant Homburger (light pioneer column). Feldwebel Langendörfer held command of 1st Company.

some smallgoods, a few knick-knacks (including a face washer), a batch of collar tabs for his imminent promotion to Major, but not luxuries like cigarettes:

> *At present we are well supplied with tobacco and alcohol. I don't have any problem with you keeping the cigarettes so that when I am next on leave – I think it will be February or March – I'll have something to smoke at home.*

Traub planned to return to the front in the afternoon but an appointment with Oberst Steinmetz in the evening delayed his departure until the following morning. Compared with the front-line, the rear – with its warm bunkers – was heavenly:

> *The weather here is quite tolerable. It's about ten below but the wind has eased, so one can bear it reasonably well. We are well supplied with warm clothing. I have not yet begun to wear my good pullover, I'll wait until it becomes colder.*

Chipping Away (15-19 November 1942)

Nothing changed when Traub came back to the factory. No sensational new tactics and definitely no reinforcements. Army Group B posed a question to 6th Army: "What is now intended for the further occupation of the Gun Factory?", to which one of Paulus' staff officers replied: "A small piece will be chipped away from the north and south every day until they join hands." The army's intentions are revealed by this small exchange. Gone were the days of throwing several battalions against the Soviet line in the hope of achieving a large-scale victory. Now, all energy would be focused on obtaining small objectives upon which the full weight of German firepower could be concentrated. Signs of Soviet desperation raised hopes: slow-flying biplanes buzzed overhead during the night and dropped sacks of bread and fat. German troops were delighted because they thought hunger would compel Lyudnikov's surrounded men to capitulate.

While Rettenmaier's and Brandt's storm groups whittled down the Soviet defences, their rear was covered by Major Braun's regiment in sacrificial fighting against a constant surge of Soviet troops attempting to break through to Lyudnikov. Grimm's company, now in Hingst's hands, lost two dead: Pionier Johann Malinowski and Gefreiter Paul Wössner. The fighting for the fuel-tank promontory became even fiercer on 16 November. At daybreak, Soviet troops broke into German positions via the riverbank but Major Braun sealed off the penetration and captured one officer and 35 men. He then ordered Hingst's company and the 294th Pioneer Battalion to blockade the riverbank with anti-tank mines and the much feared S-mines, which bounded out of the ground to waist height and tended to maim rather than kill. This was done: the beach and surrounding cliffs and gullies were thickly sewn with the deadly devices.

The foggy morning of 17 November was split asunder as Soviet howitzers pounded Braun's sector from 0500 hours, searching out bunkers and gun pits near the fuel tanks. Stalin organs joined in at 1000 hours. At the stroke of twelve, the fire shifted to the depths of the German defences. One final volley of rockets came down on the promontory and then the Soviets attacked. The fuel tanks were nothing but crumpled cylinders reeking of burnt petrochemicals. Infantrymen and pioneers had entrenched themselves as best they could. The defensive cornerstone was a thickset transformer box atop the northern rim of Appendix Gully. Obscured by milky fog, Soviet riflemen and marines pushed along the riverbank and stormed up the gullies, where mines and blindly-aimed machine-guns reaped a fearsome toll. Nevertheless, the storm groups were able to punch through the line into the fuel installation. Machine-guns sprayed the wraith-like figures as they darted through the mist. Grenades thumped. Shouts rang out and submachine-guns stuttered. The Soviet intruders were able to disappear below ground in the latticework of trenches, craters and foxhole. Braun's infantrymen immediately recognised the danger and pulled back. At great cost of life, the Soviets captured the transformer but suffered heavy losses and could advance no further. Casualties in German units were moderate and most of those had been caused by the preliminary barrage. Obergefreiter Franz Müller, father of seven and Grimm's "problem child", was in the thick of the action:

> We held a position for seven days, from 11 to 17 November, where I was employed as a squad leader under platoon leader Feldwebel Pauli. The position was taken by the Russians with overwhelming superiority from which Feldwebel Pauli, Obergefreiter Danner and myself emerged as the only survivors [of 2nd Company]… Feldwebel Pauli promised me that I'd get the Iron Cross First Class.

Müller barely escaped with his life when his bunker collapsed under a direct hit. Buried alive, he was eventually dug out and sent to the rear for medical treatment. Also wounded was Obergefreiter Hans Ruider. Despite being wounded on 2 November, he had remained up front voluntarily, distinguishing himself by his cool-headed destruction of bunkers with satchel charges before being wounded a second time on 17 November.

It did not take long for the Germans to strike back at the fuel installation. Assault groups pierced the western side of the penetration and overpowered several Soviet positions. The breach was partially sealed but the fragile Soviet line around their small foothold held. Conditions worsened throughout the afternoon as rain squalls and snow flurries drenched the battlefield and quickly froze over in the sub-zero weather.

– – –

After more than a full month in Stalingrad, the Bodensee Division was utterly exhausted and the moment was rapidly approaching when it could no longer function effectively. The arrival of a thousand replacements was timely but the division was shocked by what they had been sent: wide-eyed youngsters, hastily trained, mere human materiel for the slaughter. No junior commanders were amongst them. Oberst Steinmetz had no experienced non-coms to train them because the few not dead or hospitalised formed the core of front-line units. To be of any worth, the men needed to be drilled, not just in streetfighting tactics, but also in ordinary basic training.

At the front, the fighting dragged on in a descending circle of violence. On 19 November, Major Brandt conducted a surgical raid backed by massive power. Supported by the panzers and assault guns, two assault groups from the 336th Pioneer Battalion with a few men from Staiger's 3rd Company and Regiment 577 attacked along Pribaltiskaya Street to capture two buildings. From the safety of his cupola, a panzer officer watched the pioneers go about their work:

The pioneers used explosives to blast holes through the approximately thirty centimetre thick concrete floors. They formed the explosives into a bell shape, covered their ears and detonated them. Immediately after the explosion – in which all the force of the blast was directed into the hole – the pioneers tossed in hand grenades. The hole created by the explosives had a diameter of about 40-50 cm, so a man could not get through, only hand grenades and smoke canisters. In these last attacks in the Barrikady, flamethrowers were successfully deployed. We could see the pioneer-flamethrowers from our panzers. The pioneers poked their flamethrowers through the holes that had been blasted through. After that, the "smoked out" cellar was successfully taken by the pioneers and assault infantry.

The capture of Houses 74 and 67 were seen as victories, and measured against the Stalingrad yardstick, they had come at low cost. One of Staiger's men was wounded: Gefreiter Peter Teichelkamp was hit in the hip by a bullet. Teichelkamp later reported that at the time of his wounding, his company still had fifteen men, 1st Company seven men and 2nd Company six men. He was witness to several extraordinary events during the previous week. In his opinion, the discipline of German units in Stalingrad, which had been on operations for a long time, was no longer particularly good. He once saw some soldiers refuse to follow a Hauptmann's order and act in such an insubordinate manner that the Hauptmann started thumping the men with his rifle butt. The situation was only resolved by the fact that the Soviets attacked at that precise moment. On another occasion, Teichelkamp saw four combat pioneers hiding in a cellar when they were supposed to be attacking. Discipline and order was breaking down.

After four more days in the Gun Factory, Traub trekked back to his HQ in the rear. Operations were winding down and clerical matters demanded his attention. His adjutant took care of most paperwork but some affairs required his input. After attending to a small stack of documents, he put pen to paper and wrote home:

Early today I returned from our forward positions, washed myself properly, had a shave and am now feeling human again. I'll go forward again the day after next if nothing else comes up... In the meantime, I have eaten supper. Actually, it is only 5 o'clock and now I want to make things comfortable for myself. The shack is nice and warm and a bottle of red wine is open. This is how one can endure the war. As I wrote you some days ago, there is no shortage of liquor at present. Additionally, for commanders, there is a special allowance and I currently have an excellent cellar. I'll send you a bottle of Cacao liquor tomorrow. This will be my Christmas present to you. I wish I could give you other things but there is nothing to be had here. Anyway, you can be happy that – apart from air raids – you have been spared the war. In Stalingrad, despite the silly shooting and firing, there are still Russian civilians. They live in cellars, in sewerage pipes and potter about half-starving, pursuing their miserable existence. As my interpreter told me today and by way of an example, some days ago they found a four year-old girl in a shell crater who for two days had been lying next to her dead mother killed by a shell splinter. There are many such shocking fates. But we have become insensitive to them and are no longer touched by such events...

Traub preferred not to dwell on such tragic fates, so after discussing the weather, he returned to the subject of small luxuries available to him:

At the moment I have enough cigars and cigarettes to smoke. Other than female company, we are not starved of any physical pleasures. The night before last, we sat in our cellar by feeble candlelight and played a lively card game. The last of our personnel returned from leave fourteen days ago. They told us about getting an ample food parcel at the border station, four pounds of butter, a pound of flour and other nice goodies. If only it was my turn, but this is out of the question before February 1943.

Unfortunately for Traub and the rest of his pioneers, furlough schedules were being jeopardised by momentous events occurring far to the west on the morning of 19 November. Thousands of Soviet artillery pieces obliterated Romanian lines opposite the Serafimovich bridgehead and opened a path for an armoured stampede. Three spearheads cracked the Romanian defences within hours and speared deep into 6th Army's vulnerable left flank. Individual Romanian units offered stout resistance, but many others succumbed to tank fright and fled for the lives. Soviet armour wedges avoided resistance and raced for the Kalach bridge. German troops caught in Stalingrad's rubble had no idea of the drama unfolding deep in their rear. Soviet forces south of Stalingrad launched an offensive the following day. The intention to encircle 6th Army was clear, but there was nothing the Germans could do about it.

One member of the battalion with a floor seat to this dramatic spectacle was Gefreiter Teichelkamp from Staiger's company. Grimacing with pain from his hip wound, he was brought out of the forming pocket by car, reaching Chir railway station after an adventurous trip. He saw Romanians clinging to vehicles and screaming loudly to be taken along. Most were unarmed. An army in defeat is not a pretty sight. The Soviet jaws clamped shut on 22 November, trapping Paulus' army. A total of 35 lucky men from the battalion were on furlough when the pocket closed. Several more were strewn throughout hospitals in the hinterland. Everyone else was caught inside. Unteroffizier Theo Moller, wounded on 16 October while serving as Grimm's HQ section leader, managed to avoid encirclement by a string of circumstance and accident. He was in hospital in Konstantinovka with Obergefreiter Josef Ostermeier, also from Grimm's company. They were released on the same day, but Moller's discharge occurred a few hours earlier, so he waited for his friend. This slight delay saved his life because if he had left on the earlier train, he would have passed through Chir station and been on his way to Stalingrad. Instead, Soviet forces broke through during the night and severed the railway link, forcing Moller and Ostermeier to turn back. They met up with another pair of 2nd Company refugees – Pionier Ernst Egeler and Obergefreiter Franz Mackerle – and escaped westward in a wild flight. "I was again lucky in misfortune," explained Moller, "when I broke my ankle on black ice in pitch darkness. Evacuation to a Warsaw hospital, then a homeland one, delivered him to safety.

Encirclement and Reorganisation (20-30 November 1942)

For the moment, little changed in Stalingrad. The most active combat zone was still the Gun Factory sector. The final German offensive spasms occurred on 21 November when the attack was resumed to isolate and destroy the garrisons of two Soviet-held buildings (Houses 68 and 75). Only after an hour and a half of fighting and heavy casualties were the assault teams able to annihilate the defenders of House 75 and occupy the imposing structure. The loss of this formidable strongpoint – the largest building along Prospekt Lenina – was a severe blow to the Soviets because it now projected into their lines like a dagger. It was quickly converted into a formidable bastion by funnelling in more men, positioning extra machine-guns in embrasures and stockpiling lavish amounts of ammunition and grenades. The pioneers girdled it with thick belts of barbed wire and minefields during the night.

Closer to the river, infuriating Soviet resistance from cliff-side dug-outs provoked fierce attempts to finally eliminate it, but even the liberal use of satchel charges proved unsuccessful. Pionier Josef Zrenner clearly remembers the difficulty in attacking these swallows nests and underground thoroughfares:

> The Russians were dug in down below and we were up above, so it was not actually possible to combat them. The worst thing, however, was that the Russians were suddenly not only below us, but also behind us because they emerged from the city's sewerage canals and attacked us from the rear. The sewer system in Stalingrad was very well developed… The Russians, moving through the sewer tunnels, suddenly popped up in our rear and attacked us from behind. After that, one of the pioneer units was assigned to clean out the sewers with flamethrowers so that no further intrusions were possible.

As the fighting in Stalingrad ground to a halt, it was time for the Bodensee Division and its leaders to take stock, regroup and make plans to raise combat strengths. A large batch of reinforcements that had just arrived from the homeland brought rumours that the Russians had broken through Romanian lines south of Stalingrad and their allies were in disarray. The news swiftly filtered through to the troops, but nobody was terribly concerned. The division's main priority now was to incorporate these new men. While the grenadier regiments planned to conduct rudimentary training, the 305th Pioneer Battalion put them straight into the line. The initial allotment for each company was not great, as Gefreiter Vorherr wrote: "Today we received another fifteen replacements, mostly Austrians, young chaps, born in 1923. More will come later, a sign that we're to stay put." Gefreiter Zrenner remembers more details:

> We received replacements in November and they were parcelled out, but unfortunately I must say that they were hastily trained men. They had no experience of any sort. They were young men, and we assigned them,

proportionately, according to the individual strengths still remaining to the companies.

They were first briefed by the staff and then split up amongst the individual combat groups. They were instructed by the staff and told exactly how to behave by those men already at the front, and then at night, they were directed into the battle line, into the foremost positions. There, two, three, four soldiers were assigned to each strongpoint. Familiarisation was therefore done at the front by a man who had combat experience.

I have to say that these replacements were not a reinforcement for us. Each unit received approximately ten, fifteen new men. Some of them fell quickly, either wounded or could not be found because they'd gone astray in the positions. They had to know that our units were stationed on one side of the building and the road, Russian units on the other side, and it moved back and forth. We could no longer advance any further because we lacked forces.

It was unfortunate that these quickly trained young men were thrown into this fighting. As a consequence, which was shocking for me, after two, three days, I noticed that these young men were not up to the fighting. They came back and some were mentally stressed, they had discarded their weapons, they were insane, they spoke insanely, and I say this mildly, they had gone crazy because of the fighting, they were unable to bear the strain. We had to disarm them, take them into our care by force. Some of them could not be approached, they were totally insane, and reasoning with them did not help. One of them aimed his weapon at us, at himself, we had to disarm him; some of the others simply lay down in the snow, others ran ahead and as far as it was possible we took them back to the main dressing station and handed them over to the medics.

It was hardly surprising that Hauptmann Traub decided to give his new recruits on-the-job training. The quietening of operations offered the perfect opportunity to acclimatise the teenagers to front-line conditions, but most of all he could ill-afford the removal of any experienced men from combat duties to train them. The state of 1st Company shows why. Unteroffizier Josef Mattes, the oldest of three brothers deployed in various Stalingrad units, was in charge of the company. One of his brothers explained to a family friend that "I am always a bit worried about him. He is caught right in the crap. His company still has eight men and two NCOs. He's leading the bunch." The other companies were slightly better off because they still had officers in charge.

The battalion's command and supply personnel experienced a shock on 23 November when Soviet artillery rounds smacked down around Razgulyayevka and Gorodishche, the two small villages in which the battalion had its headquarters and rear echelon. Hauptmann Traub narrowly avoided injury:

> The Russians heavily shelled our village. We were just able to quickly dive into a bunker. Afterwards, we discovered that a few mighty shell splinters had shattered my palace, so that even my jacket, which was hanging over a chair, received a couple of respectably-sized holes. Our Zahlmeister, who was standing near the impact zone of an incoming shell, remarkably escaped with just a scratch, however his truck, which was standing in front of another one belonging to the battalion, went up in bright flames. Unfortunately the truck was loaded with a lot of explosives which, after about ten minutes, went up with a loud bang, causing a lot of damage to my shack… Now everything has returned to normal. Strange, but the Russians did not continue with their shelling afterwards. Besides the Zahlmeister, two men were lightly wounded, but they remained on duty.

Deaths continued at the front, too. On 25 November Unteroffizier Walter Mittermayer in 2nd Company, a 23-year-old Austrian from Linz, was killed.

A memorandum sent to 6th Army revealed the alarmingly low number of combat soldiers employed in the Barrikady sector: Hauptmann Traub's 305th Pioneer Battalion had 45 combat soldiers – a tenth of its original combat strength – while Major Krüger's 162nd Pioneer Battalion possessed 99, Major Weimann's 294th Pioneer Battalion was practically non-existent with 40, and Hauptmann Lundt's 336th Pioneer Battalion had 92. Maintaining these husks required more effort than benefit gained because each was supported by a full-strength supply train. Late on 25 November, the Bodensee Division was ordered to disband its attached pioneer battalions and use them to assemble one or two full-strength formations. Major Krüger's 162nd Battalion amalgamated with Traub's 305th. Some of Traub's original staff, including Traub himself, were set aside for special assignments, while Major Otto Krüger – the highest-ranking pioneer officer – took command of the composite battalion with his own staff. As Inspektor Zeller often said, "each commander gathers his own flock around him." In his letters home, Traub never mentioned to Ella that he had lost command of the battalion. He bore absolutely no animosity to Krüger. Both officers were held in high regard by their respective battalions, and soon, the men of the 305th Pioneer Battalion came to respect Krüger as highly as they did Traub. Krüger had commanded the 162nd Pioneer Battalion for over a year, and was the perfect blend of prudence and daring. He always carefully considered situations and took advice before fully committing his pioneers. His attack on 11 November was the most successful of all the new battalions, and the audacious thrust along the Volga two days later came close to collapsing Lyudnikov's bridgehead. The Bodensee pioneers learned that he had been an NCO for many years and risen from the ranks. His toughness and courage were beyond reproach: in August 1941, he discharged himself from hospital – where he was being treated for eye damage caused by a mine – in order to take over the 162nd Pioneer Battalion, and he often led from the front. His personnel file is filled with examples of bravery. Why the German Cross in Gold had not yet been pinned to his chest was a mystery. Conscientious, sturdy and brave, the Bodensee battalion could not have hoped for a better commander.

Gefreiter Albrecht Löffler from 1st Company wrote that "towards the end of November, our battalion and another pioneer battalion were dissolved and combined into one, whereby each company was about sixty men strong. These companies also contained members of the supply train and our battalion staff." The most prominent staff member assigned to the combat troops was the adjutant, Oberleutnant Max Fritz, who was given command of Grimm's old company. The man currently leading it, Leutnant Hingst, was transferred elsewhere (possibly to battalion staff). In this way, each company had an officer in charge.

The reformed battalion was quite strong and possessed five companies:

- 1st Company had 1 officer (Hingst), 5 NCOs and 75 men with 6 light machine-guns;
- 2nd Company had 1 officer (Fritz), 6 NCOs and 65 men with 6 light machine-guns;
- 3rd Company had 1 officer (Staiger), 5 NCOs and 83 men with 6 light machine-guns;
- A company formed purely from the 162nd Pioneer Battalion contained 1 officer (Oberleutnant Baransky), 3 NCOs and 29 men with 3 light machine-guns;
- A company formed from Soviet POWs had a strength of 270 men, 35 of whom were German, including its commander, Leutnant Hubert Homburger, leader of the light pioneer column.

Members of the 162nd Pioneer Battalion were spread throughout the first three companies, while a small core of the old battalion, forming the fourth company, was left intact as a "Restkommando," an administrative cadre to take care of matters relating to the original battalion. The 305th Pioneer Battalion no longer possessed a wholly Swabian/Austrian character because the men of the 162nd were taciturn Silesians, many with Slavic-Polish roots and thick accents. Oberleutnant Artur Baransky, an East Prussian company commander in the Silesian battalion, was a first-rate combat leader with an imposing presence due to his imposing size, penetrating drill sergeant stare and toothbrush moustache with goatee. He informed his wife of his new assignment:

I've been given an amalgamated company and will be deployed in the next few days. At the moment, I've been resting for three days, bonding my new bunch together. Here – in the rest quarters – it's great… I'm living in a reasonably clean "hero's cellar" in the "Red Barricade" gun factory.

Baransky was not afraid to big-note himself and take jabs at Major Krüger:

In previous battles, I have made a great name for myself here. I'm known throughout the Division. Hopefully the bosses won't forget to give me a distinguished decoration. But I'm unlucky because my childish commander is not in a position to assert himself. Besides, he doesn't like to give anyone a medal he doesn't have himself. A great egotist!

A bizarre boost to the battalion's strength were fifty Italian drivers and thirty lorries of the 248th Heavy Truck Company. After delivering the 162nd Pioneer Battalion to the city in early November, their orders were to bring the battalion back after completion of its mission. The pioneers never completed their assignment, so the hapless Italians were caught in the Soviet bear trap.

The reorganisation and consolidation of battered German units throughout Stalingrad was a prerequisite for a break-out. If they were to have any chance of surviving a gruelling march across the steppe, the divisions needed to be lean and unencumbered by excess baggage. Hitler was never warm to the idea of abandoning Stalingrad because that would mean giving up all the gains from the preceding three months, so a Führer Decree was transmitted to all 6th Army divisions on 27 November: "Stalingrad, with its positions, will be held until relieved." Measures were being initiated to break through to them from outside. For the first time, the troops were fully aware of the perilous situation. Initially, most men believed they would escape from Stalingrad, but that belief withered as the days wore on. Oberleutnant Staiger recalls:

Our morale became miserable after the encirclement. We spoke openly amongst ourselves, saying that this war would ultimately be lost even though we believed in the catchphrase: 'The army is surrounded. Hold on, the Führer will get you out!' But even that hope very soon disappeared.

One of the first tasks given to the revamped 305th Infantry Division was the construction of a new defensive position in the rear centred on the high ground around Gorodishche. Major Linden would oversee construction. Linden's labour force came from different divisional units, but of his 343 men, 305th Pioneer Battalion supplied sixty pioneers and 135 POWs. This new task signalled a shift in German planning. No longer was everything geared towards capturing Stalingrad; defence was now the highest prerogative. A week earlier, the Germans – so close to clutching the rest of the rubbled city – could not have imagined that the roles would be reversed, that it would be them building reserve defensive lines and being ordered to hold positions at any cost. Strongpoints were arranged for long-term defence, minefields thickened around buildings, extra tiers of barbed wire strung out and trenches deepened. All measures were taken to ensure the line was held. With the encirclement less than a week old, steps were also taken to ensure that the army could withstand a protracted siege. The most important was conservation of ammunition and food. Rationing was introduced on 26 November. Daily rations for each man consisted of 400 grams of bread (normally 750); 120 grams of meat or horseflesh; half the standard allocation of vegetables (normally 250); 30 grams of fat (normally 60) or three-fifths of the standard allocation of jam (normally up to 200 grams); the evening meal was still the full standard allocation, as was sugar (40 grams); salt was half of the standard allocation (normally 15 grams); three servings of beverages, thinned, if need be; and finally, the soldier's staple, cigarettes and cigars, were issued in half portions (normally seven cigarettes or two cigars). On that same day, an order was issued that left no doubt about the gravity of the situation:

> The army, surrounded by enemy forces due to events in neighbouring sectors, will hold its positions until the very last. The task of each division is determined by this: not one step back! Whatever is lost must be recaptured without delay. Relief is being initiated. Supplies will be brought in by air.

An army of 300,000 men was told to sit and wait for help to arrive. Hauptmann Traub, for one, was not particularly concerned, as he nonchalantly described the predicament to his wife:

> Some days ago the Russians attacked us with strong forces from the north and south with the aim of closing our rear, and we thereby find ourselves in a pocket... But we will master the situation.

The garrisons of Demyansk and Cholm survived months of encirclement the previous winter and emerged as shining paragons of German determination and endurance. If push came to shove, Stalingrad would do the same. The destruction of an army so large seemed unimaginable.

New Daily Routine (1-9 December 1942)

Darkness had fallen hours earlier and sentries in the fuel tank sector squinted, trying to peer through the black veil. Keen ears were the best detectors at night and one soldier heard whispering out in front of his post. Flares revealed nothing but as soon as they fizzled out, it started again. Pst-pst-pst. Then a hushed voice called out in pidgin German, "nix shiesen" – don't shoot. Two silhouettes approached under trained gun muzzles, arms raised above their heads. "Nix shiesen, nix shiesen" they repeated. Initial concerns that this was a ruse disappeared. They were deserters, trying to save their lives by defecting. It was the second day of December, a fortnight since being encircled, and the Germans were astounded that enemy soldiers were still absconding from their units, a phenomenon that would continue well into January. Ordinary Red Army soldiers had been misled so often by Soviet propaganda that they simply could not fathom that it was the Germans who were surrounded.

Life settled into a steady pattern for the Bodensee Division. The pioneers, as usual, were the most active personnel. The rearward elements, gathered under Hauptmann Traub, built new defences in the rear and kept roads clear; the combat companies deployed in the Barrikady under the command of Major Krüger were tasked with solidifying front-line positions, laying minefields and linking strongpoints with deep trenches. Fighting was sporadic and consisted of occasional gun battles and nocturnal raids. Oberleutnant Baransky, the brash new addition to the battalion, had a close call while dealing with a Soviet break-in on the morning of 3 December:

I was almost a goner on [our daughter's 4th] birthday. The Russians had broken through near my neighbours and so I received heavy flanking fire. When I rolled up brother "Ivan" from the side and then quickly took a favourable strip of terrain from him later in the battle, I was hit in the steel helmet by one of those bloody bullets. The result of this shot was a hole in the helmet into which you can easily insert 2 fingers. I got a blow to the head that took my sight and hearing. Luckily, the piece of sheet steel was only shoved backwards and the bullet went wide. But a small splinter from the steel helmet still caused two lacerations, 3cm long, on my head. So lucky again! It is already my fourth wound of such a trifling kind. Now I'm going to start reporting my wounds, even if they're only minor [...] I did, however, take my revenge many times and achieved a good local success.

On 5 December he celebrated his 38th birthday as only he could:

On the dot of midnight, my platoon leader congratulated me in the front-line trench (about 30 metres away from the Russians). I had just thrown a hand grenade into the trench because cheeky "Ivan" had crept really close to our position again. So my new year literally began with a bang. I think that's the right way for a warrior of my calibre to start a new year.

For the moment, both sides were content to hold position and let events play out. The Germans still had Lyudnikov's bridgehead in a death grip and tried everything to halt the trickle of supplies and reinforcements, but it became quite tricky on 7 December when the channel between the west bank and Volga island iced over, allowing supplies to be walked in. It could not be halted because German artillery was only allowed to fire in extreme emergencies. Small indicators like this foreshadowed future hardships, but the seriousness of their predicament was not yet fully realised. Traub was fretting about getting some smoky bacon and a can of peas to Ella for Christmas and missing out on incoming mail:

It's very doubtful if I'll get the three Christmas parcels you've sent. They must be air freighted to the front because the Russians have severed the supply lines in our rear and the ring has not yet been opened. Despite this, we are feeling confident and strong enough to break through from within, if need be. The only thing is provisions are gradually becoming less. The bread ration has been decreased several times. We will need to get used to the idea of eating horse meat. In the village we had a depot with wood and other nice objects for building our accommodation, as well as eight lovely cows, which we have had to leave behind. Lacking petrol, we could not bring these beasts with us. Other German soldiers will be overjoyed by their find. We only have two little pigs now. They have a short life expectancy, maybe one at Christmas.

Thoughts were already turning to the most festive day on the German calendar, but Major Krüger had more pressing concerns: although made up of five companies, his battalion possessed far less hitting power than a headcount suggested. Manpower needed to be shifted from supply trains to the combat units, and the most obvious targets were the three old companies from the 305th Pioneer Battalion. First to be disbanded was Hingst's 1st Company. All of its men, both in the front-line and rear, were distributed amongst 2nd and 3rd Companies and provided a significant boost in strength. Gefreiter Füssinger quickly adjusted to life in a new company:

> *Due to the disbandment of my company, I unfortunately lost my previous occupation, therefore, I am now practically in field work again. Fortunately, however, in my new company, to which we temporarily belong, I have received a lovely task: a large number of sleighs must be built for the division. There are still some carpenters who can help with that. My previous company will be re-established in the foreseeable future and then I will have my old work to do.*

Füssinger's skills as a master wainwright were invaluable for the mobilisation of the division. Almost everything with a combustion engine languished under snow drifts because of fuel shortages, so the only form of propulsion was horses. If an order came to break out of the pocket, the division planned to have the means to transport itself without relying on petrol. For now, Füssinger considered himself lucky to be assigned to a suitable task; some friends from the supply train had been sent to the front.

Softened Up (10-20 December 1942)

Machine-gunners in the Commissar's House and nearby strongpoints fired upon the Volga channel; the Soviets responded, mainly with small arms and mortars, but they brought a new weapon to bear on this day. Germans witnesses reported hearing a hollow discharge, then seeing a small sphere arching a few feet above the ground and exploding in a spectacular ball of white flame. Whatever they were, they caused great concern amongst the infantry because these orbs had the potential to fly through embrasures and gunports. The Soviets were using ampulomets, the inaugural deployment of this weapon in the Barrikady. These ampulomets – literally 'ampoule throwers', though sometimes called Molotov projectors – weighed 28kg including tripod, and were crude weapons with a smooth muzzle-loaded bore that hurled 125mm diameter spheres filled with a phosphorous-based incendiary mixture. Maximum range was 250 metres. Lyudnikov's division received two of them on 9 December. In this first test run, two German emplacements were destroyed and another suppressed. More ampulomets would appear in the following days.

At 0700 hours on 12 December, Soviet artillery commenced the first of several powerful strikes to systematically destroy German men and firing points in the salient anchored on the Volga. General Chuikov, commander of the Russian army in Stalingrad, would later write:

We marked out the area occupied by the enemy from north and south and from the Volga to the farthest point in the front-line, indicating landmarks clearly visible from the east bank. This gave us a 650-850 yard corridor occupied by the Germans. Our artillerymen, seeing this corridor clearly, could fire accurately at the enemy's firing positions. Spotters on the west bank watched the firing. They indicated and watched the targets, and errors in the gunners' aim. All this was communicated to the artillery observation posts, and then transmitted to the firing positions.

The artillery opened up again at 1300 hours. Exactly 5760 rounds were fired in the two one-hour bombardments. The same program would be implemented the next day.

It wasn't only the front-line feeling the brunt of Soviet weapons. Life in the rear was also becoming more hazardous as Soviet planes carried out nocturnal bombing runs and artillery reached deep into the hinterland. Hauptmann Traub describes these strikes and other concerns affecting those in the rear:

Everything here is much the same for us. The Russians are pretty active and frequently shell us with their artillery, and their planes are just as active at night. Just now, at 1615 hours, an ace among them dropped two bombs which lightly rattled my excellent shack. Last night, at around 2200 hours, as I stepped out, they dropped two incendiaries which caused no damage. I have also had a bunker built, which I will move into tomorrow afternoon. It isn't safe from bombs, but one has a certain sense of security when sitting below ground. Actually, I hope the bunker can be well heated. The frequent north-easterly winds are making my shack pretty cold. There are absolutely no trees here on the steppe; every vehicle that goes into Stalingrad must bring back wood. But there must be a time when it is no longer possible to demolish any more houses in Stalingrad. And I fear this may happen much sooner than the end of this winter. Equally bad is the situation with the lighting. My carbide lamp has been broken for a long time. I had collected a nice supply of candles but yesterday night to my shock I discovered there were only two left. At Christmas I probably won't be celebrating in candlelight, but in the dark. To be positive, yesterday there was a rumour that for the first time even parcels are coming through the mail. Now we hope that our Christmas parcels will arrive. Only letters and airmail are still being processed at the moment. Even official administrative mail remains unshipped and so there is little wonder my promotion to Major has not come through. From where I sit it should have come through from the [High Command's] personnel administration a long time ago.

My hut is very comfortable tonight. The wind has eased, the stove is crackling nicely and at around 1700 hours my orderly, who has just finished his watch, is coming again and I will eat my supper. I cannot have more than two slices of bread, otherwise there would be nothing for the morning. You cannot get fat on it but it's enough for me. Eating a great deal is just a habit. God be praised, I still have a carton of cigars and something to drink and these help you get through. Our paymaster confirmed today that he has kept a box of cigars in reserve, because I once told him that I could be hellishly difficult if I had nothing to smoke. We have been lucky so far, not having really experienced the bitter cold. I believe that so far it hasn't dropped lower than minus 15 degrees. Hopefully St. Peter will show his understanding and be merciful with us this winter.

The combined threat of the winter chill and bombs forced Traub to move underground, like everyone else, and after living like a mole for a few days, he was quite happy with his new accommodation:

It has become colder again but the wind has eased, so it is quite tolerable outside. The Russian planes are very active at the start of the night but I've been in my new bunker for three days now and I don't let them disturb me. My bunker has become very comfortable. I sleep like a dachshund in winter, the stove heats excellently though it uses quite a lot of wood.

– – –

Colonel Lyudnikov received an order from army commander Chuikov to advance south-eastwards, meet up with the forces of Gorishny's 95th Rifle Division and establish a solid front. Lyudnikov decided to attack towards the Commissar's House to annihilate German defences south of Taimyrskaya Street. Beginning of the attack was set for midday on 15 December. A formidable concentration of artillery units softened up the German defences for six hours, rising to a final five-minute crescendo before X-hour. Lyudnikov's division went over to the offensive at noon. German troops opened intense fire from all weapons. Artillery howled in moments later. Upon seeing the storm groups move out, mortar crews in Halls 3 and 4 laid down fierce barrages right on the Soviet flank. On the main axis of advance, the storm groups encountered minefields, craters, dug-outs, barbed wire entanglements and booby traps. The pioneers had been working on these defences for weeks and this was their first true test. Relying upon this system of man-made obstacles, Regiment 577 easily retained control of three of four fortified houses targeted by the storm groups. Regiment 578 halted their attackers after an advance of just thirty metres, one sector even preventing the Soviet soldiers from leaving their starting positions by laying down instant and fierce machine-gun and mortar fire. The German defenders also had to contend with Gorishny's 95th attacking from the south, but apart from the capture of a few dug-outs, no meaningful progress was made.

Oberleutnant Staiger's schedule was full. His pioneers were broken up into teams to carry out numerous jobs: timber from demolished houses was transformed into Spanish riders; mines were dug out of the ground in rear sectors and relaid east of the Barrikady; wire obstacles and booby traps were installed around strongpoints. This went on, day in, day out. Light but constant casualties steadily drained his company's strength. One of his men, Gefreiter Friedrich Wickers, was killed on 16 December. Two days later Staiger was making his way toward the front-line to oversee some job or other:

> Because of the snipers, we could only move to the front or towards the rear in trenches which were sometimes quite narrow. On one occasion things were moving too slowly for me, so I overtook by jumping up out of the trench with the intention of taking a few steps and then cutting back in further in front. Suddenly I was hit in the right wrist by a bullet... If I had not been so impatient and did not attempt to overtake the patient transport moving far too slowly in front of me, it probably would not have happened. But I was hasty and the projectile hit me. It was not very far to the aid station and I went there on foot. I did not take this wound very seriously but our commander, Hauptmann Traub, was of a different opinion and said: "You're flying out of here. With your injuries, you're of no use to us any more. You'll only be another mouth to feed." It was clear to me that he meant no harm because Hauptmann Traub was an officer respected and loved by all; he wanted me to escape from the encirclement in time. Thus I was flown out on a Ju-52.

Hauptmann Traub later mentioned this incident to his wife:

> Some days ago one of our battalion officers was wounded by a shot in the hand. Of the old guard who were together in France, there are now only three; myself, the adjutant and the technical inspector. Fortunately, most of the officers are only wounded. They have written to me from homeland hospitals. Anyway, our grand, proud battalion has been pretty much plucked. It must be time that we are

withdrawn from the front for a few weeks so we can be refreshed and refitted, but it is certainly not the time to think about that now.

Staiger's departure was the trigger for the dissolution of his 3rd Company; all of its men were transferred to Fritz's 2nd Company. Oberleutnant Fritz now controlled all combat troops of the old 305th Pioneer Battalion. As such, he was designated "Führer der Kampfgruppe des Pionier-Bataillon 305" (leader of the combat group of 305th Pioneer Battalion).

Pesky raids near the fuel tanks remained constant. Grenade duels raged at night and in the morning. At 1100 hours on 19 December, however, Soviet storm groups executed well-planned strikes. German observers immediately called down mortar and artillery barrages, while heavy machine-guns opened up devastating fire, but the Soviet riflemen pressed on and plunged into German positions. A hand-to-hand struggle erupted. Individual dug-outs changed hands several times. Batteries of ampulomets pelted the German defences, their dense clouds of smoke blinding the defenders and preventing accurate return fire. Occasionally, they flew straight into a dug-out: twenty infantrymen were incinerated in three separate incidents. The Soviet storm groups eventually captured three trenches and a few dug-outs, but discovered that German corpses had been booby-trapped. The pioneers of the 50th Panzerpioneer Battalion carried out this deviousness before pulling back. Unfortunately for them, they also left behind many dead comrades and two living men who were taken prisoner.

Soviet artillery was also inflicting casualties in the rear. Feldwebel Josef Kienzler, the leader of the battalion's communications section, was wounded on 19 December by shell fragments in a shoulder. Killed by the same blast was fellow staff member Obergefreiter Georg Reitle. Grimm had the utmost respect for Kienzler: "He was an extremely dependable man, modest, esteemed and popular with superiors and subordinates. His information was constantly reliable and correct." The cadre of original battalion members was tiny, so the loss of these two men had a huge impact. Among the staff remaining behind in Razgulyayevka, close to the main dressing station in Gorodishche, were Feldwebel Heinrich Bromeis, Unteroffizier Ludwig Schnöll, Gefreiter Günther Faden, Gefreiter Max Hils and Gefreiter Karl Welker. Everyone else had been sent to the front. Before Kienzler was flown out, he talked with Hils and Schnöll, as well as with Bromeis by telephone: "They were still in good health, if one can still use this word for those days," wrote Kienzler. Of the physical constitution of his comrades, the man in the best state of health in Kienzler's opinion was Schnöll. Feldwebel Bromeis, on the other hand, had to be helped along occasionally because of crippling rheumatism. Kienzler left Stalingrad in a Ju-52 on 23 December.

– – –

Over a period of four weeks, the Bodensee Division had created a web of obstacles and emplacements around Lyudnikov's bridgehead, the backbone of the defence being the rows of buildings on Pribaltiskaya Street and Prospekt Lenina. The main stronghold was the Commissar's House, the others were Houses 67, 75, 74, 84 and a transformer hut, each lavishly armed with machine-guns, some also with mortars and anti-tank guns. These redoubts were linked by communication trenches and supported by a system of dug-outs, blockhouses and pillboxes. Spaces between buildings and dug-outs were mined and secured with wire entanglements, Bruno spirals and trip-wires. The division's low combat strength forced it to base its defence on the creation of obstacles and a skilfully organised system of fire, both of the thickest density. Heavy personnel losses meant the grenadier regiments had extra automatic weapons. Despite ammunition shortages and strict limitations on expenditure, grenades were available in huge quantities. Stacked in each major strongpoint were at least fourteen cases of stick grenades – over 200 grenades in all – while every dug-out had several boxes available. The German scheme was to substitute automatic weapons for personnel and use pioneers to create defensive works. Krüger's battalion was integral to the overall plan. By providing the main line of defence with high-density fire, man-made obstacles and engineering works, the division was forced to forego the creation of defence in depth. The plan was dead simple: hold the line at any cost and prevent the Soviets from linking up. After a month of static warfare, the offensive impetus was about to be firmly seized by the Soviets. The Germans could do nothing but wait for the hammer to fall.

Lyudnikov's Revenge (21 December 1942)

Feather-light snow fluttered down from the winter sky. Silence reigned over the Barrikady, broken every now and then by a stuttering machine-guns or the snap of rifle fire. Then quiet returned. At the stroke of midnight, three red flares sizzled into the sky and Soviet artillery unleashed their customary barrage. This had been going on for several nights: three red flares followed by a ten minute bombardment. Leaving a few men to keep watch, the German defenders sought secure shelter and returned to their firing positions immediately afterwards because they knew it would be three hours until the next bombardment. Most went back to sleep. At 0240 hours a blue flare hissed into the sky, three red flares from Zaitsevskiy Island answered it and, as usual, one gun fired, followed by others, and then the overture was joined by howitzers and mortars. The Germans knew the routine. After seeing the red rockets, they pulled back from the forward line to seek cover. Two minutes in, however, the Soviet big guns switched their fire to the depth of the German defences, and a minute after that, the artillery fell silent. They had fallen victim to a Pavlovian response conditioned over the previous nights. Officers screamed at their men to get back to their strongpoints, but it was too late: Soviet storm groups had already broken into their forward line wielding spades and daggers. After slaughtering sentries and annihilating a few firing posts, they advanced along Pribaltiskaya Street, broke through thick barbed wire entanglements, seized Houses 67 and 66, blockaded Houses 74 and 75 – the most important German strongpoints – and surged toward the Barrikady's rear gates. Furious flanking fire was eventually laid down from Hall 3 and cut off the leading Soviet groups. Further east, other storm groups failed to capture the Commissar's House but did seize two buildings and other strongpoints and weapons. The lack of cellars in this sector meant the Germans weathered the artillery bombardments in their bunkers and were therefore able to react much more quickly. Determined German counterattacks began on all fronts. As the grey dawn slowly illuminated the battlefield, German observers brought down a hurricane of mortar fire and hindered the advance of the Soviet second wave.

As a result of the nocturnal attack, the German stranglehold on Lyudnikov's pocket was loosened, though all further attacks were met by fierce resistance. The German ability to rapidly mobilise reserves and launch powerful counterblows thwarted every Soviet attempt to capitalise on their initial success. Nevertheless, the capture of four buildings, each one a fearsome fortress abundantly equipped with automatic weapons and grenades, was seen as a great achievement.

– – –

After six days at the front, Gefreiter Vorherr – having somehow escaped being sent to the crisis developing behind the Gun Factory – returned to his billet in the rear, exhausted and frozen through. A pleasant surprise was waiting for him: a small pile of mail, including a one kilo package. The fact that it had been flown into the pocket was remarkable and Vorherr was ecstatic for a few moments, and then, while writing a reply, the reality of his miserable situation slowly began crushing him:

I don't want to write how I'm doing. You could go mad from hunger, one loaf of bread per day for eight men, and this has been going on for a month already! What is to become of us! Every day there is talk that things will get better, but it's all for nothing. So, slowly, I have now given up hope. We will certainly have no peace at Christmas. For the last six days and nights I've been outside on sentry duty, this evening I came back at 1900 hours to get my things in order because at 0900 tomorrow morning I'll be heading back to Stalingrad. I have not washed in the past

six days and barely slept, and no more will be had today. I haven't changed my
underwear in over a month, but that is unimportant, who asks about that here?
One should still only write "I'm going well, etc." It is miserable how we are going, a
convict certainly has it better, now we'll be forward in the positions at Christmas,
who knows if I'll even be alive at Christmas. I had diarrhoea again today but that
is due to the horse meat… I would give 25 Marks for a loaf of bread.

The Battalion's Darkest Day (22 December 1942)

The brutal fighting did not ease during the night. Gunfire and explosions echoed across the city. At 0340 hours, one of Gorishny's storm groups made contact with Lyudnikov and ended forty days of isolation. The link was tenuous but Lyudnikov now had telephone communications. German units tried to restore the position, though without success. Major Krüger's pioneers had been held in reserve but the situation forced their commitment: Oberleutnant Fritz's "Kampfgruppe des Pionier-Bataillon 305" launched an attack towards the Volga and – according to Obergefreiter Franz Müller, father of seven, now employed as a machine-gunner – it was successful. For his efforts, Müller was promised the Iron Cross First Class by Fritz. Every able-bodied man was sent to the front. One of them was Gefreiter Füssinger, former wainwright, now sleigh builder. His tenure as a front-line soldier was brief:

On 21 December I was deployed in the fighting in Stalingrad and by 22 December
had already been wounded in the right hand. Since then, I find myself back at the
supply train so I can heal. At least I will not have to spend Christmas in the trenches.

The fighting did not cease once night fell. Oberst Steinmetz was determined to recapture lost positions, so Fritz's Kampfgruppe der Pionier-Bataillon 305 again entered battle. Obergefreiter Franz Müller was wounded, his third suffered in Stalingrad, but this one was severe enough to warrant a ticket out of the pocket. Far less fortunate was his commander, Oberleutnant Fritz, who jumped into a trench filled with Soviet soldiers. Some say he was beaten over the head, others that he was shot, but all were convinced he was dead. However, there is some doubt as to whether this is true because of the following excerpt found in a Soviet combat report:

At 1730 hours on 22 December, on the sector of 241st Rifle Regiment, an officer – a
lieutenant of the sappers of 305th Infantry Division – was captured and sent to the
political department of army staff.

Records show that apart from Fritz, no other officers from 305th Pioneer Battalion were killed, wounded or missing on the front-line on this date. All German sources state quite clearly that Fritz was dead. However, the fact that he was reported killed at the same time and in the exact same place that a German pioneer officer was captured leaves some room for doubt. Indeed, Fritz's family were never convinced: in a post-war letter, Oberleutnant Schaate noted that Fritz's "next-of-kin do not completely believe it and go to fortune-tellers." Why were they uncertain? The most likely scenario is that others in the battalion doubted the official version and passed on their suspicions to Fritz's parents. Zrenner, the last known living veteran of the battalion, was convinced that the Leutnant captured by the Russians was not Fritz, but he cannot remember his name. For sixty years the disappearance of this officer has puzzled Zrenner and constantly been in his thoughts. The officer was assigned to lead one of the combat groups, but before his deployment, Zrenner advised him to remove or cover his collar tabs as they might attract the attention of eagle-eyed snipers. The next day, Zrenner went to the group to ask how it had gone:

> The Leutnant was not there anymore. "Where is he then?" I asked some comrades.
>
> "We don't know, he just disappeared!" was the reply. It remained a mystery.

When shown the Russian combat report detailing the capture of a German sapper officer, 86-year-old Zrenner exclaimed, "I now know for sure that the one they captured was exactly this officer!" For him, a six-decade enigma had been solved. Unfortunately, Zrenner does not remember the officer's name: "It could not have been Max Fritz and also not Johannes Lindner, because the Leutnant was not born in 1906 like Lindner, but was the same age as me [born in 1920]." The author has been unable to identify this officer. In any case, Fritz's death/disappearance was a severe loss for the battalion. He had been one of the battalion's original members, an adjutant without parallel, and respected by all. Hauptmann Traub mourned his loss in a letter home:

> Sadly, in the last few days, we have had some losses in the battalion, among them two officers: one through a direct hit from a bomb on his quarters, and the other my former adjutant, the commander of 2nd Company, falling in action on the front-lines during a Russian attack. I have written to the wife of the first fatality. Now I have to write to the parents of our former adjutant, Oberleutnant Fritz. These are always difficult letters to compose, but this duty cannot be avoided.

The officer killed by a direct bomb hit was Leutnant Hubert Homburger, former leader of the light pioneer column and current commander of the POW company. The night was only a few hours old when a Soviet plane released a bomb that plunged straight through the roof of Homburger's dug-out and exploded inside. His orderly was also killed in the blast. Hauptmann Traub and Inspektor Zeller were in nearby bunkers and dashed over to the terrible scene. They saw Homburger mangled beyond all recognition, just torn limbs here and there. Even though Hauptmann Traub was not the battalion commander, Major Krüger asked him to write to Frau Homburger. She received three letters at once, two from her husband and another in writing she did not recognise. Feelings of dread weighed heavy in her stomach, but she read the two letters from her husband first. He talked about Christmas preparations, including the impending slaughter of a pig they had fattened to almost 200 kilos, and the construction of an Advent wreath, yet his thoughts centred upon his "beloved sweetheart" and "terribly dear children":

> I know it will be a sad Christmas for you, as it will for me, but the war forces this sacrifice upon us… That your papa can soon go on furlough, no, that he will be able to return to you forever in 1943, is the most beautiful Christmas present and most beautiful New Year's greeting for us all. If I cannot be with you, then that is a sacrifice for you and for me, and we should gladly bear it for the ultimate purpose – victory.

With his words still audible in her head, she tentatively opened the letter from Traub:

> I must give you the sad news that your husband, our dear comrade, Leutnant Hubert Homburger, died a hero's death for his country at 1745 hours on the evening of 22 December. He was killed when a Russian aerial bomb landed right on his dug-out and was killed instantly. His remains were borne to the Gorodishche military cemetery near the church, about 8 km west of the northern part of Stalingrad, for burial with military honours.

> I know, dear Frau Homburger, that words are meaningless to console you in your deep grief for your husband and father of your three small children. I assure you of my sincere sympathy and that of the entire battalion for the heavy loss you have suffered through the heroic death of your husband.

> Your husband was a dear friend to us all and respected by every battalion member for his devotion to duty and straightforward manner. I lost in him a colleague, whom I will always miss. All of us will never forget our dear comrade Homburger.

He signed off simply as "Wilhelm Traub", no rank was mentioned anywhere in the letter. Olivia Homburger, widow, was left to care for three children, aged five, four and sixteen months.

Two founding members of the battalion, both officers, killed on the same tragic day. Also losing his life was Gefreiter Hans Schenkel from 2nd Company, long-time member and also the man who had helped Grimm to the rear when seriously ill back in November. Among the 57 wounded suffered by the division on this day were several from the pioneer battalion. The fates of those evacuated from the pocket are known. Obergefreiter Klemens Bastian, 1st Company, was struck in the right elbow by a shell splinter and flown out on 3 January 1943. Gefreiter Hermann Gamsjäger, 2nd Company, was hit in the left upper arm and both thighs by shrapnel and flown out Christmas Day. Gefreiter August Gramling, a butcher from 2nd Company's supply train drafted for front-line service, was knocked down by a bullet in the right thigh and also flown out Christmas Day. Pionier Karl Kornhuber, 1st Company, suffered phosphorous burns to the left side of his face in close combat and was flown out on 4 January 1943. Pionier Erich Zimmermann, 1st Company, was sprayed by shrapnel on his left side and flown out the same day as Kornhuber. The variety of nasty wounds resulted from the struggle with Gorishny's men. The dazzling explosions of phosphorous ampoules stunned the pioneers and generated fear, especially after the first painful wounds were inflicted and rumours were heard about the dozens of men burning to death in their dug-outs.

Another casualty was Oberschirrmeister Paul Botta from the light pioneer column, his body finally succumbing after weeks of pain. He had been living with a hernia for a long time but on this day his appendix burst and required an urgent operation that could not be performed in the pocket, so he was put on a plane the very next day and flown out of the pocket. Although not directly involved in combat, Botta had rendered invaluable service by transporting ammunition and explosives directly to the men in the front-line and ensuring the operational readiness of the flamethrowers. The illness saved his life because he would have been with Homburger when the bomb smashed into his dug-out.

Even the indomitable Oberleutnant Baransky almost met his end when his unit was pulled out of the line during the night of 22-23 December:

I was almost hit during the relief. A bloody shell landed five metres from me. The result was a slight ding to the oesophagus, the blood barely started flowing. I've been lucky, lucky and lucky again. Other officers go to the front once and either fall or fly home to Germany with a Heimatschuss [home shot]. In any case, I'm once again the last man standing in a larger group, which I'll probably take over during the night.

Grimm's Odyssey (23 December 1942)

Still suffering mildly from an intestinal infection but feeling vastly better compared to a month earlier, Oberleutnant Grimm gathered his kit and headed to Stuttgart's central railway station on 18 December. Lotte and little Gisela accompanied him. What can be said at such a time? Mostly small talk and practical matters, laced with moments of sad silence, but nothing profound was uttered. Despite the visceral desire for him to stay, at least for Christmas, they were helpless in the face of duty and circumstance. The frigid maw of the Eastern Front awaited. The conductor's shrill whistle brought on teary farewells and rushed professions of love all along the platform. Lotte tried to maintain a brave face, but tears welled up as her husband kissed them goodbye and climbed into an officer's compartment. The train pulled slowly out of the station in clouds of steam and disappeared from view. The sobbing crowd dispersed. Would Gisela see her daddy again?

The sombre mood on the train gradually lifted as the officers started talking. They were all members of 6th Army and everyone was trying to find out if the rumours were true. Was Stalingrad encircled? As military men, they had sources unavailable to civilians, so collective knowledge was pooled until a vague picture formed. The truth would only be revealed when the destination was reached. The journey took five days. What met Grimm upon his arrival on 23 December was recorded in his diary:

Furnished with march rations, the train journey proceeded via Berlin – Kovel – Kiev – Lozovaya to Yassinovataya. The forested stretch of line between Kovel and Kiev was plagued by partisan activity, but our train crossed this sector without incident. All men returning from leave had to get off the train in Yassinovataya. It was known that the rail connection to Chir had been disrupted by the enemy. I joined several other officers of 6th Army and together we tried to get back to our units in Stalingrad by plane. From the closest airfield at Stalino, only officers ranked battalion commander and higher were being flown into Stalingrad. Despite this, different officers flew in with transport planes by fulfilling certain functions, for examples, as gunners, all with the objective of reaching their units and their comrades. Our small circle [of 6 officers] reported to the railway station's security officer. He gave us the order that all officers of 6th Army should head to the Führerreserve [reserve pool of officers] in Rostov and enlisted men towards Lichaya. The journey to Rostov could not be continued on 23 December and we spent the night in an officer's hostel – without the normally required delousing certificate – in one room, so we could cluster around an oven by candlelight. There was snow, the cold wind whistled, but not into us. Overnight camp was assembled, two slept on the table without any sort of underlay. The room was free of bugs and lice, what a blessing. On the way to the latrine, you had to leave the house and go to the train station, where news of the changing situation reached our ears, but everyone was convinced that sooner or later he would return to his men in Stalingrad.

Our train was due to depart at 0300 hours on 24 December. We learned in time that the train was full and we should not clear out our quarters. We were very glad to know that we would not have to spend Christmas Eve on a train. A fir branch was organised, decorated with a few small candles and mounted. We shared whatever we had and everyone had a swig of cognac. During the distribution of rations, there was an unexpected bonus at the train station in the form of "chicken noodle soup", naturally with horse meat – I noticed it by its sweet flavour – as well as a handful of biscuits for every soldier. Stomachs did not growl, so the Christmas spirit resurfaced. By candlelight, everyone talked about his family, but nobody was able to separate that from his war experiences, plus everyone wanted to know to which unit you belonged, why he had been allowed to go on leave, who was a convalescent coming from a replacement unit, amongst other things... We turned in at midnight and slept soundly.

No possibility existed of getting a train on Christmas Day. A fellow officer managed to get hold of some tickets for the Ukrainian theatre. The German officers were impressed: "The amateur actors and actresses made a real effort and only the curtains showed that this was a theatre." After another plate of noodle soup and instructions from the town's garrison officer to report in the next day, the group turned in early. Their new assignment was revealed on 26 December: each one of the six officers was ordered to establish a 200-man strong "Urlauberkompanie," an ad hoc company formed from men returning from furlough:

I pulled 200 men from the train, loud guys who wanted to get back to Stalingrad. Horse-grooms, artillerymen, medics and so on, but no pioneers or infantrymen. Without winter uniforms, camouflage clothing, automatic weapons, binoculars, first-aid dressings, field-kitchens, without anything.

Grimm was fortunate to find a couple of efficient and well-trained NCOs amongst his lot. He would need all the help he could get. The company shifted to Makeyevka and occupied abandoned Russian barracks. An exchange of soldiers took place inside the company as men from the same divisions found each other. Winter clothing, albeit grey, not white, was issued to those who lacked it. Weapons were short because members of several divisions had been called out, equipped with all available rifles, and flown into the pocket. All remaining soldiers were only given light duties in the freezing cold weather. This period of limbo continued for almost two weeks, but it did afford Grimm the opportunity to recover from his illness:

> My chronic intestinal catarrh from Stalingrad had still not completely eased and I turned to an Oberarzt of the Luftwaffe who was housed with us. He had long since observed that I did not smoke and only drank tea. I gave him my collection of cigarettes and the garrison commandant authorised a trip to a Luftwaffe pharmacy. I was treated with tablets for three days, but I had to lie in bed. After this treatment, I recovered noticeably quickly, took myself off the bland diet, could eat everything again, drink and even enjoyed something to smoke. I put on weight and soon weighed over 50kg. This medicine was apparently not able to be prescribed by the Wehrmacht "army" in Stuttgart.

For the moment, he bided his time, often pondering the fate of his comrades in Stalingrad. Snippets of information about the general situation reached his ears, but nothing concrete about his own battalion because his current posting in an ad hoc unit, which lacked a Feldpost number, meant mail – both official and private – did not reach him. In this case, ignorance definitely was bliss, as the men of the Bodensee Division were experiencing frozen purgatory.

Holiest Days of the Year? (24-26 December 1942)

In Stalingrad, gun battles raged all day, every day. Huddled inside their brick citadels, the German defenders peered out of embrasures and loopholes, searching for any sign that a Soviet attack was about to erupt. Every movement was hosed with machine-gun and small-arms fire. Relentless pressure, however, forced the German's to cede House 83. Next in line was House 81. After two hours of brutal hand-to-hand combat, Soviet sappers hauled in hundreds of kilos of explosives and detonated a massive charge that prompted the Germans to abandon the ruins. Blocking the Soviets now was the Commissar's House. It could not have been better designed as a stronghold. Stout walls, narrow windows, compartmentalised cellars and an advantageous siting, coupled with German improvements like barbed wire, mines, booby traps and machine-guns. Only a well-planned attack carried out with audacity stood the remotest chance of success. The Germans were fully aware of the building's importance: it anchored their entire defensive line. Without it, they would struggle to hold their position east of the factory. According to Rettenmaier, "the Commissar's House was defended with characteristic Swabian pigheadedness." Nevertheless, its garrison was not safe because Soviet heavy mortars bombarded firing points while anti-tank rifles took out soldiers behind narrow embrasures.

Some of the German casualties on this day, Christmas Eve, belonged to the 305th Pioneer Battalion. All were from 3rd Company. Stabsgefreiter Ignaz Kainz was shot through the right forearm and struck in the right shoulder by shrapnel. He was flown out of the pocket on 4 January 1943. Gefreiter Max Lattner suffered phosphorous burns in close combat and was flown out on 30 December. Gefreiter Gottfried Riegler was wounded in the left upper arm, left shoulder and face by shrapnel and flown out on 4 January 1943. All three were replacements that had arrived just before the Soviet counteroffensive on 19 November, but while Lattner and Riegler were young, inexperienced Austrians, 30-year-old Kainz was an old hare.

– – –

The importance of Christmas to Germans cannot be over-emphasised, and even more so during the war. It was the year's most important feast day and everyone began thinking about it and their absence from home. It was a time of introspection for most soldiers and each perceived it in a slightly different way. Despondent and consumed by misery, Gefreiter Vorherr experienced a Christmas "like never before in my life. No mail, no presents, nothing, on the contrary, it was a day exactly like any other that has passed in the last month. The company is in action up front, I reported in sick, yet the doctor wrote that I'm fit for service. My old bladder trouble is back, every night a wet shirt, underpants and pants, and in the almost minus thirty degree temperatures, they never dry." Gefreiter Füssinger lamented a full month without mail and the non-appearance of parcels, including one with honey from his hives back home:

> Only empty stomachs, as during the past five weeks. Two hundred grams of bread
> a day. Still, anyone who gets out of Russia with their skin intact will be happy. The
> hope for better times is never given up... The memory of festively covered tables
> from years past is the only thing that can bring joy this Christmas.

Inspektor Zeller was able to celebrate in grander, tastier style. He met up with Oberinspektor Schneider from the 50th Panzerpioneer Battalion, an instructor at the pioneer school where Zeller trained. Schneider had a hunting rifle, so the pair went on safari, looking for anything edible on four legs. Luck was on their side: a dog was bagged by crackshot Schneider. It was roasted and eaten with great relish. "It tasted delicious," Zeller recalled over half a century later.

Oberleutnant Baransky convened his celebration in a gun factory cellar:

Christmas! Relatively well supplied with chocolate and other specialities and thankfully not deployed in the front trench. I spent Christmas Eve with my boys in a deserted cellar (similar to a foundry). I was able to quickly fetch my radio so that the evening was quite entertaining and festive... I am so glad that we were able to spend the night in the old filthy cellar.

Hauptmann Traub had it best of all, snug in his bunker and far from the fighting. Tacked to the wall were drawings of Christmas trees sent by his sons. A shimmering candle and open bottle of wine stood on the table. Tucked safely out of sight were some tins of meat and peas, as well as some bacon, presents for his wife that remained unsent because of the blanket ban on heavy parcels. Ignoring the obvious temptation, he refused to open them, despite his deteriorating condition:

I have become very thin, there is no trace of my stomach and my cheeks have become noticeably hollow. Despite this, I feel quite well and cheerful, after I overcame a light tonsil infection.

As well as being a self-sacrificing husband, Traub was a caring father. Thousands of kilometres from home, he had organised a bicycle as a Christmas present for his son, Jürgen, courtesy of his battalion comrades. Feldwebel Dittenhöfer, Spieß of 2nd Company, had been instrumental in procuring it and getting it sent to his parent's house in Nürnberg. Because Dittenhöfer did not trust Reichsbahn freight to care for the bike on the long haul to Wesermünde, Traub arranged for a soldier on furlough to bring it to his wife. Content in the knowledge that everything possible had been done for his family, but almost ill with a dreadful longing to be with them, he celebrated the day quietly, as described in a letter to Ella:

For Christmas Eve I invited over our doctor [Wirtgen], who has been transferred to an infantry regiment. I got something to drink and his refreshing presence helped remove my sense of loneliness. We talked about home. The doctor has a practice in East Prussia and also two children. There were many subjects of common interest.

Naturally there was no mail. Who knows where it is stuck. There was not even a letter for me. Well, maybe next time. Hopefully you at least received a letter from me. How did you enjoy Christmas Eve? I hope the boys were satisfied with the Christmas spread. I barely dare to hope that the cacao liqueur arrived for you after today hearing that large quantities of parcels have not even been shipped by our postal service.

Despite your parcel not arriving, I still had a small exchange of gifts here. From one company I received a sausage and two pork cutlets, from the other a packet of cigars, a fried piece of filet and a bottle of cognac. I was very happy with these presents. From this experience one can see the bond between us old comrades. In addition, I have the feeling that the companies are of the mind that they must do something to help keep me fed. They are probably scared that one day I might go into hospital for malnourishment. I sometimes hear various remarks that I have become fearfully thin, but I feel very well, even in my becoming very slim. Otherwise, there is little news to report. We still have hard combat with the Russians and, sadly, casualties. Who knows when it will be different.

Up front, periods of icy silence were broken by the crack of rifle shots, energetic machine-gun fire and crackling flares. It would die down, silence returned, and the performance repeated a short time later. Shells warbled in occasionally. German mortars could be heard coughing out rounds somewhere in the bowels of the factory. Trouble was brewing near the fuel installation. At 0100 hours, after a brief flurry of artillery, Soviet storm groups captured House 41, one dug-out and a pillbox, and in the process captured a small arsenal and an Austrian Obergefreiter from the 305th Pioneer Battalion.

Thus dawned Christmas morning, frosty and cold, the snow stained by soot and brick dust. Mercifully for both sides, it was to see little ground combat, though gunfire was continuous. With so much metal flying around, casualties were inevitable. Obergefreiter Johannes Jäger from 1st Company was killed, while Unteroffizier Kressner, a member of battalion staff, was hit in the right thigh and left hand by shell fragments. At 1700 hours, Dr. Dopfer – former battalion doctor but currently attached to a divisional aid station – operated on Kressner, removing most of the metal shards. Despite the horrendous situation, Dopfer was in high spirits, making jokes, conversing with one anaesthetised patient with amusing results, all to the general delight of the other wounded men. Such lightheartedness had a restorative power all its own. Nevertheless, the influx of new cases never stopped. Amongst the division's sixteen wounded on 26 December was Gefreiter Johann Breimschmidt from 2nd Company, struck in the right knee by shrapnel. He was flown out on 30 December 1942.

Introspection (27-31 December 1942)

"The vermin torment us terribly," Rettenmaier wrote, "as soon as you lie down on the bunk to rest, the crawling begins all over your body, from your feet right up to your neck."

I am very sensitive to this and am barely able to sleep. To 'hunt' them in my present surroundings is impossible, at least while in the dug-out. If we step out into the open we can certainly take action, but at twenty degrees Celsius below zero, this also has certain difficulties. I changed my clothes but the lice infestation returned in a very short space of time. One has no peace from these monsters.

Given the squalid conditions, the proliferation of vermin is hardly surprising:

Personal hygiene was no longer thought about. The incessant combat activity allowed no time for it. Beards sprouted. In the cellars, the cylindrical iron stoves were fuelled by oil-soaked wood paving taken from factory buildings. These produced dreadfully thick clouds of smoke and a layer of black soot covered everything. The entire garrison looked the same. On top of that, faces became more and more distorted by hunger. Eye sockets were larger and the look in one's eyes took on a peculiar expression. Vermin gained the upper hand and did not allow exhausted bodies any rest. Relief could only be had when one sought out a quiet spot somewhere in the open and – despite the icy cold – removed clothes and scraped off the pests with the flat of one's hand. This helped for at least an hour.

The reason why the Germans endured these conditions – and continued to do so – can be found in a simple explanation proffered by Hauptmann Traub: "We must hold out, otherwise Germany will endure a terrible disaster. Hopefully our children will have it better than we do and experience their own lives without war." In some ways, Traub was fortunate because he was spared the bloodletting. A multitude of jobs kept him fully engaged deep behind the line, and he received a new one on 28 December: expansion of the Gorodishche position to the south. This second major line of defence had been under construction for weeks, but a threatening situation to the south, at the Krasny Oktyabr Factory, necessitated the prompt extension of this fall-back position. At Traub's disposal were all non-employed elements of 305th Pioneer Battalion, the staff (without officers) and unemployed elements of 50th Panzerpioneer Battalion, and all alarm units within the zone of Gorodishche's local commandant. Traub's mission was multifold: remove snow from previously constructed positions, install tank-proof foxholes next to strongpoints, construct positions for heavy infantry weapons, accurately designate observation posts and command posts for battalions and regiments, and finally, specify potential gun emplacements for artillery and rocket-launchers. All of this while simultaneously fulfilling his previous duty of keeping the roads clear.

Traub's road clearing detail contained the bulk of his old battalion staff, men like forager Unteroffizier Anton Ell and cook Unteroffizier Helmut Schäfer. The fact that these two were shovelling snow rather than preparing hot meals speaks volumes. And despite his bladder problem and perpetually wet trousers, Gefreiter Vorherr was also assigned to this detachment: "At the moment I am working on a road that leads to Stalingrad. But what is called work is just so that we don't freeze." He summed up his daily routine, which understandably revolved around mealtimes:

> It is now 0400 hours, bitterly cold outside, there's about fifteen centimetres of snow, exactly like a month ago. For a few days I've been working on the road, which is a good hour away from the company in the village where I live together with another man in a bunker that can accommodate just two. At night, when the fire is out, it is cold. The food is supplied to us once a day, we heat the tea and coffee ourselves, we get up at 0540 hours in the morning, an hour later we go to work until 1100 hours, then there's a hot meal and afterwards we keep working until it gets dark. That is at 1430 hours. Then there's tea and the evening portion, which is swallowed up in a jiffy. Just once I would like to eat until full, like we used to at home during afternoon tea.

Nobody was left idle. Inspektor Zeller was also assigned a new chore:

> We had vehicles, but no petrol. In connection with this, I recall a tough dressing-down from Steinmetz. At the end of December I had to prepare a number of vehicles and deliver them in accordance with his orders. However, to him, the camouflage did not seem good enough and he yelled at me about it.

The New Year was welcomed in a humble and introspective way. Those with access to a radio listened to Goebbels' speech, some lit candle stumps on Christmas trees and others toasted with whatever alcohol was available. "How we celebrated Christmas and New Year will not be shown to you in a weekly newsreel!" Gefreiter Vorherr explained to a girlfriend. "Bread must be saved, even when it is scarce like ours is. Barely 200 grams per day, the right amount for breakfast, but then it's gone. What then in the evening? To bed on an empty stomach? There is nothing edible to be found in this area, everything is gone. The Führer once said, 'No one should go hungry and cold.' We are doing both, in fact, we're doing more." Hauptmann Traub still had a small cellar to keep hunger pangs at bay:

I invited the gentlemen from my staff for a glass of schnapps. They left around 10 pm. Then I read a little until 12 pm and went peacefully to bed. Around 1 am there was loud shooting. The Russians had attacked, but were stopped and slaughtered, suffering heavy casualties.

A snapshot of the battalion on New Year's Eve would have been unrecognisable to the men who belonged to it in France. The character of the battalion had been altered by its amalgamation with the 162nd Pioneer Battalion. Very few original members were left because practically every combat soldier in the battalion had been killed or wounded. The few remaining veterans, like Feldwebel Pauli, kept a tight rein on the inexperienced reinforcements. Things did not fall apart. On the contrary, the battalion was still a potent force, kept in reserve to fight fires. After settling the situation around the fuel tanks, Krüger's combat pioneers were pulled out of the line, ready for the next emergency. The battalion still possessed a high headcount because its ranks were swollen by newcomers. Most were involved with construction and road maintenance, improving positions in the front-line and erecting secondary defences. Krüger led the combat elements, Traub was overlord in the rear. The dual system worked well. Personnel passed back and forth between them: Krüger drew reinforcements from Traub's work detachments and sent men there to recover from minor wounds or illness. With the failure of Manstein's relief attempt in late December, the writing was on the wall. Nevertheless, the abandonment of an entire army was still unfathomable.

Waiting (1-9 January 1943)

Gefreiter Johann Bonetsmüller from 2nd Company was happy; news of his promotion to Obergefreiter had come through. He cared little about the advancement in his military career but was pleased with higher wages, in this case, an extra thirty Reichsmarks per month. Vorherr had received a similar promotion a month earlier as recognition of three years of service. Although fighting for their country, soldiers were still heads of their families and provided for their wives and children. Money was useless at the front, so every soldier sent it home, married men to their wives, bachelors to their parents. These funds were used to buy luxuries required by the soldier at the front, such as stationery, facewashers, film, anything not provided by the army. Encirclement and the subsequent breakdown of the postal service disrupted this lifeline of care between the soldiers and their families. Vorherr had a wad of money in his bunker, ready to be mailed home, but weight restrictions prevented him from sending it. Lack of mail from home was the greatest morale sapper. Some men, like Bonetsmüller, had not received any for seven weeks. His last letter home was dated 2 January; after that, the Bonetsmüller family was met only by silence. Vorherr's final message went out on the same day. Gefreiter Wilhelm Füssinger got his last letter out two days later:

I have not received mail since 20 November and the Christmas packages did not reach any of us. We have now known for over six weeks what hunger really means. And the hope for better times still remains a long way off. Lent has arrived early for us! And so, at midday, we receive a watery soup with horse flesh, 100 grams of bread in the evening and coffee with cigarettes for breakfast. My birthday [1 January] was spiced with the black humour that dominates at the moment. Now we hope that everything will soon get better.

This was the last time Füssinger's family would ever hear from him. No more letters arrived from Willi and Johann. The first week of January was the last period in which mail officially left the pocket. After that, days with no news turned into weeks, and only with the announcement of the fall of Stalingrad did families learn about the catastrophe that had claimed their loved one.

- - -

German defences around Lyudnikov's bridgehead were so tough that the Soviet leadership gradually shifted its offensive emphasis south to the sector held by Oberst Krüder's Regiment 576. Attacks along this axis offered a better prospect of success because a deep thrust into the Barrikady Factory from the south would endanger the entire German position opposite Lyudnikov's bridgehead. Krüder's men held a line that zigzagged from a large concrete oil reservoir, through a vast cratered district of trenches and flattened shacks to the Bread Factory. Soviet storm groups nibbled away at the line, storming a dug-out here, probing there. Major Krüger's pioneers were active in this sector most nights, strengthening the defences and occasionally restoring a breach with a counterattack.

Nuisance fire continually fell upon the Gun Factory and nearby settlements; it was part of the daily rhythm and nobody thought much of it. Every now and then, however, a painful blow was dealt. On 4 January, Oberst Steinmetz, accompanied by his adjutant, an orderly and Oberstleutnant Brandt, visited a command post in Hall 3c. On the way back, a single mortar shell sailed in, exploded right next to the quartet and seriously wounded all of them. Shrapnel peppered Steinmetz's right side, striking his shoulder, arm and fingers, but in spite of these numerous wounds, he staggered to a nearby aid station to fetch medics. The other three were carried back to a dressing station but died of their wounds the same day. Steinmetz was flown out on the night of 8 January and required nine months in hospital before he was back on his feet. A replacement division leader was quickly found in the form of Oberst Dr. Albrecht Czimatis, commander of Artillery Regiment 83 (100th Jäger Division). The monocled artilleryman was described as an "outstanding personality, of high intellect, versatile, very skilful, energetic and cool" and his south German roots and many years with an Austrian division provided a decent fit for the Bodensee Division.

- - -

Continual fighting throughout the pocket drained combat strengths, yet ration returns showed ample manpower to adequately replenish depleted units. Even at this stage, most of the army's men were in non-combat roles, so new measures were being constantly implemented to remedy this situation. Individual corps and divisions were encouraged to raise combat strength by any means. Battalions and companies were consolidated, dissolved or reinforced with men freed from other duties. The tooth-to-tail ratio throughout the army needed re-adjustment: in the current situation, twenty or thirty men in a supply train supporting four or five front-line soldiers was obscene. The solution pursued by 51st Army Corps on 6 January was to free up soldiers by amalgamating staffs and supply trains, dissolving unused supply units and reducing those not operating at full capacity. The released men then underwent fourteen days of infantry training. The number of soldiers freed up for infantry deployment in the Bodensee Division was 11 officers, 131 NCOs and 577 men.

The division received another massive increase in strength when General von Schwerin's 79th Infantry Division was disbanded. Each of its three infantry regiments formed one battalion, these were then consolidated into a new Regiment 212 and subordinated to the Bodensee Division. Other windfalls to the division included the artillery regiment, three labour companies, some signals troops, a training staff, a few supply men, but most significantly, control of two ammunition dumps. Two units that did not come the division's way were the pioneer and anti-tank battalions. At 1300 hours on 8 January, 305th Infantry Division assumed tactical command of 79th Division's sector and all the men in it.

Koltso (10 January 1943)

General Rokossovsky, commander of the Soviet Don Front, had sent Paulus a surrender ultimatum on 8 January and it was turned down flat. Hitler made it clear that surrender was not an option. And so, on the morning of 10 January, Operation Koltso (Ring) was launched by the seven armies of Rokossovsky's Don Front. A tremendous preparatory barrage from seven thousand artillery pieces, mortars and rocket-launchers left the encircled Germans in no doubt that the end was nigh. The Soviet objective was complete liquidation of the pocket. Their main effort concentrated on the weaker western and southern faces. The first day brought gains of four or five kilometres, which was disappointing for Rokossovsky but an alarming development for Paulus. His units managed to prevent an outright breakthrough over the next two days.

– – –

Back in the ruins of Stalingrad, most Soviet units launched small attacks to tie down the German divisions, but not Lyudnikov's men, for they were being replaced. The German strongpoints, particularly House 79 and the Commissar's House, were just too formidable. Attempts to crack the German defences east of the Gun Factory had failed, so Lyudnikov's division was transferred in order to launch attacks on softer sectors where progress was being made, grudging recognition of the defensive prowess of the Bodensee Division. Czimatis' men would continue to resist until the very end but their defensive efforts now focused on their right flank, along the main battle line manned by Regiments 576 and 212 that ran from the fuel tanks on the Volga bank, encompassed the Bread Factory and meandered south to the outskirts of the Krasny Oktyabr Factory.

The men of 305th Pioneer Battalion were unaware that a significant milestone in the battalion's life – and death – took place on 13 January: this was the day that the last man from the battalion escaped the pocket. Gefreiter Paul Nuoffer, Grimm's orderly and driver, wounded on 10 January, was flown out, though under conspiratorial circumstances. A rumour persisted that he had purposely injured himself. Zrenner recalls:

> He reported to me after the "wounding" but he was not even in the front-line or in battle and it sounded funny to me. The gunshot wound was in his left arm. I sent him to the hospital at Gumrak and thought to leave this determination to the medics (discovery of possible gunshot residue). However, everything was in the biggest shambles and no-one said anything. Anyway, Nuoffer got a place in a plane and made it out. If this is true, then I have to say it was dishonourable and should have been severely punished.

Nuoffer was fortunate that the surgeon who treated him was Grimm's good friend Dr. Dopfer, a man who lacked malice and took the Hippocratic Oath seriously. If he had blown the whistle on Nuoffer, his vow to "do no harm" would be broken because a firing squad was the penalty for self-inflicted injuries.

Grimm: So Close, Yet So Far (15-21 January 1943)

As the plane carrying his orderly landed at Salsk airfield, Grimm was being drawn into his own desperate situation. It began fortuitously enough. On 15 January, collection points were established to gather 6th Army members based on their divisional affiliation, the one for Bodensee Division being dubbed "New York." That same evening, however, Grimm was unexpectedly ordered to ignore this order and instead get to Stalino in all haste with a new 200-man strong company for deployment as pioneers. Permission was given to search out suitable men from the other companies, and by dawn the following day, Grimm had assembled 198 men of all ranks. Their only weapons were a few German rifles, but after discussions with the local administration, about 60% of the company were soon equipped with Russian firearms. A platoon of prisoners from a punishment battalion was attached to help lay mines. Grimm was ordered to reach Voroshilovgrad as quickly as possible to create defensive positions. Their lorries departed in snowstorms and blistering cold on the afternoon of 16 January, but fell far short of their objective. In darkness, they sought shelter in a school in Nikotovka. The thermometer showed thirty below zero Celsius (-22°F). Grimm's new charges were miserable and restless. Atrocious weather thwarted all attempts to move forward over the next three days, each time being forced to turn back to Nikotovka.

Motors were left running overnight but that did not prevent the fuel lines of some lorries from freezing. Under sunny skies and pushing upstream through throngs of retreating Italians, Grimm and his motley band finally reached Voroshilovgrad just before noon on 20 January. The Red Army had reached the nearby Donets, so Grimm's mission was no longer valid. The city was in unrest. Grimm learned that the Soviets had relaunched their offensive on 13 January and destroyed the Hungarian Second Army and, despite some valiant resistance, the Italian sector had also collapsed. The Red Army was now flooding south. Any available unit was being loaded onto lorries and thrown towards the enemy. For the moment, Grimm's company was split up and deployed to defend the city outskirts, while he was put under the direct command of the defence staff. Leadership of an important outlying strongpoint was thrust upon him. With six of his men, he drove through the darkened city and witnessed scenes of drunken Ukrainian policemen plundering a storehouse, occasional buildings in flames, pitiful civilians fleeing west with overloaded wagons and random outbursts of gunfire. After an uneasy and bug-ridden night in a Ukrainian hut, Grimm and his tiny group were thrilled to see regular German troops retreating into the city with panzers, flak and anti-tank guns. They did not feel so alone now.

Battle for the Bathhouse-School Strongpoint (15-31 January 1943)

Holding attacks initiated in Stalingrad on 10 January by Chuikov's 62nd Army slowly gained ground in the flattened settlements west of the Krasny Oktyabr and Bread Factories. Small raids stabbed at the line each day; no major objectives were achieved, no deep penetrations forced, they were just relentless small-scale operations that pecked away at the defences, capturing a single house or dug-out, occasionally causing a section to crumble or compelling minor withdrawals. Sectors without solid buildings were less defensible and it was upon these that Soviet storm groups focused. The Soviet divisions operating in this area (39th Guards, 138th, 45th and 95th, from south to north) had orders to reach the city's western edge and made steady progress until 15 January when Sokolov's 45th Rifle Division was held up by small-arms and machine-gun fire coming from the only solid buildings in their sector: the four-storey School No. 20 and nearby T-shaped Bathhouse. This pair of stout structures was a formidable German strongpoint, their impregnable basements converted into fire-spitting pillboxes and their upper floors bristling with machine-guns. Commandant of this strongpoint – known to the Germans as the "Weißhaüser" (white houses) or "Teehäuser" (T-shaped houses) – was Major Krüger, its garrison drawn from the remnants Pionier-Battalions 162, 294 and 305, plus a smattering of men from other formations. Unit integrity was a rock bottom priority at this stage of the battle. Krüger was fortunate to have several combat officers with him, including Hauptmann Artur Baransky and Oberleutnant Alfons Schinke (both from 162), and Oberleutnant Gerhard Menzel (294), as well as seasoned cadres of survivors from all battalions. The bulk of the men, however, were cannon fodder, hastily trained, drawn from all branches of the service.

Not one official German document exists to show the courageous last stand of Krüger's battalion. The only sources available are brief snippets, a few eyewitness accounts and Russian records, but these bear testimony to the ferocious showdown that capped the short-lived but proud history of 305th Pioneer Battalion. By using the war diaries of Soviet units to examine this battle in detail, the grit and astounding fortitude of the pioneers is revealed.

The Bathhouse blocked the advance of Colonel Sokolov's right-hand regiment. His two other rifle regiments continued west. The central regiment tried to capture School 20 but was stopped twenty metres from its eastern wall. The regiment on the left barely moved but it did capture two immobile tanks being used by the Germans as bunkers. Sokolov recognised a troublesome strongpoint when he saw one, so he decided on a risky manoeuvre to bypass it so that he could reach his objective along the city's western outskirts. After that, he would allocate a group to liquidate the Germans in the Bathhouse-School complex.

In the pitch-dark night of 15-16 January, Soviet storm groups bypassed the Bathhouse by stealing through the deep east-west ravines and establishing an all-round defensive position on the western outskirts of the settlement, deep in the German rear. Krüger's men in the Bathhouse-School strongpoint were distracted by enemy blockade groups and two artillery batteries positioned on a slight rise that had lowered their barrels and were firing directly at both buildings. When daylight came, he found Soviet units in his rear. His garrisons laid down methodical fire that isolated the advanced Soviet groups and then launched three counterattacks, all of which were repulsed.

Colonel Sokolov was counting on the Germans in the Bathhouse-School to follow the pattern established over previous days: realise their predicament, abandon the position and fall back. He even extended an olive branch; Major Krüger was astonished when two Red Army soldiers approached under a white flag and, in halting German, proposed that the garrison surrender in order to avoid unnecessary bloodshed. After long parleying, Krüger's men asked for bread. The Soviets took pity on the hungry Germans and handed over a few loaves. Having received the bread, and presumably feeling refreshed, they ungraciously started firing again. Their distrust of Soviet promises was too ingrained. After such "diplomatic negotiations" the enraged Soviets drew up a number of artillery pieces and blasted away at the strongpoint at close range.

Taking advantage of darkness once again on the night of 16-17 January, small storm groups closed in on the Bathhouse-School strongpoint. Alert sentries heard them scuttling about and began sweeping the approaches to both buildings with dense beads of fire. A latticework of tracers and wheezing flares lit up the death zone, forcing the Soviet attackers to hug the ground. They could be seen wriggling forward, probably trying to get within grenade range, but the insurmountable volume of fire stopped them cold. They fell back, leaving behind several crumpled forms, some lifeless, others still moaning. Although soundly beaten, the Soviets had identified several firing points in the buildings and preceded to zero in on them with heavier weapons.

Every German officer near the Soviet intrusion recognised its dangerous potential. If this small Soviet foothold was reinforced, they could push northward into the rear of the Barrikady position and precipitate a collapse of the entire sector. Local commanders launched counterattacks on their own initiative. Krüger's garrison in the Bathhouse-School launched two of their own, but retracted into their buildings under deadly fire from snipers, machine-guns and mortars. In return, Krüger's machine-guns easily controlled the long gully, nicknamed "Baumschlacht" (Tree Gully), north of the Bathhouse during daylight hours and prevented the Russians from maintaining contact with their units embedded in the German rear. As Tree Gully was the obvious route for a Soviet relief attempt, Major Krüger set his pioneers to work, thickly sewing the ravine with mines and coiled wire. Defences were also bolstered in other areas. And not a moment too soon. A fireworks show exploded at 0200 hours on 18 January. Mines erupted in the gully, followed by anguished screams, and then gunfire broke out around the buildings. Machine-gunners were quick on the trigger and sprayed likely approach routes while mortars began dropping volleys into the gully. Several storm groups felt out the German line, but the main focus was on Tree Gully where Russian porters, weighed down by crates of ammunition and food, had been the first to stumble into the deadly minefields and trigger the gunfight. Every Soviet attack around the Bathhouse was bloodily repulsed.

Fighting flared up further south, too. On the other side of the open expanse south of School 20 was the western end of another gully through which a Soviet group was trying to advance. An Unteroffizier from the 294th Pioneer Battalion, now part of Krüger's composite battalion, crept into the gully and directed fire against the encroaching enemy, but a pair of Red Army soldiers sneaked up from behind and took him prisoner. Under interrogation, he revealed to his captors that pioneers from the 162nd, 294th and 305th Battalions were defending the Bathhouse-School complex. Nevertheless, this attack also failed and sunlight halted Soviet ground attacks, though artillery and machine-guns continued to relentlessly pound the German fortress.

Although the Red Air Force possessed absolute aerial supremacy, ground operations near the Bathhouse-School strongpoint were confined to night-time, a tacit acknowledgement of the omnipotence of Krüger's formidable bastion. For the fourth night in a row, on 18-19 January, Sokolov undertook a nocturnal operation to bolster and resupply the group holed up in the German rear. The Bathhouse-School was once again attacked, but this time the garrison also faced a solitary T-34 which emerged from the shadows and sent round after round into identifiable embrasures. Soviet troops ran forward under its fire support. Krüger's soldiers adroitly switched to alternate firing slits. The twin-pronged Soviet attack crashed against the binary strongpoint. One storm group charged towards the Bathhouse but were machine-gunned to a halt forty metres from the building, while the other stormed across exposed terrain towards School 20 and was massacred. The T-34 churned forward helplessly until it stumbled into a shell crater between the Bathhouse and the School. Its engine howled like a wounded beast as the driver tried to reverse out. Spotting an opportunity, a few pioneers armed with T-mines and grenades tried several times to knock it out, but were scattered by a fifteen-man Soviet group sent to defend the tank and protect its dismounted crew.

The following night (19-20 January) a second tank arrived. First it hammered away at the façade of the Bathhouse and caused a fire to break out, then carefully crawled over to its brother lodged in the crater, apparently to try and recover it. This distraction did not prevent Krüger's men from hosing down small Soviet groups attempting to infiltrate the line in order to reach the encircled group. They wiped out every last man in several groups. Sokolov tried a new tactic on 20 January: a daylight assault. At midday, storm groups tried to capture the Bathhouse-School strongpoint, but were bowled over by vicious automatic fire and sustained ghastly casualties. The ripped-up ground was littered with Soviet bodies.

Krüger's strongpoint was reaping a bloody toll, but more importantly, it was preventing the disintegration of the entire German position in the Barrikady settlement. Nothing was denied to him. Reinforcements, albeit hastily trained, replaced those who had fallen. Machine-guns were available in lavish quantities, as was ammunition to feed them. Hand grenades were getting short but Inspektor Zeller ensured a continual supply of home-made bombs manufactured from the battalion's own stores of TNT. Zeller, gregarious as ever, was a true asset, both in terms of morale and martial knowledge. If he wasn't buoying spirits with a joke or creating new means to arm the garrison, he was leading counterattacks. On one occasion, he avoided obliteration purely by luck:

It was here, at the so-called white houses, where I once again narrowly escaped death.
I was supposed to lead a supply train out of a bunker during an action. However,
shortly before that, someone else was appointed. This bunker then took a direct hit
and all the men were killed. The fighting was hard, the Russians attacked relentlessly,
and we kept defending ourselves. But to surrender, we actually did not want that
because it was said that the Russians immediately shot prisoners, especially officers.

At some point during the first week of the siege, Major Krüger was wounded, not
severely, but enough for it to be detrimental to his leadership. He was helped back to
the battalion's rear area to recover. Hauptmann Gast replaced Krüger as commander of
both the 305th Pioneer Battalion and the Bathhouse-School strongpoint. Gast was a
strong personality and proof that the combination of youth, ambition and ability got
you a long way – fast – in the Wehrmacht. Many thought him aloof and arrogant, but
Zeller preferred to view these traits as "very confident and decisive. If others felt
differently, I will say nothing about it." Gast's tenure as strongpoint leader would be just
as circumspect and skilled as Krüger's.

During the night of 20-21 January, the encircled Soviet group abandoned their dead
and stealthily made their way back through German lines, bringing with them a small
arsenal of captured weapons and ten Russian prisoners, almost certainly Hiwis who
defected after realising they had picked the wrong team. At this stage of the battle, these
"volunteers" knew justice would be swift and brutal if captured by their countrymen.
The 305th Pioneer Battalion had hundreds of Soviet prisoners in its ranks. After
practically the entire POW company was destroyed when caught in the open by a Soviet
bombing attack, the survivors were distributed wherever manual labour was required,
some being used to haul supplies to the front, but most going to Traub's road clearing
details. One of these teams was led by Unteroffizier Ell, a former member of battalion
staff. As the battle neared its end, Ell was treacherously murdered by his Hiwis. Zeller
believed that these Hiwis deserted and wanted to provide an alibi in order not to be
denounced by their own countrymen and shot as traitors.

For most of the daylight hours of 21 January, the Bathhouse and School 20 were
clobbered by artillery and mortars. The Germans responded with methodical fire by
solitary rockets fired from six-barrelled Nebelwerfers. The combat journal of Sokolov's
45th Rifle Division summed up the situation: "The enemy, with the remnants of the
294th, 305th and 162nd Pioneer Battalions, continues to stubbornly defend the
Bathhouse-School strongpoint and adjacent structures." Soviet ground assaults resumed
at 0800 hours on 22 January. Two storm groups tried to capture the Bathhouse and

nearby structures to the north but made no headway and sustained heavy casualties. Another group had the same result at the school.

Gast's men were exhausted. Day after day the Soviets threw themselves at the buildings. When the storm groups stopped, artillery fire recommenced. Both structures, with their stout cellars and thick walls, were natural forts and kept the German soldiers relatively secure, but they offered few comforts. The Bathhouse was a stark place, with tiled walls and floors, and colourful yet mocking frescoes. The desk and chairs in the school had disappeared into fireplaces long ago. Manning guns at embrasures was a hazardous task because the enemy made a point of identifying these firing points and taking them out with 45mm guns or anti-tank rifles. Gast's wounded had little prospect of survival because medical supplies in the pocket were depleted and the possibility of aerial evacuation slim. The only airfield available for Luftwaffe planes on 22 January was the tiny Stalingradskiy airstrip, and even that would be in Soviet hands the next day. Any type of wound was now tantamount to death.

The greatest challenge so far for Gast came on 23 January. Under the veil of darkness, Soviet storm groups exploited artillery cover to close in on the Bathhouse, overcame stubborn resistance and for the first time broke into the building. Vicious grenade battles and murderous scuffles echoed through the tiled rooms. Pistols, daggers and entrenching tools were the weapons of choice. Part of the redoubt was soon in Soviet hands. Attack and counterattack ebbed and flowed all day through the corridors and communal shower rooms. The clash ended in a stalemate. Meanwhile, other storm groups stormed the school four times, always resulting in bitter hand-to-hand combat, but they were beaten back every time by heavy automatic fire from the school and flanking fire from the Bathhouse. They dug in 20-30 metres from the school's southern and western walls. Dozens of Red Army soldiers lay sprawled on the sooty snow, dead. Storm groups tried to finally capture the Bathhouse on 24 January but Gast's obstinate pioneers stood their ground, went over to the counterattack and squeezed the Soviet intruders out. The Bathhouse was wholly back in German hands. With their own men clear, Soviet artillery and mortars walloped the building. No attacks were forthcoming on 25 January; it seemed the Soviets had had enough. The pioneers slumped into an exhausted silence behind their weapons.

– – –

On 17 January, after reducing the pocket by half, General Rokossovsky called a temporary halt to the offensive so that his forces could regroup. The final stage of Operation Koltso began on 22 January. Rifle units of 57th Army pressed in from the south-west, broke through at Voroponovo Railway Station and marched towards

Stalingrad-South. German units were incapable of closing the yawning gap. Ammunition was low and troops could not be transferred from other sectors. The end was coming and Paulus knew it. That night he sent a grim message to the High Command:

> *Rations exhausted. Over 12,000 unattended wounded in the pocket. What orders should I give to troops who have no more ammunition and are subjected to mass attacks supported by heavy artillery fire? The quickest decision is necessary since disintegration is already starting in some places. Confidence in the leadership still exists, however.*

Hitler's reply was blunt:

> *Surrender is out of the question. The troops will defend themselves to the last. If possible, the size of the fortress is to be reduced so that it can be held by the troops still capable of fighting. The courage and endurance of the fortress has made it possible to establish a new front and begin preparing a counteroffensive. Thereby, 6th Army has made an historic contribution to Germany's greatest struggle.*

The western defensive front began to crumble and the remnants of once-proud divisions retreated eastward into the chimeric safety of the ruined city. Tanks of the 21st Army, coming from the west, linked up with 62nd Army on 26 January and cleaved the pocket in two. This event prompted a hysterical reaction amongst some German units, particular the rear echelons. First, individuals and small groups moved from the west into the Barrikady Factory area. By midday it was motor vehicles with trailers, quickly followed by chaotic movement of German infantry through Skulpturny Park and "Ilyich" Hospital towards the Barrikady Factory. Shortly before nightfall, columns with thousands of men flooded in, like rats leaving a sinking ship. This inrush of bedraggled and desperate newcomers overwhelmed the rear areas as they sought sanctuary in overcrowded cellars and tried to acquire any form of sustenance. Most of them found shelter in the buildings of the Schnellhefter Block, a vast complex located in the centre of the northern pocket. Amongst the refugees were the staff and supply train of the 305th Pioneer Battalion, forced to abandon Razgulyayevka and Gorodishche on 22-24 January. Oberzahlmeister Max Keppler led the remnants of the orderly offices with cooks and supply personnel to Stalingrad-North and prepared bunkers in the Schnellhefter Block. Hauptmann Traub was there, too. Defence of the Gorodishche locality, which he had been preparing for the past month, was nullified by the Soviet approach from the west. After the loss of the Stalingradskiy airstrip on 23 January – the pocket's last connection with the outside world – it was suggested that an airfield be laid out in the northern pocket. The job fell to Traub and his construction detachment. Before he started, however, the proposed site fell to the advancing Red Army and Traub's new mission was to erect obstacles.

The arrival of whole echelons of unemployed soldiers proved beneficial for the strongpoints still holding out along the Volga front. Gefreiter Zrenner was one of the men sent to the Bathhouse sector:

In the middle of January, after we had burnt our written matter and destroyed everything in our staff, the staff members were assigned to individual combat groups, and I went to a combat group in a northern direction, through the workers settlement that was the housing estate for Red October. I marched on with two comrades, towards the forwardmost strongpoint that had been indicated to us on a staff map, and there were still four infantrymen to whom we were allocated as reinforcements. We then had to hold a sector. On the left was a main road and on the right were large buildings, actually masonry buildings, designated by our units as the white houses, even though they were actually apartment blocks. And there, for the time being, we were housed in the cellars, everything was bombed out, everything, the entire street, everything bombed out. We were in a bunker from which we could survey the entire area. I then went from the shelter into the cellars of the white houses and spoke individually with comrades because I wanted to inform myself about the situation directly on the Volga and what was happening there. There were many wounded men in the cellars, half of all the men were wounded, without medical personnel, nothing, they screamed because of their wounds, they were just left behind. To the left was another unit with a field-kitchen, they still had things to cook so I grabbed a meal from them and they also still had a radio. I knew the unit from earlier times and they gave me something to eat for my people. In the evening we had to stand guard because the Russians were on the other side of the road. They would sing and stoke up fires at night and get drunk on vodka and because they had snipers, they would fire over at us. Whenever one of us stirred, a shot rang out.

– – –

General Chuikov shifted the axis of attack of Sokolov's division northward as part of a concerted effort to encircle and destroy the Germans in the Barrikady Factory. The

Bathhouse-School complex would be screened. Sokolov launched his fresh attack at 1500 hours on 26 January. By pivoting in place, the direction of his advance was now northward, away from the troublesome buildings, so only his left-hand regiment had to deal with it. On 27 January this regiment, together with a regiment from Lyudnikov's 138th Rifle Division, took School 20. Seizing the initiative, small storm groups immediately launched an attack against the Bathhouse and actually penetrated its gloomy innards. Gast's garrison prevailed and forced the Soviets to retreat, but they returned a short while later and broke into the Bathhouse once more. The defenders scrambled and evicted the Soviet intruders. Nevertheless, the school was in Russian hands. Incredibly, Gast's plucky pioneers even managed to recapture it on 28 January. The 305th Pioneer Battalion, a Frankensteinish entity stitched together from different parts, had now held the buildings for two weeks, an incredible feat of arms under current conditions, though the stubborn defence came at an exorbitant cost. Reinforcements were scraped up from wherever they could be found. Hauptmann Gast sent Inspektor Zeller to the Schnellhefter Block to collect the last dregs and bring them back to the "white houses." This task gave Zeller a chance to do the rounds and speak to some men from his old 305th Battalion. He met up with Major Krüger and Hauptmann Traub:

> *They were very good officers, but at this point they were just poor dogs who had no way out for their soldiers and themselves. They were in bad shape, firstly, they were hungry and malnourished, although every day there was still a few corners of processed cheese, but without bread or anything else. Second, death was before their eyes, both as defenders of Stalingrad and in captivity. And thirdly, they were faced with the decision: suicide or Russian captivity? Should I shoot myself or not?*

During his long conversation with Zeller, Traub affirmed that he would never go into captivity.[1] Zeller then visited his close friend, Dr. Wörner, the battalion doctor, who took him down into one of the cellars filled with wounded men. Vast numbers congregated to seek food, shelter and treatment for their wounds. Upon seeing all his untended comrades, Wörner burst into tears: "If only I had something to help, but there are no drugs, no pain-relieving injections, no dressings!" Looking around at the miserable wretches, Zeller spotted the familiar face of Gefreiter Hermann Raissle, former driver of the battalion commander. They had known each other for years and even travelled to Russia in the same train carriage. Now Raissle lay in this "hospital", his

---

1   "Under no circumstances must you pass on this comment to his wife," Zeller told Grimm in a 1948 letter

last breaths wheezing through a hole in his chest. The predicament of the wounded weighed heavily on those still fit to fight. Front-line soldiers witnessed hell on earth, but they dreaded going down into the first-aid cellars. And it became worse for those unlucky enough to be struck by enemy metal: to preserve the strength of its combat troops, the army stopped issuing rations to all wounded men on 28 January. Zeller was almost glad to head back to the front-line. He gathered the last levy of thirty soldiers, leaving behind only the most essential men. "I no longer remember who went and who stayed and I never again saw anyone from our old battalion, except Dr. Wirtgen." The forlorn group trudged across the battered settlement toward the Bathhouse, backlit by explosions and tracers, occasionally halting in craters and ruins until sporadic fire died down. "They were just cannon fodder," Zeller recalled, "but I did not check whether all thirty had come along. It was terrible. The 305th Battalion now consisted of the remnants of the actual battalion, the 162nd, 336th and 50th Pioneer Battalions, artillerymen, musicians, altogether two platoons, and they fought at the white houses. I was also there with my remaining equipment and ammunition."

– – –

Chuikov's army had turned its front to the north so that it could face the German pocket in the industrial district. The Soviet attack began at 1100 hours on 28 January with three divisions. Left to right, they were Lyudnikov's 138th, Sokolov's 45th and Gorishny's 95th. In many places progress could only be made with flamethrowers. Gorishny's division charged into the Gun Factory with its three regimental groups. Two storm groups, totalling 31 men in all, rushed Hall 6e and managed to seize it. German counterattacks relentlessly battered the storm groups and as a result, just one Red Army soldier returned to friendly lines unscathed.

Soviet forces continued their attacks late on the morning of 29 January. "Enemy defence remained strong and tight," wrote Chuikov's chief-of-staff. "The Germans clung to every fortified house. To break the desperate resistance of these condemned men, heavy guns had to be used in direct fire, flamethrowers employed and bombers sent in." Artillery was the main source of Soviet firepower. With the battle so close to its end, Soviet soldiers knew they had a chance to survive, whereas the desperate Germans, with nothing to lose and expecting no mercy, kept fighting. Their system of strongpoints in the Bread Factory and the Barrikady was almost unbreakable. If a workshop or masonry building was lost, a counterattack was immediately initiated to retake it. Somehow, one of Gorishny's storm groups captured Hall 6e once again and created a strongpoint into which four 45mm guns were laboriously rolled. These fired directly at German dug-outs and firing points. With a firm foothold in the factory, all units of Gorishny's division then tried to press deeper into it, but were halted every time.

Sokolov's division had battered itself senseless at the Bathhouse-School strongpoint, now it was Lyudnikov's turn to take on his old foe, the pioneers of the Bodensee Division. His 138th Rifle Division advanced at 0200 hours on 29 January. Fifty minutes later his storm groups broke through to the southern approaches of the destroyed school and for the next eight hours were embroiled in nasty scuffles that relied upon bladed weapons and grenades. German counterattacks were repeated and intense. At the end of the day, Lyudnikov had nothing to show but heavy losses. Whole storm groups were wiped out. Such was the devastation wreaked that the Bathhouse-School strongpoint was left in peace the following day.

– – –

In an effort to prepare the German people for the forthcoming disaster and convert the catastrophic loss of an army into an heroic epic, Reichsmarschall Hermann Göring delivered what was effectively a eulogy, except in this case the corpse was still alive to hear its deeds and sacrifice being distorted and mythologised. Göring compared Sixth Army's death throes to the "heroic sacrifice" of the Spartans at Thermopylae, and went even further with this thread, referencing the battle of the Nibelungs: "They too stood in a hall full of blazing flames, slaking their thirst with their own blood, but they fought to the very last." The starving soldiers in the pocket were disgusted by Göring's insensitivity and everyone now realised that there was absolutely no hope of being rescued. Gefreiter Zrenner clearly recalls how he felt at that moment:

I heard the radio and stood there because someone was speaking, and it was Göring's speech. Göring said something like future generations will remember the historic battle of our soldiers in Stalingrad, almost like the Spartans during the battle of Thermopylae. 'Wanderer, if you come to Sparta, say you have seen us lying here'. That is the meaning of what Göring said. [It was] terrible that we had been given up on by our highest military leadership. That we already no longer existed, that we had been abandoned. We had already been written off as lost.

Zrenner looked around his dark cellar to gauge the other men's reactions. Many were wounded and displayed no outward emotion, and it was impossible to see what the few healthy men were thinking because their filthy, bearded faces were sullen masks. The "Stalingrad heroes" in Zrenner's cellar absorbed the fact that they had been abandoned and contemplated the gloomy future.

– – –

The killing continued in Stalingrad on 31 January. Gast's garrison faced yet another attack with fatalistic composure. A composite group from Lyudnikov's division attacked the school with the help of a tank, but an order from army commander Chuikov halted the assault. The division handed over the line to another formation and moved to a new sector.

A few hundred metres further east, in the Gun Factory, another Soviet unit received a startling reminder not to underestimate the Germans, even though they were starving and at the end of their tether. Shortly after 0700 hours, a tremendous explosion rang out and shook the factory to its foundations. The ceiling trusses and thick masonry walls at the eastern end of Hall 6e crashed down upon Gorishny's riflemen, causing grievous casualties. The Soviet commander in the factory building, a major, was severely contused when dragged out of the rubble, but his chief-of-staff fared much worse, losing his eyesight. Taking advantage of the confusion, combat groups from Regiment 577 rushed into the workshop from several directions shouting "Russians, surrender!" The German plan had worked to perfection: detonation of large demolition charges accompanied by a simultaneous attack. The shell-shocked Soviet survivors abandoned the building and pulled back a hundred metres.

Incredible commitment was still displayed by the men of the Bodensee Division. Their demise was in sight but they kept attacking. Before noon on 31 January they launched a four-pronged counterattack from the fuel tanks and Hall 6e, each assault wedge formed from thirty to forty soldiers. Their courage was remarkable. Sustaining losses, they fell back to their starting positions.

Flickering Out (1-2 February 1943)

German opposition incensed the Soviets. It was senseless and would change nothing. Couldn't they see that further resistance was pointless? The Soviets brought out sledgehammers to finish off them off; an artillery bombardment of incredible strength, over a thousand guns and mortars concentrated along a six-kilometre sector west of the industrial district, rattled the crisp morning air of 1 February. Bombers and ground attack aircraft roared overhead. The pocket seethed from countless explosions. In some sectors white flags were displayed after only fifteen minutes of drum-fire, but most German troops resisted stubbornly, at least initially. The few remaining howitzers, flak and anti-tank guns shot up tank after tank and repelled the first attack. The wrath and hatred on the Soviet side rose immeasurably. A new inferno was unleashed, even stronger and more powerful than the first, and when the attack recommenced in three hours time, barely any resistance was offered because German units had fired off all ammunition in the morning. Entire sections of the line surrendered. Red Army soldiers ruthlessly shot severely wounded Germans, prompting some bandaged men to remain in position and defend themselves. Because of the stiff resistance offered, German prisoners taken on this day suffered much more physical abuse than those the following day.

The last morning of the Stalingrad battle dawned. It was clear that it would all be over on this day. An hour of artillery preparation ended at 0800 hours and Soviets units went on the offensive. For the first one and a half hours the Germans returned quite strong fire, though they never once tried to launch a counterattack. And they could in no way hold off the advance. Soviet storm groups pushed forward resolutely and the entire German defence collapsed. What was a sheer impossibility barely two days earlier now took only a few hours. Some storm groups attacked right through the Gun Factory to its northern boundary while others veered east and linked up with units holding Lyudnikov's old bridgehead, taking dreaded German strongpoints, like the Commissar's House, Apotheke and House 79 from behind. Hauptmann Gast abandoned the Bathhouse-School strongpoint, and together with most others, pulled back in a northern and north-western direction. Thousands fled into the opposite corners of the increasingly narrow pocket, right into the clutches of Soviet formations attacking from the north and west.

– – –

Just as the life of the 305th Pioneer Battalion was flickering out, Grimm's ordeal was only beginning. Since 23 January, he and 80 men had constructed four strongpoints near Voroshilovgrad. Also incorporated into the defences were an old Russian anti-tank ditch, dug by the civilian population the previous summer, and so-called Czech hedgehogs, tank obstacles made of iron angle. Small-arms were scavenged from abandoned Italian billets and close-combat weapons improvised from explosives and grenades. Sporadic attacks by Russian planes had little effect as most of their bombs proved to be duds. Frequent alarms were sounded because of marauding tanks but none came close to Grimm. This state of uncertainty continued until 3 February when Grimm was ordered to immediately assemble a one-hundred man company, including men from his Urlauberkompanie:

I showed the order – stamped with the word "secret" in red – issued by Stalino's commandant, according to which my Urlauberkompanie could not be deployed as infantry on the front-line. I was lectured that I should not be carrying this secret order. I did not let go of it and it was not taken from me. I accurately reported the lack of equipment and weapons: half the men had German and Russian rifles, two Russian light machine-guns, one of which was not fully functional, one semi-automatic Mauser, hand grenades, pistols. No helmets, no binoculars, only a few camouflage uniforms. First-aid dressings were found and one sled for wounded men was organised. The enemy had broken in, I heard.

Grimm's protest was in vain. His company was sent into the line. They set off on foot at eleven the following morning without vehicles, without field-kitchens, just cold rations in their packs. The destination was reached at twilight, where fresh orders sent them on to their allocated sector in a strenuous nocturnal tramp across snowy plains:

I occupied the defensive line from left to right and unexpectedly ran into a slender, tall lieutenant with three men behind a tent-quarter fastened to a tree for protection against wind and cold. This was my left wing. Still scared from shock, shivering from the cold, he reported on the situation and from his post I overlooked the position with its deep transverse cuts, in some places dropping straight down below us to the frozen and snow-covered Donets, while beyond it, as if on a platter, were woodlands running alongside the Donets. The view over the whole hinterland was unimpeded. No fighting could be heard, but with the naked eye I could see individual Russians on the far bank dashing about in short bursts and slipping into snow holes. He reported that my defensive line was previously occupied by a well-armed march company that was unexpectedly attacked from behind and rolled up with heavy losses. The enemy was suddenly standing in the trench in front of him and at night it was not apparent who was friend or foe. I do not know whether he had more pity for us because of our weapons than I did for him.

Combat started the following day, 5 February:

Enemy patrols and storm groups were taken under fire and driven back. Our right neighbour supported us with flanking fire from at least three machine-guns as soon as the enemy moved onto the Donets' ice. I observed that one attacker, who was struck down roughly in the middle of the Donets, blew himself up. He must have been carrying a lot of explosives. A large hole appeared in the ice. A similar case on our riverbank was reported to me and it left behind a deep impression. Our strongpoint on the left had no flanking fire support and had our only captured machine-gun of Russian make, which was instantly recognisable by its slow rate of fire. [Strongpoint leader] Unteroffizier Süßbrich prevented a storm group from breaching his strongpoint. After that, he came to my command post, all agitated, and reported that his hand grenades were not detonating. To test this, I immediately had one thrown, and true enough, it only partially detonated. These were grenades we had collected from the Ukrainian police battalion. They were all destroyed. Both of these incidents bothered the company more than second and third degree frostbite.

Uneasy days followed as Soviet troops and even some tanks slipped through gaps and occupied a pair of hills behind the German line. These enemy intruders were dealt with by Grimm's neighbours but it was apparent that the line could not be held against a serious attack. A ski patrol brought a written order to fall back. Grimm's company left their positions at 0300 hours on 11 February, successfully reached the designated rallying point and were given fresh orders to occupy a new line. That same evening he received more instructions to pull back two more kilometres. Exhausted and frostbitten, they trudged along snow paths. The fighting was getting fiercer. A few pieces of German artillery helped hold back their relentless Soviet pursuers. By falling back, Grimm and the other defenders were absorbing more and more assets. A few more men, some heavy weapons, but most significantly, a communications network. All forward positions were interconnected by phone.

Rocket-firing planes were a nasty surprise for Grimm. Because the planes were coming in low, about 300-400 metres altitude, he permitted his men to open fire. He mistook the launching of a rocket from beneath a plane's wing as a hit by one of his men, and called out, "It's burning!" The whistling and cracking of the projectiles instantly highlighted his error. "Such rockets were unknown to me," he remembers.

Another withdrawal was ordered on 13 February, this time through Voroshilovgrad. Grimm and his men walked through the abandoned city, past demolition teams awaiting the order to trigger their charges. At the end of their draining 14km march were warm billets, water and hot food. The next day, Grimm was told to keep marching and reach Aleksandrovka as quickly as possible. The Russians had broken through north of the city and every soldier was mobilised for its defence. The enemy penetration was sealed off and Grimm's rag-tag group was soon relieved by a well-equipped infantry company. More march orders followed and a new defensive position occupied at night. This pattern repeated itself day after day. On 20 February Grimm was ordered to disband his company and report for other duties. Two days later he took command of the military police unit in Voroshilovsk and on 1 March assumed control of position construction in Debaltsevo, a task that occupied him for almost an entire month.

It was during this quieter period that he heard the devastating news that Stalingrad had fallen. Mixed emotions swayed his mood: naturally he was happy to be alive and able to see his wife and child again, but the guilt of not being with his men gnawed at him. He was their commander, their father, and he had a responsibility to look after them. "I should have been with them," he would say. Still, maybe it was not as bad as he thought. He heard that men had been flown out of the pocket. Surely some of his soldiers had escaped. In his current posting, Grimm had no way of learning more because he was in the doldrums when it came to communications: outside the battalion grapevine, not privy to any reports from official sources, and far from home. A few letters from his wife started to trickle in, but her focus was on the welfare of her husband rather than his battalion. And then the best news of all was delivered to him on 17 April: the 305th Pioneer Battalion was being reformed and he had been summoned. But first, he was headed home for some leave. He boarded a train the next day and was back in Stuttgart within a week. Lotte and Gisela were waiting for him.

Furlough permitted Grimm time to contemplate the fate of his men. To those outside the Stalingrad pocket, it seemed like the entire 6th Army had just disappeared. Once the Soviet pincers clamped shut around Paulus' men, the chances of making it out alive were slim. At least a few men from the battalion must have survived. Grimm made it his job to find out what happened.

CHAPTER 11

At 0700 hours on 2 February, the senior unit commanders of the Bodensee Division received their final order: "All fighting will cease; weapons will be destroyed; expect the Russians in about an hour!" Major Krüger was in the Gun Factory when he received word of this order. He had visited the headquarters of his old 162nd Pioneer Battalion in the Schnellhefter Block the previous evening to talk to his staff, but there was no way he could get back there now, so he decided to stay with the staff of Regiment 578. "Assemble down in the Ziegelofen[1] (brick kiln)," he was told. As he walked through the dead factory, a weird atmosphere hung over the city. Weapons were quiet; the stillness seemed eerie. About 35 men gathered in the Ziegelofen, including Krüger and several other officers. The cease-fire order was announced. Those hearing it for the first time were struck dumb. Even the most optimistic knew the end was coming, but to hear those words! There was not much more to be said, though the situation presented one last opportunity to fulfil a basic need, one that had been denied to them for so long. Everyone was invited to a last supper. The iron reserve of a field-kitchen, consisting of Zwieback,[2] sausage, meat, tinned vegetables and coffee, was opened and everyone packed as much as they wanted to carry. Nobody fell upon this delicious manna like a ravenous animal: they ate tranquilly, to strengthen themselves for the march into a dark fate. With stomachs full for the first time in months, conversation started up and inevitably turned to suicide. The senior officer present spoke up:

Rubbish! Should we fail in the last few minutes? Here are the men who did their
duty in all the heavy fighting and – always willing to make sacrifices – would fulfil
any order and stand faithfully by us in each and every situation. Are we to desert
them now and escape in this last and crucial moment by shooting ourselves? No, I
deem it to be our role to see to it that our men go respectably into captivity and that
we remain with them for as long as the Russians permit it.

---

1   Ziegelofen = brick kiln. Contrary to what the name suggests, the Ziegelofen was not actually a blast furnace, but was some sort of storage depot. It was a long narrow building, approximately 30 to 40 metres in length with a semi-circular vaulted roof about 2.5 metres high. It was 4 to 5 metres wide and was lined with shelves. The entire building was covered with earth.

2   Zwieback, literally translated as "twice-baked," is a type of crisp bread.

He told everyone to remain calm and keep their heads on their shoulders. When the Russians turned up, he would go out first.

If I'm not shot out of hand, then everyone else will also go into captivity unharmed.

It is important that you have no weapons or ammunition. Everyone check their baggage once again.

These words calmed anxious nerves. A short while later noises were heard outside. Everyone stood still and waited. Would grenades come tumbling in? Instead, a gruff voice hollered "Raus!" with the characteristic rolling Russian 'r'. The senior officer replied immediately, "Ja", grabbed his satchel, and said to the group, "It's happening, comrades, keep your heads high, and good luck!" He walked out the door. Krüger waited inside with the others, staring at the sunlit doorway. Silence. No shots rang out. He had not been gunned down. It was an odd sensation, standing in a gloomy room as a free man just moments away from crossing a luminous threshold into captivity. Minutes later, the same loud voice told them to come out. They all stood there, rooted to the ground, unsure of what to do. Nobody wanted to be first. A shadow darkened the opening, then the senior officer called out: "Comrades, come out, the Russians are here!" The trance was broken and everyone started filing out. As he emerged into the dazzling sunlight, Krüger saw the officer standing calmly next to the Soviet soldiers. "Davai!" said one of the Red Army soldiers, "Move!", and the small column shuffled off.

The Russians – loud, well-fed, strong young men – were happy and danced about with pure joy. The contrast between victor and vanquished could not have been greater. Ruddy clean-shaven faces beamed from white uniforms while bedraggled German prisoners filed past, dirty and grey. The survivors of the 305th Pioneer Battalion were swept along in this tide of misery, all semblance of unity broken, just small groups of friends clinging together.

– – –

Gefreiter Zrenner knew his time was up when Red Army riflemen began closing in from all sides. With some regret he recalled an earlier conversation that may have spared him from his forthcoming fate: "In December, I once asked Hauptmann Traub if my detachment for studies, which had already been approved, was granted, and whether I could not head off [to Germany] now. He said no: 'That's not happening now!'" On 31 January Zrenner ran into Traub again: "Well Zrenner, what do you say now?"

"It doesn't look good!" Zrenner replied.

"Remember one thing," Traub said, "either we all come home, or no-one does!"

Zrenner thought hard about Traub's words and concocted his own plan to get home:

We intended to break out at night on 1 February, but that did not succeed because we barely made it a hundred metres before we took machine-gun fire. Then came the morning of 2 February and we were calm, internally calm, about what lay ahead of us. There were five of us, and we conferred as to whether we should voluntarily emerge from our shelter so that there was no danger of the Russian soldiers approaching and throwing hand grenades into the bunker. Then someone said he would not go into captivity, he would shoot himself beforehand. And it is still firmly in my memory, my exact words were that I would not shoot myself, if the Russians shoot me in captivity, that was different, because I would be killed by enemy forces, enemy action, but I would not shoot myself. And then came dawn on 2 February, and at eight o'clock we heard the sounds of battle, and I said to my comrades that it would be better if we went outside, stood in front of our bunker and waved to the Russians. But we already saw movement in our immediate vicinity, it was Russian soldiers, and we came out as they approached. They were Siberians [...]. And these Siberians came up to us and took us prisoner. We admired them for the equipment they carried. They were wide-eyed at seeing German soldiers. It was like a dream. German soldiers seeing the unimaginable, slit-eyed men with such things as fur caps, fur gloves, fur coats, everything, the best, and I simply had to say to a comrade, look at how they are equipped, and we still have no winter clothing. They treated us humanely, all five of us, and the first thing they did was to take our watches and the second things was our medals and decorations. They opened my coat and reached for my chest, where my medals were, took them from me, and my watch, then allowed us to get ready before leading us off along the beaten path upon which they'd arrived. The snow was thick, a metre deep, and they led us to a command post, down on the Volga... Someone from the German side had told us that deserters were immediately shot and for that reason none of us deserted. We therefore noticed that it was not as we'd been led to believe. They treated us decently. We were not struck, we were marched off, down the steep cliff to the Volga, and there we were led into a Russian command post that was built into the steep bank. We were interrogated, and all my notes – I maintained a diary – that I kept in my left breast pocket, and my Soldbuch in the right, all of that was taken from me.

After interrogation by German-speaking officers and their female assistants, Zrenner and his comrades were forced to stand out in the open, hour after hour, watching as German soldiers from other strongpoints were brought into captivity:

Eventually, there was a group of about 200 of us and we had to stand there all day, it was very cold, not snowing, it was sunny, but very cold. And then we formed up in the evening and were herded over the Volga, across the ice. And then we had to keep walking, mostly at night, for about four weeks, through a few villages. Men who died from exhaustion and those that could not keep marching were just left behind in the snow. Nobody took care of them. The Russians shot no-one, and we had to walk every night, during the day they put us into collective farms where we slept in cattle stalls. We were so tired. The main thing was we were glad we could sleep soundly again, without the noise of battle, without planes, without anything. They had to literally beat us with clubs to get us up, we were so tired... During this trek, my feet froze, and my fingers froze. After four weeks we arrived in a town called Katyar. It was not destroyed, it was on the other side of the Volga, where no German troops had set foot. We went to an old school in the middle of the city, it had a courtyard, and in the middle of the courtyard was a well. We were able to draw water and had shelter without barbed wire, without anything, without guards.

The worst part was that we were only there a few days when we all caught typhus, and of the 200 men housed there, only about 50 made it. I got typhus on about 19 March. That is my saint's day, St. Josef's Day, that I know for sure, and at this time I got typhus. One comrade who was with me in the strongpoint got typhus two or three days earlier and died in front of me. It was delirium with high fever, and could not be avoided. And I got over it, I had a high fever, and wanted some water, and once a day a Russian woman came with a barrel which had soup in it, and dished it out. Once a day and a piece of dry bread. And I believe I was delirious for about 14 days, then I was able to rouse myself, and I remember very well that I went outside and washed myself with snow, I was glad to feel any sort of sensation again.

– – –

Hauptmann Gast, Inspektor Zeller and two other officers (Oberleutnant Schinke from 162nd Battalion and Leutnant Hingst from 305th Battalion) were glad to have escaped the Bathhouse-School strongpoint. All were in little doubt that the bloodied

Soviet besiegers would seek retribution on the officers leading its defence. Despite their wounds, both Schinke and Hingst were ambulatory and refused to go to an aid station. Their best chance of survival was to stick together. Zeller was in a feeble state:

I got jaundice because of the poor diet, I could only eat a bit of jam and sugar, so I was already very weak. Many talked about shooting themselves. I also thought about it, but a comrade next to me said something like: "God has given me life and I won't meddle in his affairs!" I knew that I wanted to trust in God. One of our comrades, however, shot himself with his last bullet to avoid captivity. Then the moment finally came. A few Cossacks took our guns from us. That was an incredibly humiliating feeling.

The four of us were separated from the soldiers and taken to the command post of a Russian engineer regiment, interrogated there, and spent the night in the anteroom of the shelter of two Russian captains, who were quite companionable to us. We were then incorporated into a group of prisoners [the following day].

There was still nothing to eat, but we had not been killed immediately. They made us march, four days and five nights, always just outside the city of Stalingrad. We wondered why. I think there was no camp and they could not find a place for us to stay. They sent us from one place to another. There was no organisation for this vast number of prisoners. The Russians themselves occupied all habitable shelters. So what to do with us? By moving and running, death from freezing could be avoided. Nevertheless, our limbs of course froze during the icy nights, especially feet, because our shoes and clothing were not warm enough, like the felt boots and fur coats of the Russians. In addition, they could also be relieved of guard duty. But to where and how could we flee?

The miserable throng shuffled south, the weak and forlorn being winnowed to the tail where they collapsed into the snow and were finished off by a bullet to the head. The Russian sentries were rarely malicious; some prisoners even perceived the fatal shot as a merciful coup de grâce. What it did do, however, was underline the fact that those who wanted to live should not succumb to fatigue and despair. The column eventually stumbled through the gates of a transit camp set up in Beketovka. At least 60,000 prisoners were eventually herded in. Hunger and disease reaped a fearsome toll: over 45,000 men would perish there.

Everyone was awoken very early on 1 March and told to prepare for transportation to permanent POW camps. Everything would be better there, they were told, but the prisoners were sceptical as they had already been deceived so often in the first few weeks of captivity. After they had all assembled, an announcement was made: "Everyone who is feeling particularly ill and cannot survive a long rail journey, step forward. Also, all pioneers!" Being singled out could improve chances of survival if one's skills were of use, but on the other hand, it might lead to a quicker demise; there was never any middle ground. Zeller fell into both categories:

We officers were being sorted into the sick and the "healthy." Pioneers should present themselves separately. I was heavily wounded and went to the sick group. Next to me stood an officer whom I had never seen before and whom I also didn't know, even though I had been a technical supervisor to four battalions, and he said to me: "Don't step forward. I think they only want the pioneers to clear mines." Others, however, presented themselves without suspicion because they thought they might not be sent to Siberia.

Hauptmann Gast, Oberleutnant Schinke, Leutnant Hingst and many other pioneer officers stepped forward. Zeller's cautionary friend proved to be correct. The mines laid by the pioneers throughout Stalingrad would now be cleared by them. Plans of the minefields had been drawn up when they were laid, but they no longer existed, so detection by probe and clearance by hand was the only method. Needless to say, the work was tremendously dangerous. Zeller's destiny headed down a different path:

I was to go in the officer's train to the Yelabuga camp, but since I had frozen feet, they took me off the train and I went into some kind of hospital. There, according to several doctors, my feet would be amputated. There were two Russian, two Austrian and one German doctor working: the German doctor was Dr. Iffert. He was able to ensure that the amputation did not happen and with his help my feet remained firmly attached to my legs. How grateful do you think I was to him! Unfortunately, I soon completely lost track of him. I was still so weak and thin that a female Russian doctor carried me in her arms to examinations. The many lice, which crawled in masses in the dressings of the wounded men and tortured them, were pure evil. They also caused typhus, which almost cost me my life. The Russian doctors were mostly good to us wounded Germans and did what they could. I saw that almost everyone was a doctor first and a politician second.

When I was more or less able to stand on my own two feet, we went to another
camp. It was Camp No. 75 at Kazan, Arsk. Here, every bed was occupied by two
men. A single blanket was available for them both. In the same room as me was
Dr. Wirtgen, who had received a single bed.

Zeller observed that "everyone who was younger than 25 and older than 35 died of
typhus if they were not lucky enough to receive a blood transfusion." This fact was
noted by the Red Cross after studying all the figures.

– – –

Hauptmann Traub went into captivity at 0830 hours on 2 February with the last
seven members of 2nd Company: Feldwebel Franz Dittenhöfer, Feldwebel Adam Pauli,
Schirrmeister Martin Holler, Obergefreiter Johann Bonetsmüller, Gefreiter Alfons
Thanisch and two others. Everyone else had been drafted into a combat group or been
scattered by the final chaotic days of the battle. Traub and his small cadre were in a very
bad state. All were emaciated, unshaven and lousy. Pauli was nursing an arm wound but
otherwise remained in tolerable health. Their column marched northward. Snow
flurries whipped around them and bitter cold ate away their reserves of strength. A
snowstorm set in during the night of 2-3 February but they kept walking. Traub's group
clung tightly together, helping each other, providing unspoken but palpable
encouragement. Brief rest stops were granted but, as one soldier remembers, "sitting
down meant danger (or deliverance) because many fell asleep in the snow and froze to
death. Most of the soldiers who stayed behind because of exhaustion were shot. Time
and again the quiet of the night was split by rifle shots." No food was distributed. After
two days of marching in a roundabout way, they reached a transit camp in Dubovka,
60 km north of Stalingrad. They were housed in buildings and huts destroyed by the
Luftwaffe, a fact repeatedly pointed out to them. Not one window pane was intact, all
openings had been plugged with stones and rugs, shrouding their interiors in perpetual
darkness. The prisoners were packed in like herrings. Everyone was lousy; lice were
scraped off by the handful. Eventually, about 15,000 men were crammed in, and in mid-
February a large proportion of the prisoners were transferred to the ruins of a nearby
monastery. Available as shelter were the church, an open courtyard, crumbling
buildings and earthen bunkers, all with no protection from the vicious cold. And so in
mid-March began the epidemics and skyrocketing death-rates. Dysentery, typhus and
typhoid fever ravaged the weakened inmates, killing twenty to thirty men every day.
Empty beds in the makeshift hospitals were continually filled: the dead went out, more
sick men came in. Somehow, Traub and most of his pioneers avoided the pestilential

diseases. The inmates were split up on 8 April, officers going to Moscow for "re-education", lower ranks to work camps in central Asia. Before they parted ways, Traub's band promised to honour their pact made on the eve of capitulation: whoever got home first would notify everyone else's next-of-kin. And so Traub bade farewell, as did Thanisch and Bonetsmüller. Left behind in Dubovka's dark, foetid rooms were Dittenhöfer, Pauli and Holler, utterly exhausted and not up to the physical strains of a long journey.

– – –

In Stalingrad-North remained thousands of dead and severely wounded men, their names unknown because the constant indiscriminate assembly of new combat groups separated friends. Their comrades remembered them, knew that they had been commandeered for an urgent defensive task by such-and-such an officer. If their names were not to fade into obscurity, however, those who marched into captivity needed to survive, retain their mental faculties, remember names and dates without the aid of notes, and write down their memories when they made it home. Simply staying alive was a battle against the odds. Few would manage it. Hundreds of ordinary men from the 305th Pioneer Battalion vanished in the vastness of the East. Details regarding the fates of individual soldiers are vague. What is clear is that battalion arrived in the Soviet Union nine months earlier with over 700 men, and on those fateful days in early February 1943, it was scrubbed off the Wehrmacht's order of battle.

Countdown to Destruction: the Death of the 305th Pioneer Battalion

In May 1942, nobody in their darkest nightmares could have imagined the vaporisation of the full-strength battalion. Grimm's company was actually overstrength, with 210 men instead of the authorised 191. The other companies were the same. A full allotment of officers filled all command positions. The pioneers were in a can-do frame of mind – and retained this determination – despite piecemeal deployment to thwart the Soviet spearheads south of Kharkov. The green companies performed well, especially the 1st and 3rd of Hauptmanns Klein and Häntzka. Klein's company took a battering on 19-20 May, losing 15 dead, including 2 officers, and 44 wounded, 59 men in all. In just two days it had lost half of its officers and a third of its combat soldiers. Hauptmann Beismann's presence calmed the nervy company and enabled it to attain its objectives in the face of withering resistance. Häntzka's company also incurred losses, while Grimm's avoided them altogether.

Having a combat-hardened commander in charge of the battalion compensated for the overall lack of experience. The Wehrmacht was outstanding in this regard, basing appointments and promotions on merit and past performance. Beismann was part of the new generation of officers, full of fire and always ready to push forward. Gefreiter Zrenner states that Beismann instilled true Prussian discipline and that "during combat, he was always in the lead. An exemplary officer." A few units of the Bodensee Division succumbed to tank fright. Not the pioneers. Grass roots strength in a unit is important, yet a cool head in charge is vital. Panic is infectious, but so is calmness and confidence, and these traits were exuded by Beismann. Battle-tempered NCOs and enlisted men will look upon a novice officer with indifference until he has proven himself and gained their respect. In Beismann's battalion, nobody came close to having as much combat experience as him, so everyone looked up to him and tried to prove themselves. This was the crux of the battalion's admirable debut and justification of the wisdom of replacing old-timers like Oberst Hertel with medal-encrusted combat veterans like Beismann. When tank fright was nullified, inexperienced battalions performed just as well as in combat – sometimes better – than those that had entered the Soviet Union in 1941. "In general," wrote Beigel, "I have established that the discipline in the units who were here over the entire winter is quite excellent, at least as far as I could tell." Whether new or seasoned, German infantry divisions were equally reliable in mid-1942.

One battle down, one victory. Operation "Wilhelm" was next. This minor envelopment commencing 10 June, a preliminary offensive for the main summer offensive, enabled the pioneers to enjoy the thrill of the assault. Confidence surged as their foe fled before them. It was all over within a week. Another triumph chalked up. Casualties were light, although 3rd Company mourned the death of Leutnant Hupfeld. More serious was Operation "Fridericus" in late June. Until 29 June, Grimm's 2nd Company had lost three dead and six wounded, light casualties for almost six weeks on the Russian front, but the black day of 30 June – including three men wiped off the face of the earth by a mine accident – caused the tally to jump to twelve dead and ten wounded, one of whom succumbed the following day. Twenty-two men, or 10% of ration strength, gone. This was the moment the stars fell from Grimm's eyes; fighting for Germany's future was neither heroic nor glorious. Nevertheless, another victory was chalked up. The scoreboard stood at three-zero.

Death stalked the unit at every step during the advance across the steppe. Like a bird-of-prey circling overhead, he would swoop down and nab a soldier from the land of the living, usually in an unexpected way. These ongoing casualties gradually induced fatalism. Even the most prudent soldier who eschewed foolhardiness could be picked off. In overall numbers, however, July and the first half of August were not very costly for the battalion. Most days passed without any blood being shed and the men were either busy marching or holding position, sitting on the edge of foxholes, sun on their backs, as they wrote to their loved ones. Belief in themselves as a combat unit and in overall victory reached a zenith after three months on the Eastern Front. The front-line had shifted from Kharkov all the way to the eastern tip of the large Don bend.

Death and wounding accelerated from 15 August onward. Back in May, Grimm recorded the name and details of every man who became a casualty and continued this procedure until mid-August. The names of the dead were still put on paper, but impersonal numbers now represented those who were wounded. This change, forced upon Grimm by circumstances, was tantamount to depriving these men of their individuality, their humanity, and reduced them to mere figures, a resource to be accounted for. There was no malice or forethought on Grimm's part, it just happened, his schedule was too hectic, operations were constant, Beismann was on his back, and a thousand other worries crowded his mind. In this last two weeks of August, Grimm's company lost 9 dead and 25 wounded. The toll stood at 22 dead and 38 wounded since its arrival in Russia, almost 60 men in all, or just under a third of his ration strength, but close to half his combat soldiers. Precise casualty figures for the other companies do not exist but they were similar, if not worse, especially 1st Company. Grimm was stripped of his two officers (Beigel and Hepp) so that they could fill leadership roles in the other companies, an indication that Grimm's company was probably the strongest.

Construction of winter positions commenced in September and while proud to have played its part in the summer campaign, the battalion was happy to settle down in warm, fortified positions to weather the cold months ahead. The initial dread of being pierced by enemy metal was replaced by the hope of being wounded just the right amount, a so-called "Heimatschuss" (home-shot): enough to be sent to a German hospital, but without crippling injury or disfigurement. Grimm witnessed a prime example of this phenomenon when, during the din of battle, one of his men gleefully held up his hand with the index finger shot off and shouted elatedly: "Now I can go on marriage leave!" This was an early sign of a decline in morale, when the individual relied upon chance rather than the combat prowess of his unit to emerge alive from battle. And this was when the battalion had attained every objective and suffered no defeats, no setbacks. September was quiet in the static positions, yet Death claimed another four of Grimm's men, as well as one of the battalion's stalwarts, Hauptmann Häntzka. On 5 September Grimm's company had 135 men (1 officer, 13 NCOs and 121 enlisted men), but after losing another 4 dead and 14 wounded, a head count at the end of the month showed 117 men. In reality, however, this figure was too high because ten of those men were undergoing medical treatment in hinterland hospitals for various illnesses. Grimm's 2nd Company therefore possessed half the number of men with which it had entered the Soviet Union. Schaate's 1st Company was slightly stronger: two officers, 13 NCOs and 107 enlisted ranks, a total of 122 men.

And then came the battle that would scar them forever. Beismann was still in control, yet even his cool head and combat acumen were futile when faced by the daunting task of subjugating a fanatical foe prepared to defend every building in Stalingrad until the bitter end. Three days in, the god of war liquidated Beismann with a mortar round as he rode into battle on an assault gun, a reckless move from a man who knew that armoured vehicles drew gunfire like a magnet attracts metal filings. Hauptmann Traub, possibly the oldest man in the unit, took charge of the orphaned battalion. Beigel, Schaate, Hingst and Hepp were wounded, leaving Grimm and Staiger as the only combat officers in the battalion. The fighting was unrelenting. Every day men died in different and revolting ways, yet this was now commonplace and rarely elicited comment. As company strengths dropped into the teens, and then single digits, the hollow survivors grew despondent when, instead of being withdrawn for a rest, they were reinforced by men harvested from the supply trains. Grimm was crippled by illness, as were many others, but he stayed on, fighting in the terror-filled Gun Factory, watching his subordinates die and collecting a lifetime's worth of nightmares. The Gun Factory was a monstrous holding cell, converting its living occupants into dead ones, and the only way out was the golden ticket of furlough or some type of wound, and it could not be a light injury either: that just got a man sent back to the supply train to recuperate, if he was lucky, while a moderately serious wound warranted admission to a hospital in a Ukrainian city with immediate return to the front upon recovery. Stalingrad was transformed into a pocket on 22 November, preventing the return of about a dozen or so men in rear-area hospitals and 35 men on furlough. With hindsight, they would thank their good fortune.

The battalion was in a sad state. Each company was roughly the size of a squad. Combined, they could muster no more than a platoon's worth of combat soldiers. The battalion's ranks swelled after amalgamation with 162nd Pioneer Battalion. Remnants of other units, including fifty Italian drivers, were also bundled into the hybrid Swabian-Silesian battalion. After the appointment of Major Krüger as commander, Hauptmann Traub was allocated other tasks, which essentially split the battalion in two: combat elements under Krüger and a labour force under Traub. Hunger, exhaustion, frostbite, illness, injury and death whittled away the battalion over the following months. If a man was not wounded or very ill, he did not leave the pocket, simple as that, there was no ambiguity. Gefreiter Nuoffer was the final battalion member to escape on 13 January. The Soviet offensive progressed through Gorodishche and pressed the battalion's clerks and supply personnel into the ruins of Stalingrad-North, where most found shelter in the Schnellhefter Block. When Krüger was wounded, Hauptmann Gast took over and maintained the unit's combat integrity, so even at the very end of the battle, when all hope was lost, the remnants of the battalion held the "white houses" for a fortnight against relentless Soviet attacks.

The battalion's decay is not unique in the annals of military history, especially for units fighting on the Russian front, and more particularly those engaged in battle at Stalingrad, but the totality of the unit's destruction in just nine months when viewed from the perspective of each man and his next-of-kin is shocking. At least 28 of the battalion's men, including one officer, escaped their battalion's fate by being flown out of the Stalingrad pocket due to wounds or illness. Everyone else was killed in crumpled buildings, died of hunger and exhaustion, or was marched into captivity. Only six men returned from Soviet camps, three of them officers, none of them front-line soldiers. All the others were listed as "Vermißte" – missing in action.

CHAPTER 12

Sun-drenched Brittany in France was the site of the battalion's reformation in summer 1943. Only three of its 19 officers belonged to the original battalion: Grimm, Beigel and Zorn. All others came from different battalions and replacement units. Grimm was the last officer to turn up, so upon arrival he found all command positions already filled. Attachment to battalion staff as a "special purposes officer" solved the problem. Grimm handled any matter relating to the old battalion and its men. This official mandate granted access to documents that would help him research the fate of his old battalion. The big problem was that most paperwork – war diaries, orders-of-the-day, personnel files, Wehrpasses,[1] award recommendations – almost everything had been lost at Stalingrad. In fact, Grimm's personal archive was the most comprehensive source of the battalion's activities in 1942.

Like any good investigator, Grimm set about gathering material and interviewing eyewitnesses. A request to the Wehrmachtsauskunftstelle für Kriegerverluste und Kriegsgefangene (Wehrmacht Information Office for War Losses and POWs) resulted in a list of names and addresses of all company members present in September 1942. In July he compiled a roster of the battalion's officers and administrative officials who remained in the east, either confirmed dead or still missing. Seven killed, six MIA. The list was far from complete because it lacked the names of those who had joined the battalion during the campaign, but these details would emerge over time.

Enquiries made through official government channels by families of missing men invariably found their way to Grimm, who thereupon questioned soldiers in the reformed battalion and made enquiries of wounded men in replacement units. Grimm wrote hundreds of letters in his off-duty hours. He harnessed the soldier's grapevine: one man often knew the addresses of several others. He visited unit after unit, talking, probing, asking for information. For example, to ascertain what happened to his friend Oberarzt Dr. Dopfer, Grimm personally questioned over one hundred men, not just from the pioneer battalion, but from all divisional units. The best witnesses, that is,

---

1   A Wehrpass was an ID book issued to all personnel when they registered for military service. It was administered by the soldier's current unit and contained such details as units, dates of assignments, promotions, awards, battles, major injuries or illnesses.

those with the most valuable information, were survivors of the pocket, men evacuated by plane. Six wounded pioneers had been flown out of the pocket on 4 January, followed nine days later by Gefreiter Paul Nuoffer, officially the last man of the battalion to make it out of Stalingrad. His statements were crucial in determining who was still alive in January, who probably went into Soviet captivity and who had earned medals, yet even the date of his escape, three weeks before resistance ceased, meant many questions went unresolved. It truly felt like the battalion had been swallowed by the unfathomable depths of the Soviet Union.

A matter with better prospects of success for Grimm was awards, specifically who had already won them but not received the physical medal and certificate, and who had earned them but never been recommended. Without proof, each case required sworn statements from witnesses, and fortunately, because the leaders of all three companies survived the battle (Schaate, Beigel and Grimm), they were able to attest to the deeds of most individuals from their respective units. An instance of the lengths to which Grimm went to ensure the right and just allocation of medals is shown in the following letter sent to battalion HQ in early June 1943:

> As leader of the former 2nd Company, I submit the attached recommendations for bestowal of the Iron Cross Second Class, Assault Badge and Wound Badge for men who belonged to the former 2nd Company and now find themselves back with the battalion. I took over this company on 26 March 1942 in France and led it without interruption until the start of my convalescence leave on 14 November 1942 in Stalingrad.

> Through the fall of Stalingrad, all of the company's written documentation was lost with the exception of the diary for the time from departure from France until the end of September 1942.

> The Iron Cross Second Class could, in most cases, not be handed over because in the first three days of the attack from 14 to 16 October 1942, about 70% of the company became casualties. A divisional order regarding the extraction of manpower from the supply train and MuMa (ammunition and machine troop) enabled the company to be strengthened. Iron Crosses that were awarded could not be sent to wounded men afterwards because their addresses were lacking, and in any case, correspondence of any form was impossible because the entire company remained continuously deployed in the northern part of Stalingrad, contrary to all

predictions at the time. In addition, a number of Iron Cross recommendations with the battalion's combat- and command section were lost on 16 October when the command section was ripped apart by the enemy and the battalion commander killed on the same day. Furthermore, Iron Cross recommendations at the divisional command post in Gorodishche were lost in a direct bomb hit which killed the IIa [division adjutant] Hauptmann Göhner and the successor of the Ia [division chief-of-staff] Oberstleutnant Kodré. An investigation immediately initiated by me regarding Iron Cross recommendations forwarded by the battalion, especially the proposal for the Iron Cross First Class for Oberfeldwebel Platzer (killed on 15 October 1942), remained unsuccessful. – On 22 or 23 October 1942, at the command post of III./Inf.Rgt.576 in the Gun Factory, I was informed that every man who had fought in the forward line since the first assault day in Stalingrad should be given an Iron Cross Second Class. I thereupon submitted to Battalion recommendations on a page from a message pad. As far as I can recall, this was for six men. In the meantime, a number of Iron Crosses Second Class and two First Classes were allocated to the company without prior recommendations. They were to be submitted later. In the first instance, these same men were decorated, those wounded during daring surprise attacks on bunkers and resistance nests.

The <u>Assault Badge</u> was issued to the company just one hour before the departure to assembly positions in Stalingrad-North. Bestowal had been delayed mainly by the late acknowledgement of assault days by Infantry Regiment 576. Before deployment in Stalingrad, the company had accrued a total of 7 assault days acknowledged by the regiments and the former battalion. Most of the Assault Badges awarded could not be presented or sent afterwards because the company suffered over 30 casualties while bridging the Don and during the defensive battles between the Don and Volga. All award certificates for the Assault Badge remained with the company supply train. The battalion was prepared for a three-day battle. – On the basis of a telephone enquiry to Infantry Regiment 576, I was informed that for the time being, only the first three days of the attack, namely 14, 15 and 16 October 1942, were acknowledged as assault days. I made no use of this because Oberstleutnant Gunkel, leader of Infantry Regiment 576, told me that only two blocks still needed to be taken and the speedy relief of the division could be counted *on. The following day, the company was slated for the storming of Hall 6. The processing of recommendations was no longer possible for me.*

Nominated in the attached recommendation list for the Assault Badge are only men who participated in close combat in the forward line on acknowledged assault days.

This letter and its attached lists led directly to the bestowal of ten Assault Badges, a dozen Iron Crosses and 16 wound badges, with several men receiving one of each. Word spread about Grimm's deed, and soon he was inundated with letters from ex-company members, a few permanently crippled by their wounds, some recovering in hospital, others biding their time in replacement units. They were not glory hounds eager to display trinkets of courage upon their chests, they just wanted something to show for their sacrifice and service on the Eastern Front. A blanket bestowal never took place: each man was carefully evaluated on a case by case basis, a painstaking procedure of interviewing witnesses, gathering facts and composing sworn declarations. A determination by the German High Command did make things easier: any man in the Stalingrad pocket after 20 November 1942 and then flown out was automatically entitled to the Iron Cross Second Class if he did not already possess it.

Although loath to admit it, Grimm too felt he had been denied recognition of his Eastern Front service, yet there was no superior officer to whom he could turn for help, with one major exception. Grimm sent the following letter to the army's personnel office with the request that it be forwarded to General Oppenländer:

Esteemed Herr General!

As leader of the former 2./Pi.Btl.305, I would like to humbly request that you, Herr General, as commander of the victorious former 305th Infantry Division, submit an assessment of my active service on the basis of my remarks below.

Herr General would probably best remember me from the occasion of your inspection of my company during a river crossing and bridge construction with inflatable boats over the Plavet near Gouarec in France. I led this company continually until 14 November 1942, on which date I was permitted to depart on convalescence leave with the division's authorisation. As far as I know, I am the only company commander that took part without interruption in the hard and victorious fighting of the division from Kharkov until the complete conquest of the Gun Factory.

For this not insignificant period of action in the east, there exists no assessment about me. My commander at the time, Major Beismann, was killed on the third day of attack in Stalingrad, while his successor, Hauptmann Traub, is missing in Stalingrad. All superior officers in the infantry to which I had been repeatedly subordinated, Oberst Winzer, Major Brandt and Major Braun, are either dead or missing. I'm therefore left with no other option than to humbly request that Herr General assess me for the period from the beginning of May until the end of October 1942.

It is by a fateful chain of peculiar circumstances that I have received no higher decoration than the Iron Cross Second Class. In addition to having no assessment about me for my period of operations in the east, and my combat service coming up for discussion time and time again, I cannot avoid the constant and depressing impression that a question mark still stands over my unrelenting deployment which – as one of the very few – I was able to survive with a lot of blessings and soldier's luck. My relevant impression is further substantiated by the fact that younger comrades who are also reserve officers have been promoted to Hauptmann der Reserve [reserve captains] which, on the basis of the latest regulations, are not currently applicable to me.

After devoting several paragraphs to his operations in the east and emphasising moments that Oppenländer may recall, Grimm wrapped up his appeal:

I consider it worthless to report my commander's messages about promises, particularly during the fighting in Stalingrad. Even when I think over it again and again, then I must say that I truly have no reason to bicker about my fate. After many considerations, however, I decided to make my current request to Herr General and I hope I haven't taken an unauthorised or characterless step.

A response is not recorded, but whether army authorities declined to forward the letter because it was contrary to regulations or the General decided he simply could not help – assessments [Beurteilungen] were taken very seriously – Grimm gained nothing. He remained an Oberleutnant and emerged from the Eastern Front with just the Iron Cross Second Class, an Assault Badge and little proof that he was actually there.

Stalingrad Heimkehrer - Returnees from Soviet captivity

Of the hundreds of men of the 305th Pioneer Battalion abandoned to their fate in the pocket, just six are known to have emerged from years of hardship in Soviet camps: three officers – none of them combat leaders – and three enlisted men, all from the supply train.

Although not part of the pioneer battalion when captured (he was working in a divisional aid station), Oberarzt **Dr. Werner Dopfer** is regarded as such because he was an original member from January 1941 until mid-1942 and maintained close ties throughout the battle. Captured in northern Stalingrad, he endured the death marches through the steppe and was held in Beketovka until 1 March, when officers were transported to an officer's camp at Yelabuga. Grimm established contact with Frau Dopfer in October 1943:

> *I first met your husband during the formation of the battalion in Riedlingen under the leadership of Oberstleutnant Hertel, and while both on battalion staff, we lived together during the following times in France: Libourne – St. Georges – Vannes – Gouarec. Your husband and I had an intimate, comradely relationship and were on first name terms. We always kept in contact during our deployment in the east.*
>
> *I have not learned anything in particular about your husband from Abwicklungsstab Stalingrad.[2] Because I found myself back with the reformed battalion, I undertook research into all my missing comrades. On 30 September this year, as is well known, the "Stalingrad-Aktion"[3] concluded and you won't be able to receive any more information from that agency. I'd therefore like to inform you of the results of my investigation.*
>
> *I last spoke to your husband in September 1942 at the main dressing station in Vertyachii-on-Don. Among other things, we discussed your home in Ludwigsburg… Healthwise, your husband was feeling good again and he supervised the entire operation of the dressing station. As far as I can recall, your husband had previously been stricken by jaundice. We also talked about our little ones, during which I saw a whole series of photos.*
>
> *Now, my former driver and orderly, Obergefreiter Paul Nuoffer, was able to provide more information about your husband. [Nuoffer] was heavily wounded on 10 January and flown out of the pocket, and as he tells me, only with the help of your husband. I'm corresponding with Obergefreiter Nuoffer and on the basis of my enquiries have learned that Nuoffer has already written to you…*

---

2   Abwicklungsstab Stalingrad was a deactivation staff established to wind up the affairs of 6th Army units lost at Stalingrad.

3   "Aktion Stalingrad" was a broad inquest into the whereabouts of members of the former 6th Army.

Otherwise, I cannot locate any other witnesses. Men that got out after 20 November tell me that the area in which the main dressing station was set up – Gorodishche – fell into enemy hands without a fight in the second half of January 1943. Your husband surrendered to the enemy with the dressing station. But I must point out that this news is hearsay and I urge you to treat it accordingly. Despite my best intentions, I could not locate a witness for this story.

Intermittent contact was maintained between Grimm and Frau Dopfer, who in the meantime had received a few sparsely worded postcards from her husband. Grimm wrote in March 1948 with some significant news from Karl Binder, an administrative officer from the Bodensee Division's staff, recently returned from the gulags:

Herr Binder was with your husband for a long time. As I had supposed, things were a bit better for the doctors. The fortunes of the prisoners now seems to be more bearable. Of the Stalingrad officers, it can be said that whoever wrote home is alive, whoever did not is dead.

Grimm correctly prophesied that Dr. Dopfer would be home that year. The following letter reached him on 21 August 1948:

Dear Richard,

At last I now have my five and a half years of captivity behind me and returned to my family a few weeks ago in good health. It's best if we discuss this time and everything else face to face because the distance between Ludwigsburg and Degerloch is not too great. I have seen from the letters you exchanged with my wife the self-sacrificial way in which you have concerned yourself with the fate of former members of our battalion. I met nobody else in captivity, but I did once hear that Zeller is to be found somewhere else in captivity, I've been unable to find out any more. At the end, all that was left of our little group was Hauptmann Traub and Zeller, our good Max [Fritz] was killed in the pocket shortly before Christmas 1942. And about those who got away earlier thanks to their wounds, you no doubt know more than I. At the moment I am being bombarded by so many new sensations after such a long time in captivity that I am hiding away in my house and only slowly re-introducing myself to the outside world.

If your travels bring you to the Ludwigsburg area any time soon, I hope that you'll visit us. Otherwise, let me know how you think we should meet.

To Grimm, Dopfer's letter was like a voice from the grave. He was the first man he knew that had returned from Soviet captivity. Grimm wrote back immediately:

An indescribable joy overwhelmed me. My heartfelt congratulations for your return. I am and was with you all, those who remained in Stalingrad, more viscerally linked than foreshadowed by our earlier camaraderie.

Unfortunately, Dr. Dopfer was unable to help Grimm with any of his MIA cases. Circumstances forced the doctor to relocate his family to Esslingen in 1949 and then Ulm in 1952. After several years working in a local hospital he opened his own gynaecological clinic. Upon retirement in the mid-1970s, he and his wife moved to Langenau-Albeck, where he passed away on 6 March 1990, aged 84, survived by his wife and married daughter.

The second officer to survive Soviet imprisonment was another doctor who did not belong to the unit when he surrendered. Oberarzt **Dr. Paul Wirtgen** had cared for the battalion's wounded soldiers throughout the summer campaign until transferred to Regiment 576 in November 1942. Captured on 1 February 1943, he was transferred around the Soviet camp network until, incredibly, he landed in the same tiny compound as Inspektor Zeller, a satellite camp of Lager 75 in a forest near Uva, in the Udmurt Republic. German prisoners slaved away in the republic's industrial complexes and logging camps. Zeller remembers that "[Wirtgen], six other doctors and I shared a room from 1944 to 1947. The camp was closed in 1947 and the prisoners distributed to other camps in Izhevsk, Donets, Sverdlovsk. I don't know where he was sent." Wirtgen was released on 23 June 1948. Because his East Prussian homeland was in Soviet hands, he was summoned to the Niederrhein, where his wife had set up the family in Oberhausen. His two sons gained a brother a few years later. "He wrote to me," Zeller recalls, "but his return address was written in such doctor's handwriting that nobody was able to decipher the name of his village." Wirtgen was able to conquer his nightmares and record his experiences in Stalingrad and captivity in a thick manuscript for eventual publication. He passed away on 23 January 1982 in Scharbeutz. Just over two decades later, his manuscript was destroyed by a family member before it saw the light of day.

The third surviving officer was Inspektor **Georg Zeller**, officially the only officer of the 305th Pioneer Battalion to return. Typhus and frozen hands and feet were endured in the first months, but he somehow walked out of the mephitic POW hospital under his own steam. Anyone not laid up in a hospital bed needed to work to earn their meagre daily ration. Zeller's skills as a civil engineer were put to good use by his captors:

I went into a labour camp with 100 men and four officers. The four officers were then sent to an officer's camps and I remained behind and led the construction column. The camp was closed in autumn 1947. I then worked in a machine factory in a camp in the city of Izhevsk, where I fell ill for a third time and was then released home in mid-May 1948 because I was too weak. I arrived home on 2 June 1948. In October 1948 I met my future wife, became engaged at Christmas and married her on 28 April 1949.

In September 1949, Grimm was informed of Zeller's return by an MIA's wife. He could scarcely believe his eyes, reading it over and over: "Georg Zeller came home in summer 1948". If true, that would make Zeller the first officer of the battalion to emerge from captivity. "Should I believe it?" he wrote in his introductory letter to Zeller, "I have become very distrustful." Zeller answered immediately with a long handwritten letter detailing the battalion's demise in Stalingrad and his own experiences in captivity. This was Grimm's first real evidence of what happened: "Almost every word of your letter is valuable to me because I have concerned myself with the collapse of the old battalion in extraordinary detail." Regrettably, Zeller's incarceration in a secluded camp far from the bulk of other Stalingrad prisoners precluded him from resolving the fates of those still missing.

By relegating this ghastly experience and only looking forward, Zeller resumed a normal life. Employment in his parent's construction firm got him back on his feet and his wife soon gave birth to the first of several children. After the death of his wife and flight of his children from the nest, Zeller lived by himself in his own home, but he took advantage of a standing invitation from the nursing home next door to come over for dinner every day. This same Altersheim staged various events, including a "Fasching", a catholic festival of fun and food before the Lent fasting period began. It was a cheerful and boisterous affair, with song and dance, and Zeller was a hit with all the ladies. "I always told my dance partners at the end of the dance, 'Now you know what it's like to dance with a 90-year-old', which nobody wanted to believe." Even though a nonagenarian, he was in good shape and his mind razor sharp. When presented with a copy of the author's previous title "Island of Fire", he did not let the fact it was in a foreign language deter him: he asked his daughter to buy an English-German dictionary. Zeller suffered a serious stroke in early February 2007, but his tough old body held on, and after a few weeks in hospital, still without regaining consciousness, he was transferred to palliative care. Death took him on 22 February 2007, aged 91, in Alzenau, the town in which he was born and had lived his entire civilian life.

Incredibly, the three enlisted men who survived all belonged to Grimm's company. Obergefreiter **Johann Bonetsmüller** went into captivity with Hauptmann Traub's small group. Gefreiter Nuoffer informed Frau Bonetsmüller that her husband was still alive in January 1943. After that, she endured years of silence. Finally, on 12 December 1945, she received a letter from a man freshly released from Soviet captivity which stated that her husband was in a camp behind the Urals, was "relatively well" and had been put to work as a bricklayer. In 1947 another comrade told Frau Bonetsmüller that her husband had been due for release but the Soviets unexpectedly extended his sentence. Johann Bonetsmüller finally made it home in July 1948, a broken man. His wife would later write to Ella Traub:

> You have no concept of how wretched he looks, wasted away to a skeleton. After 14 days he was admitted to a sanatorium and lay there for half a year. He received over 250 injections and they did everything they could for him. To this day [July 1949] he still cannot work for one hour. The doctors say it will be at least two years before he'll be able to get up again. He was completely malnourished.

His 100% war disability entitled him to a meagre pension. Over the years, as he slowly regained his health, the degree of war disability was downgraded. Still in great pain, he gained work as a bricklayer foreman, a job he held until retiring at 65 years of age. Amazingly, he lived to the ripe old age of 90, dying on 23 January 1997.

Gefreiter **Alfons Thanisch**, a cook, also entered captivity with Hauptmann Traub. In April 1943 he was transferred from Dubovka to a camp in central Asia. Hunger drove him to steal some bread, he was caught and sent to a punishment camp. One of the penalties was the inability to write home, so his family had no idea he was alive. After five years of silence, he was released in 1948, arriving home unannounced, much to the utter astonishment and rapture of his family. He ran the family wine bar in Lieser an der Mosel until his death.

Obergefreiter **Josef Zrenner** and the few survivors of the death march into the steppe east of the Volga were eventually sent to Beketovka in mid-1943. Reckoning his captivity would last for many years, Zrenner learned enough Russian to be able to communicate with guards and civilians. He was then shipped up the Volga to Stalingrad to help rebuild it: there he knocked down furnaces, carried away broken oil pipes, cleared rubble from the streets, all by hand, with no equipment at all. "You destroyed everything in Stalingrad," the prisoners were told, "therefore you must rebuild it before anyone goes home." From Stalingrad he was sent to a recovery camp for the ill in Uryupinsk, 340km to the north-west, where he harvested sunflowers, cleared fields, and shovelled grain in massive elevators. In September 1945 the prisoners heard that the war was over and they all looked forward to going home soon. Hopes were raised when the first transport of ill men from Uryupinsk hospital headed back to Germany that same month. Hopes were dashed in November when every man fit to work was hauled by train back to Stalingrad. Zrenner was set an onerous task: cleaning individual bricks for reconstruction of the city, then assembling prefabricated Finnish wooden huts sent as part of war reparations. During the cold days and nights of winter 1945-46, in snow and ice, he oversaw the construction of these wooden buildings. Due to sleeping in earth bunkers and being perpetually exposed to the elements, he came down with pleurisy. Every deep breath and cough caused severe stabbing pains in his chest. A lung infection soon followed, as did admission to a hospital and confirmation by a female doctor that he was unfit for any sort of work. Release and a long train ride to freedom took place in 1946. For many years he was the last living member of the battalion present at Stalingrad. He died 23 October 2012 in Neustadt/Waldnaab.

Those Who Did Not Make It

The first man home from those who made the pact in Dubovka was Alfons Thanisch. Establishing contact with families in post-war Germany was difficult, but he tried his best. When he found himself in Nürnberg on business, he called in on the family of Feldwebel **Franz Dittenhöfer**, fully expecting to talk to Franz because he believed he was already home. It turned out that the trio left behind at Dubovka – Dittenhöfer, Schirrmeister **Martin Holler** and Feldwebel **Adam Pauli** – were still in Russian hands. And that never changed. Holler died on 19 July 1943 near Saratov, while Pauli and Dittenhöfer are still listed as MIA. All of their families pursued promising leads and hopes rose when Thanisch contacted them, but as the years passed, so did the possibility they would come home.

Hauptmann Erwin Gast was last seen by Inspektor Zeller in Beketovka as he unwittingly volunteered for mine-clearing duties. Gast's wife recalls that "sometime in the 1950s, an Oberst, a noblemen from Bad Godesberg, who was with Erwin in captivity, told me that Erwin had to help clear mines and probably died doing it." Officially, Gast is still listed as missing in action, but a determination on 18 March 1950 by the district court of Hanau declared him legally dead. Almost certainly killed in the same high-risk mine clearance operations were Oberleutnant **Alfons Schinke**, Leutnant **Erwin Hingst** and many others. The use of pioneers in this way, both enlisted men and officers, explains why so very few combat pioneers survived captivity.

As the Red Army approached Germany's borders in early 1945, Gast's beautiful young wife Ilse was forced to hurriedly abandon their home and, taking nothing except a few photos, flee to her parent's place in Hanau. This house was subsequently bombed and burned out completely, consuming all the family's possessions. But Ilse held on to the photos. As she had only been married a few months and never fallen pregnant, it was all she had of her husband.

Major **Otto Krüger** managed to see out the fatiguing death march, bitter cold, starvation rations and the first outbreaks of disease in Beketovka camp. Winter slowly turned into spring. A member of his 162nd Pioneer Battalion staff, Unteroffizier Hans Krauss, entered the camps in Krassnoarmeisk and Beketovka in July 1943 and immediately asked about his commander:

I learned in conversations with several officers I knew that Major Krüger had already died in April 1943. About five officers told me this. They did not belong to the battalion commanded by Major Krüger [but] the details provided by these officers were unambiguous.

Krauss' statement was crucial in declaring Krüger dead. Although still officially listed as missing in action, the district court of Goslar on 11 January 1954 set the date of his death as 20 April 1943. His wife never remarried and died, childless and alone, in a nursing home in the 1990s. With no recorded next-of-kin, her possessions and documentation were disposed of.

Hauptmann **Artur Baransky**, a sharp critic of Krüger, simply disappeared. His last letter home was dated 5 January 1943. Perhaps he was killed in the final days of fighting, or maybe he died anonymously while trying to break out or on a death march.

Hauptmann **Wilhelm Traub** was last seen on 8 April 1943 in Dubovka as he bade farewell from his small group. When Bonetsmüller went to Moscow in 1946, he met a major from another battalion and asked whether he knew the name Traub, which he did not, but he stated that officers were not required to work and received marginally better rations there. Oberstabsintendant Karl Binder, the divisional quartermaster, kept detailed notes of who died in captivity, but his secret journal was confiscated a few weeks before his release by a turncoat German officer. Binder remembered that Traub was on the list but Frau Traub doubted the accuracy of his recall because Binder also mentioned that Max Fritz – the battalion adjutant killed on 22 December 1942 in Stalingrad – had died in captivity. Frau Traub continued to conduct the heartbreaking search for her husband until 1954, writing letters to every conceivable government department and questioning those who had returned from captivity. All to no avail.

Johann Bonetsmüller and Alfons Thanisch were the last people to see him alive. Traub almost certainly died en route to Moscow, his body unceremoniously dumped somewhere along the railway line. He is still officially listed as missing in action.

While her husband endured the privations and anonymous death of Soviet captivity, Ella Traub was looking out for the best interests of their two sons. In January 1943 they stayed with Traub's parents in Helmstedt and listened every day to radio news about the battle at Stalingrad. In February they returned to their Navy-built apartment in Wesermünde, but later in the year the Traub boys were evacuated to the countryside to escape Allied bombing raids. Here, Jürgen witnessed for the first time the downing of an Allied plane: "Hundreds of planes in formation came back from an air-raid somewhere in Germany. One plane was trailing smoke and all of a sudden it plunged down from perhaps 20,000 feet, one wing came off and floated down like a leaf, no crew members parachuted. It was shocking and I never forgot it." The Traub family was on the receiving end of a bombing raid in late 1944. One night, they were in the shelter under their Wesermünde residence when a massive bomb came down nearby with a tremendous crash. Every window was blown out. Two days later the navy had new windows installed. Ella moved the family back to Helmstedt to stay with relatives and remained there until the war was over. Tragically, 9-year-old Hans, her youngest son, was killed on 14 May 1945 when he and another boy were playing with discarded ammunition. The Traub's Wesermünde apartment was occupied by American forces, so Ella and Jürgen were only able to return in 1948. She found a job as a secretary for a large fishing company while Jürgen returned to school. He went on to a very successful career in the shipping industry, including several positions as president of container shipping lines, mostly based in the US. Ella Traub retired in the late 1960s and moved to her hometown of Hamburg. She passed away on 21 January 1993. Jürgen Traub was killed instantly when hit by a van on 12 December 2013 while visiting a local farmers market north of San Francisco. His widow sold up in the US and moved back to Germany to live out her days surrounded by friends. The couple had no children.

The mother of Obergefreiter **Friedrich Vorherr** despaired but never relinquished hope her son would come home. She was electrified by two events. The first was when her daughter had a vision of Friedrich "standing in the doorway" one night during a kind of waking dream and he said everything was alright, he was doing well. The other was the sworn testimony of Emil Kistner, a late returnee from Soviet captivity, who stated that he had met the stonemason Friedrich Vorherr from Freudenbach during a soccer match in July 1953 at Prison Camp I in Stalingrad and stayed with Vorherr until 14 October 1953 when he, Kistner, was sent to hospital. Frau Vorherr immediately informed the Red Cross of this latest development:

> *He brought me news that he was with my son from June to October 1953 in Stalingrad*
> *Camp 1. Kistner corresponded with his next-of-kin, while my son has never given a*
> *sign of life from captivity. Kistner presumes that my son went to Camp 4 in October*
> *1953. Kistner also said that all prisoners in Russia have been registered since 1948.*

She sent this letter in November 1955, when even hard-core high-ranking prisoners, typically serving 25-year terms of imprisonment after being branded war criminals, were being released under the "Adenauer amnesty."[4] Three other returnees from the same Stalingrad camp told Frau Vorherr that they had never heard of her son, and moreover, since 1951, everyone there had been able to write home. "Why your son has not written is incomprehensible to me," wrote one of them diplomatically. Kistner passed away in 1956, so was unable to clarify the situation. In March 1958, the Red Cross in Moscow informed Frau Vorherr that they had failed to find any trace of her Fritz. Still, she kept a room ready for him and brusquely dismissed any suggestion to have him declared dead for legal reasons. Only upon her death in May 1959 did her other four children take this step, agreeing amongst themselves to redistribute the inheritance if Fritz returned. He never did. As the oldest son, he should have taken over the family's quarry and stonemasonry business, a tradition that stretched back 300 years. Instead, it went to his younger brother, a very talented sculptor. Today, the family business is still in operation in Freudenbach, where the impressive Vorherr family home is a local landmark.

Oberzahlmeister Max Keppler embarked on a death march through the barren steppe. On 25 February 1943, exhausted, and without an ounce of strength left, he collapsed into the snow and was shot by the guards. The other staff officers of 305th Pioneer Battalion who did not return home were **Stabsveterinär Dr. Rudolf Pompe**, **Stabsveterinär Dr. Erich Scheffel** and **Stabsarzt Dr. Donatus Wörner**. Pompe survived the hellish winter journey in cattle cars to Yelabuga POW camp only to die on 16 April 1943. Scheffel was last seen by Nuoffer in a gully near Gorodishche and was still in good health, but like Dr. Wörner, he simply vanished, lost amongst the faceless figures that stumbled into captivity and collapsed by the roadside or died of disease. Scheffel's three sons and Wörner's daughter grew up without fathers. It was even worse

---

4   German chancellor Konrad Adenauer flew to Moscow for talks with Nikita Khrushchev. By offering
    economic assistance, he successfully secured the release of the last German prisoners-of-war from the
    USSR, which later become known as the "Return of the 10,000". Releases under this program began in
    the first week of October 1955 and continued into 1956.

for the Scheffel boys because their mother relocated the family from Stuttgart to Leipzig in her home state of Saxony, both for familial support and to find a teaching job. This area was within the Soviet zone and behind the Iron Curtain in the 1950s, and as an employee of the Communist state, it was impossible for her to tell her sons about the war, and by extension, their father. He was, and remains, a stranger to them.

The amount of sadness heaped upon Grimm in his self-imposed duty to resolve the many MIA cases was overwhelming. Letter after letter from the wives of missing men, each more heart-rending than the next. One of the most tragic was Thilde, the wife of Feldwebel **Heinrich Bromeis**, part of Grimm's commissary staff in France and Traub's labour force in Stalingrad. His fate was one of Grimm's high priority cases and as such, he maintained an upbeat correspondence with Frau Bromeis. His efforts to distract her with friendly banter sometimes backfired, such as the time he described his family life. She wrote back: "I have yet to see your daughters, I hope they're doing well and that you have better luck than I did with my little son. That is still the worst, that he died after one and a half years. So I'm basically without any hope. If only news comes soon from my dear husband." Grimm was unaware that she had lost a son. Karl had come into the world on 7 June 1942, while Heinrich Bromeis was in Russia, so he never laid eyes on his first-born. Thilde received the devastating news in February 1943 that her husband was missing, and then little Karl died on 31 August the same year, aged just 14 months. How she went on is unimaginable. The only glimmer of hope was the return of her husband, but even dreams of his homecoming were overshadowed by the expression on his face as he looked around for his six or seven-year old son. The dread of telling him ate away at her. That is if he even came home. She did not know which scenario was worse.

In 1947, Frau Bromeis obtained a pass to visit Grimm in Stuttgart, "but do not be alarmed," she wrote, "my visit will be very short." She accepted Grimm's offer to stay in an attic apartment in his house, even though she had a relative – a gloomy aunt – in the city: "I'm always happy when I can see a healthy family." A year after this visit, Grimm sent her a long letter with his latest discoveries, including a list of the battalion's missing officers now reported as dead in captivity. "Many thanks for your letter," she replied,

I can almost say it made me happy, if not for all the sad news. I remember some of those names well and would be overwhelmed by the horror of this catastrophe were it not for the small consolation that you and others are going to so much trouble.

It was around this time that small packets of men began returning from Russia, their homecoming reported in local newspapers, and Grimm held this out as a branch of hope for Frau Bromeis. It did hot help:

Although you have achieved so much in the year since I stayed with you, I have much less hope of a message [from my husband] than you think. It's unfortunate, but I have become a dark pessimist. My hope is gone, and many other things with it, because I am taking all of this very badly.

The remorseless stress and worry weakened her immune system and she fell prey to one illness after another. A routine tonsillectomy required a four-month recovery, and any anxiety at all caused painful neuralgia that could lead to a severe heart attack at any moment. Despite a genuine risk to her life, she asked Grimm to write to her if he learned anything new. He never did. Bromeis had simply vanished. On 9 September 1950, Frau Bromeis had her husband officially declared dead. Several years later, overcoming her crippling grief, she met another man and they married. Heinrich Bromeis is still listed as MIA to this very day.

Dozens of stories just like this could be related here, but the sad case of Frau Bromeis is representative of them all. The Deutsches Rotes Kreuz (German Red Cross) maintains a list of 271 missing men from the 305th Pioneer Battalion. Ordinary men, not automatons, with families and loved ones.

Postwar Reunions and Camaraderie

After the war, the veterans bonded together, forgot war-time issues and looked out for each other, especially those down on their luck in the post-war chaos. The first divisional reunion on 2-3 May 1953 in Geislingen was attended by about 2,400 people. An important function of these types of meetings was to research missing soldiers. Page after page of MIA photos were perused in an effort to determine fates and provide hope/closure to families. Weeks before the reunion, the veterans were told that "the Suchdienst [tracing service] of the Red Cross will visit individual units during the reunion to clarify the fates of the numerous MIAs. Documents still in your possession, like photos of comrades, notes and such like, should be brought along. It is the duty of every one of you to support the efforts of the Red Cross and find a place for the Red Cross helper at your table."

Grimm brought along his files and spent a lot of time talking to families of missing soldiers, particularly Frau Keppler. This led to a friendship that endured for decades. "My father had good contact with Frau Keppler," recalls one of Grimm's daughters, "and in the years after the war, we got together with the Keppler children, who were a bit older than us. We lived in the next street."

The second reunion on 26-27 May 1962, again in Geislingen, was attended by about

900 veterans and a few hundred wives and guests, roughly 1,500 people all told. The Red Cross team was again in attendance and thanks to new information, they gained insights into 39 cases, 14 of which were eyewitness accounts about the fates of missing comrades. Compared to the first two, the third reunion in June 1965 was much smaller, with just 124 veterans present. Only five were Stalingradkämpfer, including Hans Zorn, former Leutnant and leader of 1st Company in Stalingrad, but he was the only pioneer present that had set foot in Stalin's city. The poor attendance was due to life: the war had finished twenty years earlier and the veterans were in their forties and fifties, the height of their business responsibilities and earning power. Edda Grimm recalls:

My father did not have much time in the years around 1960 because he had a lot of work in the company and at the same time applied for a patent, which meant more homework for him and also for us. In that time, he did not bring up the subject of war very often. Also, nobody wanted to hear about it.

– – –

And what of those who left the battalion in 1941-1942 or escaped Stalingrad before it became a pocket? A surprising number survived the war. Avoiding or escaping Stalingrad definitely improved the chances of making it through the war.

Oberleutnant **Hans Oster**, the inaugural commander of 3rd Company, was killed near Staraya Russa on 23 February 1943 while leading a company in the 254th Pioneer Battalion. The first commander of 2nd Company, **Hans Bloch**, was more fortunate. He led a company in the same battalion as Oster for almost a year before attending a training course and being transferred to the 198th Pioneer Battalion as its commander in September 1942. His performance over the next six months was so impressive that his division commander obtained him a promotion to Major. A severe case of jaundice in mid-December 1943 hospitalised him for several months. After several close calls in 1944-45, he returned home to his wife and young son. Another officer to survive the war, but only barely, was Hauptmann **Hugo Neddermann**. He passed away on 4 October 1947 in his hometown of Wuppertal, leaving behind his wife Erika and three infant children, the oldest 4 years old, the youngest just 14 months.

Ten months after being wounded in the dreaded Ventsy woods, **Peter Buchner** was discharged from hospital and transferred to a reserve unit. His hip wound greatly hindered him, but manpower of any type was needed, so Buchner returned to action in March 1944. Marriage to Annaliese in summer 1944 spared him from some catastrophic battles, but as a career officer, front-line duty was unavoidable, even if his limp was pronounced. Back to the east, this time Poland, where he assumed the role of adjutant of Einweisungs-Abteilung 1024.[5] Officially listed as missing in action since February 1945, a returnee informed his wife in November 1949 that heavy mortar rounds took his life on 7 February 1945 during the Soviet Vistula-Oder offensive.

After losing command of the 305th Pioneer Battalion, **Hans Hertel** was shifted around various assignments, mainly command of constructions troops and electrical cable maintenance regiments, but at one point he did find himself deployed outside Leningrad. Grimm crossed paths with his old commander in the west in January 1945. Hertel told him the east was in flames and to think of comrades whose homes were now under Russian control. Hertel's birthplace of Eibenstock, a stone's throw from the Czech border, was under threat. After the war, he took up residence in Stuttgart. On of the first person to visit him was Grimm, but that revealed the aspects of Hertel's character that caused so much aggravation in France. Others also called in to greet their old commander, as Grimm related in an amusing anecdote to Zeller:

> *Funny old Hertel lives in Stuttgart. The following very characteristic incident made me happy to pay a second visit. Heinz Schaate told me about his visit. Good old Hertel brought out a bottle of VSOP containing exactly the right amount to half-fill two glasses. There was no more. Now, in 1946, there is absolutely nothing wrong with welcoming someone without VSOP. But this same thing happened to me, down to the smallest detail, even with the same bottle! That is too much for us old Hertel connoisseurs!!!*

Hertel died 9 August 1967 in Stuttgart.

Hermann Klein, former commander of 1st Company, spent the remainder of the war with reserve units, first in Czechoslovakia, later in the homeland. Wounded on 29 March 1945 by multiple shell fragments in his face and skull, the most serious leaving a gaping hole in his left cheek and damaging his lower jaw. He ended the war in a Kempten hospital. Due to his wound, his stay in US captivity was brief and he returned to his civilian calling as a bailiff for the enforcement bureau of the Stuttgart district court. He died 24 January 1966 in Stuttgart.

The injury to his left arm prevented **Paul Nuoffer** from holding a rifle, so he rode out the war in reserve units. He remained famous amongst the battalion's survivors as the last man out of Stalingrad. Expectations of resolving MIA cases fell upon his

---

5   Einweisungs-Abteilung = A pioneer unit that defined and constructed fortifications but possessed little manpower of its own. It deployed unskilled labour from any source to build fortifications based on plans drawn up by its highly knowledgeable staff.

shoulders. After several failed attempts, Grimm finally traced Nuoffer in late 1946 and Nuoffer replied in January 1947. A small car rental business kept him away from home for long periods, but promises to visit Grimm were repeatedly made and broken. Contact was lost in 1948, which seemed to coincide with the homecoming of Dr. Dopfer, the man responsible for getting Nuoffer flown out of the pocket and the only witness to the nature of his wounding. Nuoffer's young wife Gertrud passed away in August 1957, aged just 38, while he died on 1 November 1966 in Münsingen at 47. Their only offspring was a daughter who sidestepped all contact by the author's research team.

Gefreiter **Franz Müller**, father of seven and probationary pioneer, recovered from his wounds and rejoined the battalion in Italy. There he was severely wounded and processed through the medical system until arriving at a hospital close to his home town of Bruchsal. He died of illness (panmyelophthisis and pneumonia) on 6 March 1944, leaving behind his wife Friederike, who raised their large family on her own.

Berthold Staiger was the last officer of the 305th Pioneer Battalion flown out of the pocket. After initial treatment for his hand wound at Mobile Field Hospital 772, he was transported by hospital train on New Year's Eve to Reserve Hospital I in Lublin, arriving on 6 January 1943. Another train journey began two days later and admission to Reserve Hospital Graz, Austria followed on 11 January. There he remained for over three months before transfer to a hospital in his home town of Reutlingen on 27 April 1943. The remainder of the year was spent in hospitals, on convalescence leave and performing light duties in reserve units before transfer to the newly formed reinforced Grenadier-Regiment 1021 to command its pioneer company. In March 1944 he was appointed commander of 1/177th Pioneer Battalion, part of the first-rate 77th Infantry Division, and deployed to the Normandy coastline. Combat against US paratroopers and the advance by forces landing on Utah Beach took its toll on Staiger's pioneers. Positioned to block the US advance across the base of the Cotentin Peninsula, Staiger fought hard, but his service in the Wehrmacht came to an end on 17 June 1944 when captured near St. Sauveur le Vicomte. He was sent across the Channel to Windermere POW Camp. At the end of July, Staiger and 700 other officers were shipped across the Atlantic to Seebe Camp 130 near Calgary, Canada, an idyllic spot in the shadow of Mount Baldy. Facilities included a billiard hall, tennis courts and ski trips in winter. Understandably, the inmates caused no trouble and there were no escapes. In January 1945, Staiger was moved to Wainwright Camp 135 set up in Buffalo National Park, and while not as nice as Seebe, conditions were still very good. April 1946 saw Staiger transported back across the Atlantic to Camps 17 and 296 in England. Prisoners were allowed out to work for local families in exchange for daily payment of a shilling or five cigarettes. Repatriation occurred on 23 September 1947, and after spending three days in Munsterlager, he was processed and released on the 26th. During his trip back to Germany, news reached him that his father had died a week earlier. The bachelor commenced the second stage of his life. With his diploma in architecture and years of pre-war work experience, he opened his own design bureau and ran it well into his eighties. He neither attended reunions nor remained in contact with his comrades. Handsome, eligible and charming, he had his pick of local women, finally marrying Ingeborg, seventeen years his junior, in 1960. He was 46. Their union resulted in six children. Upon the death of his mother in 1982, Staiger inherited a grand house built in 1932 by his father, also an architect, on a large elevated plot in the middle of Reutlingen. With all his children married and his wife tragically dying of cancer aged 65, he was alone in his large house. He moved into the ground floor and rented out the upper floors so the house had life in it again. A malignant tumour developed on his left wrist, right at the site of his Stalingrad wound, as well as on his chest, and he underwent treatment at 89 years of age. A minor stroke in December 2005 forced his admission to a nursing home in mid-January 2006, but the stubborn old pioneer refused to stay in bed, and while walking about he took a tumble and cracked a femoral head. An operation and titanium screws fixed the problem, although the weeks of bed rest left him weak. Nevertheless, his good spirits were unaffected. A wheeled Zimmer frame granted him some mobility and while going for a short stroll, he said "I'll be amongst the fallen soon." The next day, 21 February 2006, he died, three weeks shy of his 92nd birthday. His ashes were interred in the family grave next to his wife and parents. The sale of his multi-million dollar home ensured a comfortable future for his offspring.

Hans Zorn recovered quickly from his Stalingrad wounds, rejoined the new 305th Pioneer Battalion in April 1943 and remained with it throughout the Italian campaign. Apart from Grimm, Beigel and Hepp, he was one of the longest serving members of the battalion. After the war, he ran a building company in Eutingen with his father. Married, with several children, he died 16 September 1989 in Frankfurt, aged just 68.

Grimm, Beigel, Schaate and Hepp remained in contact and met up occasionally, jobs permitting. Retirement made the get-togethers easier and more frequent. Even though they all possessed their own networks of comrades, men they had grown close to during battle, Grimm was the hub of the 305th pioneers, mainly because he was interested in gathering all information and recording the history of his unit. War-time rivalries were forgotten, especially when the perceived ingratitude and hostile attitudes of young Germans forced the ex-officers to become tighter, a closed circle where they could discuss the war. As Beigel once said:

Of course I think about these times every now and then. A nice but also a terrible
time. But with whom can we talk about this today? Old comrades rarely get
together. Youngsters have no interest in these things. They are completely re-educated.

Alfons Hepp rejoined the 305th Pioneer Battalion in Italy and remained with it
almost to the end. Shell fragment injuries to an eyelid and right hand in March 1945
were shrugged off, but a bullet in his left shoulder a month later broke his scapula,
immobilised the joint and sent him to hospital. V-E Day found him laid up in bed.
Discharged in July 1945 and cleared by Allied authorities for immediate release, he soon
returned to his pre-war job in Stuttgart as a clerk for the Baden-Württembergische
Bank. He built up his career there and retired a very wealthy man. He owned a beautiful
house in Bad Waldsee and a large apartment in the centre of Stuttgart, but never
married or had children. Mentally and physically fit, he moved into a retirement home
in Ravensburg at the age of 85, but suffered two strokes in quick succession and passed
away there on 5 August 1999.

Despite being blind in one eye, **Heinz Schaate** returned to combat late in the war.
He was hit in the top of the head in 1945. Presuming he was dead, authorities sent his
tattered hat to his mother as a keepsake, together with the death notice. Incredibly,
though seriously injured, Schaate would recover from his head wound. He married a
few years after the war and began a family. His pre-war friendship with Grimm
continued and they met every 6 to 8 weeks, keeping each other abreast on the latest
news from their comrades. Schaate worked as a structural engineer for a construction
firm in Stuttgart. The little piece of glass that pierced his eye in the Gun Factory was
never removed and over the years it migrated around his body; doctors were of the
opinion that this tiny shard caused his death on 30 October 1988.

While recovering from his leg wound, **Ludwig Beigel** married his girlfriend. He
rejoined the new 305th Pioneer Battalion in May 1943 and took command of 2nd
Company. His wife gave birth in January 1944, almost the same time as Grimm's second
child, and a battalion order of the day congratulated both men on the birth of their
daughters. He was promoted to Hauptmann in April 1944. After almost 22 months of
unbroken service with the reformed battalion, he was sent to a fortnight-long battalion
leader's training course in March 1945, found to be suitable and given command of the
36th Pioneer Battalion. His pioneers were part of the effort to halt Patton's advance into
the middle of Germany, but were pushed back steadily, and in April 1945 withdrew into
upper Bavaria. Together with the remnants of his battalion, he went into American
captivity within walking distance of his home. Immediately after the war, he commenced
studies at a technical college and gained employment as a civil servant. However, the

lure of military service beckoned and he joined the newly created Bundeswehr, eventually reaching the rank of Oberstleutnant (Lieutenant-Colonel) and retiring in 1975. He gave up his civilian job the same year. "Cardiac disease was a signal for me to call it quits," he explained. Beigel suffered his second heart attack in 1978. Despite these heart problems, Beigel outlived Schaate and Grimm by a couple of years. He died in Traunstein – the site of his surrender to US soldiers – on 25 January 1991.

The reformation of the 305th Pioneer Battalion was an ambivalent period for **Richard Grimm**. While bathed in glorious French sunshine, dark images of icy Russia and missing comrades occupied his mind. His indifference and lack of energy went unnoticed, but he could not shake it off; traumatic events were still too fresh. His world brightened on 11 July 1943 when Lotte informed him she was expecting their second child. In late July the battalion relocated to the Italian Riviera for four months. Frolicking in the palm-fringed Mediterranean, witnessing a lunar eclipse on 16 August and relaxing in a garden-enclosed villa complete with a swimming pool and tennis courts were just what Grimm needed. The Russian cold was finally driven from his marrow and he felt recharged. The Allies invaded continental Italy in early September but the Bodensee Division remained in situ. As the threat grew, the division shifted south. Unforgettable for Grimm were operations on the Maiella Massif in the central Apennines, over 2000 metres altitude and with lots of snow in winter, a sector of such difficulty that he was given permission to wear the Edelweiss badge – symbol of the Gebirgsjäger (mountain troopers) – on his cap.[6] Furlough in January 1944 was timely because he arrived home a few days before the birth of his second child, a daughter they named Edda. Grimm only returned to his unit in the last days of February 1944 and was given the news of his long-desired promotion to Hauptmann. An unexpected transfer in May deposited him into a job he truly loved: a Stopi (Stabs Offizier Pioneer), a pioneer staff officer at corps level, a position equal in responsibility and respect to a battalion commander. "I was happy to remain in this position for the rest of the war, but my joy was short-lived." The massive Allied offensive at Cassino had begun and casualties in the 305th Pioneer Battalion drew Grimm back from his beloved assignment. He realised that combat deployment was imminent. Sure enough, he took over 3rd Company, and when his commander fell ill with dysentery on 2 June 1944, he assumed leadership of the battalion. A retreat through the mountains towards Subiaco by various elements of 44th and 305th Divisions was saved by a tenacious rearguard

---

6   This honour was retracted on 27 June 1944, probably because only soldiers with approved mountain infantry training and who had also completed an "Edelweißmarsch" (high-altitude training hike) could wear such an emblem on their cap.

action, and especially by Grimm. He had two vital roads rendered impassable to Allied tanks, a task made difficult because of the lack of explosives, but he improvised by using artillery rounds, aerial bombs and anti-tank mines. The division was able to occupy its designated defensive line on 5 June unchallenged, and the kudos went to Grimm:

> It it thanks to the swift and decisive handling by the leader of the pioneer battalion, Hauptmann Grimm, and his pioneers, but also the bravery of the reinforced Grenadier-Regiment 132, that the division, under the most difficult conditions imaginable – without rations for days on end, continually without supply trains and almost without tank-busting weapons – was able to fulfil its mission. Naming this brave regiment and its affiliated units in the Wehrmachtbericht (Armed Forces Report) has been requested.

Grimm was decorated with the Iron Cross First Class on 6 June. At long last, the revered cross was his, although he was perplexed because only four days of combat in Italy was required to earn it, whereas it had escaped his grasp in Russia despite six months of constant operations. Grimm relinquished command of the battalion when the commander returned on 12 July. He was proud of his accomplishment but glad to give up the responsibility. Apart from garnering accolades, his six-week stint as battalion leader attracted the attention of his superiors, and it was decided the time was right to advance his career. His direct involvement with the 305th Pioneer Battalion came to an end. First he attended a month-long supplementary course for pioneer leaders at Dessau-Rosslau, and then six weeks of intensive instruction in the art of commanding a battalion. He lived in the barracks with the other 37 attendees, all grizzled veterans covered in medals, at least four of them wearing the German Cross in Gold. A renewed sense of purpose and belief in ultimate victory surged through Grimm's veins. Once the course was over, he went home for almost two months and then returned to duty, missing Christmas with his family again for the sixth year in a row. This time he headed west, to the so-called Invasionfront, and assumed a role on the staff of 74th Army Corps. His superior was an old Oberst "who had organised nothing." On Christmas Day he witnessed a V1 fly overhead and felt confident the war would end with Germany the victor. This attitude gradually crumbled as he withdrew steadily, demolishing historical bridges and yielding German towns to the enemy. A letter dated 6 March 1945 was the last one received by Lotte. Anxious months followed. Grimm had been captured in the Ruhr pocket, first being interned in a large cage at Remagen, and then an American camp at Attichy, France "where we experienced a three-month long starvation diet. We were so weak that we were unable to walk around." The hunger was so great that prisoners boiled left-over coffee grounds and grass just to have something to chew on and get into their stomachs. This deliberate starvation – some claim on Eisenhower's orders, others that it was an unavoidable consequence of food shortages in post-war Europe – took the lives of tens of thousands of POWs across the network, but there was little sympathy for them because the ghastly excesses of the Nazi regime had been exposed. Grimm returned home after 13 months and started rebuilding his family's life. Because of fierce Allied bombing raids, Lotte had abandoned their Stuttgart residence and moved in with her parents in Kirchheim. In the weeks immediately after the war, four families, including Grimm's, were evicted from their large private residences to make way – ironically – for UNRAA (United Nations Relief and Rehabilitation Administration), an organisation whose mission was to care for displaced persons. Lotte Grimm, her two daughters and her parents travelled back to Stuttgart but found their apartment occupied. They took over two small rooms. This was the situation Grimm found upon his return, yet despite obtaining an eviction order, the unwelcome guests took over two years to move out.

When Grimm applied for his old job at Hahn & Kolb, the new managers – all strangers to him – said they had nothing, citing the firm's reduced output and recent retrenchments. However, old work colleagues informed Grimm that management did not want to employ ex-soldiers for fear it might attract the attention of the American occupation force. Fortunately, an opportunity presented itself when the Robert Gutekunst Fabrik, specialising in the production of bolts, nuts and screws, relocated to Kirchheim. The combination of his past experience and the shortage of skilled workers made Grimm a shoo-in, so he was soon working there as the main agent, controlling the purchase of raw materials and sales of the finished product. Over the years he rose to assistant manager, remaining there with much success until his retirement in 1976. With employment taken care of, Grimm attended to a deeply personal matter: he undertook legal action against his brothers because during his absence the two of them – both of whom never bore arms for their country – had smoothly fleeced him of his legacy in the family business. Every Deutschmark to his name was poured into the lawsuit. The district court found in his favour, but several more years were required before Grimm was awarded his rightful inheritance. This ugly matter caused a schism in the Grimm family that has never healed.[7]

---

7   The author originally wanted to call this book "Grimm's Brothers", a not-so-subtle play on words of the Brothers Grimm, the well-known German folklorists. The name also captures the essence of the soldierly brotherhood formed during combat operations. When I mentioned this title to his daughter, she said, "I'd really prefer you didn't because of the trouble my dad had with his brothers." The connection had not even occurred to me. I abided by her wish, even though I believe it is the perfect name for the book.

The processing of his war experiences weighed upon him until about 1955. After that, he wanted to hear nothing more about the war. Newspaper reports were a constant source of annoyance to him because he always said that most of them were filled with lies. He never read books about World War Two. He had his own opinions and shared them often with family and friends. Ramifications of his traumatic experiences, particularly Stalingrad, were reflected in his post-war personality and altered temperament, as described by his daughter Edda:

> *With press reports nowadays of soldiers traumatised by deployment in Afghanistan, I now realise that he was very badly affected. However, as a man and soldier, he could never admit it. He was extremely anxious around us, his family, while at the same time being exceptionally strict with us as children. My aunt once said: "Richard, they are children, not soldiers!" We had to do everything he wanted right away, no exceptions. He did not tolerate tardiness or insubordination and would never enter arguments. He never hit us, but shouted so much that we ran crying from the room. My sister, as she still says, suffered greatly from it, myself actually less so; I was sometimes able to go back in an hour and get my way. We also had a good education and the social environment was alright. We often met up with other war widows and their children and had good times.*

When time permitted, Grimm visited his old comrades and one time, to his great satisfaction, he took his wife and daughters to the Cassino battlefields, to show them where he fought and to lay wreaths on the graves of fallen comrades. He also delivered eulogies for many comrades, including his faithful Oberfeldwebel **Lorenz Locher** who died a haunted man on 16 June 1966. Old comrades also turned to him when they required evidence of their service and wounds in order to receive government benefits.

The stomach problems that plagued Grimm during the war came back to haunt him in his twilight years. While holidaying in 1988, a section of bowel irradiated in an earlier operation became inflamed and he was taken to hospital, where an invasive procedure set things right. Confinement to bed gave him a lot of time to think and he decided to get his war-time files in order. He had started typing up his manuscript years earlier but the recent health scare renewed his gusto. Letters and questionnaires went out to his old comrades, photos were identified, small mysteries solved. "He was always at the typewriter," remembers Edda, "and the memories caused insomnia. That was a tough time for my mother." Try as he might, he could not finish his Stalingrad memoirs because any attempt to do so aggravated painful memories and induced hideous nightmares. Every time he closed his eyes he relived the most awful moments of the battle and his mind's eye involuntarily focused upon the most disturbing deaths. Nevertheless, he did complete the main manuscript and experienced a catharsis that changed his character. He devoted himself to friends and was an affectionate grandpa. His grandsons have fond memories of him instructing them on how to fire air-rifles properly, using the long hallways of his Kirchheim residence as a shooting gallery. He fully committed himself to his wife, spoiling her with small gifts, new furniture and other things that women want but that held no interest for him.

Prostate cancer ended his life on 24 May 1990, a week after his 77th birthday. As per his wishes, his daughters donated a copy of his manuscript to the Bundesarchiv in the new millennium, though it took well over a decade to be catalogued and made available to researchers. Grimm was survived by his wife, two daughters, five grandchildren and an impressive archive that has enabled the sacrifice of his battalion to be immortalised. The men of the 305th Pioneer Battalion – Grimm's war brothers – are not forgotten. Their memory has been rescued from oblivion.

Oberstleutnant Hans Hertel, first commander of Pi.Btl.305.

Major Friedrich Beismann, second commander of Pi.Btl.305.

Hauptmann Wilhelm Traub, leader of the battalion in Stalingrad.

Hauptmann Hermann Klein, Chef 1. Kompanie.

Oberleutnant Richard Grimm, Chef 2. Kompanie.

Oberleutnant Friedrich Häntzka, Chef 3. Kompanie.

Oberleutnant Heinz Schaate, staff officer and later company leader.

Leutnant Ludwig Beigel, platoon- and later company leader.

Leutnant Peter Buchner, platoon leader in 2. and later 1. Kompanie.

Leutnant Bernhard Staiger, platoon- and company leader.

Leutnant Hubert Homburger, light pioneer column leader.

Oberfeldwebel Lorenz Locher, leader of company HQ.

Feldwebel Wilhelm Platzer, platoon leader in 2. Kompanie.

Unteroffizier Adam Pauli, squad leader in 2. Kompanie.

Unteroffizier Heinz Rinck, squad leader in 2. Kompanie.

Gefreiter Friedrich (Fritz) Vorherr, Grimm's motorcycle driver.

Gefreiter Willi Füssinger from 1. Kompanie. MIA at Stalingrad.

Gefreiter Hans Bonetsmüller was one of the few to make it home.

Battalion HQ in "maison rouge", a seaside mansion in Saint-Georges-de-Didonne, became a gilded prison in summer 1941. Initial pleasure at laid-back lunches dissolved when Hertel insisted his staff officers remained seated until dismissed, usually for two or more hours. From left: Gabelmann, Zeller, Hertel, Fritz, Haan, Grimm and Dr. Dopfer. Grimm's "office tan" reveals the extent to which he was chained to his desk.

Lunch at l'Oasis, a Breton cottage nestled amongst pines, after a horse ride past lakes and quaint villages, 20 April 1941. Schaate and Grimm had known each other for almost eight years and were good pals.

Endless marching, May 1942. Oberleutnant Grimm and Oberfeldwebel Locher march at the head of 2nd Platoon. Its leader, Leutnant Hepp, is on the left. The platoon is divided into squads. On the right, with an MP-40 over his shoulders, is Unteroffizier Pauli. On the left, between Grimm and Locher, is Unteroffizier Bub.

Grimm (squatting) and Hepp (standing, with binoculars) oversee the recovery of the battalion HQ's bus after its right rear wheel broke through a bridge deck. May 1942.

Oberfeldwebel Locher reads his mail while Grimm takes care of his personal hygiene. He had to head off every night on horseback or motorcycle to receive orders for the next day. Early June 1942.

Hauptmann Beismann bestows the Iron Cross Second Class to some of his staff members on 10 June 1942. From left: Oberleutnant Heinz Schaate, Leutnant Max Fritz, Gefreiter Franz Rapp, Gefreiter Ernst Köhle, unknown, Hauptmann Beismann and unknown. The Second Class medal was worn on the day of bestowal; thereafter, only the ribbon remained, sewn into the buttonhole.

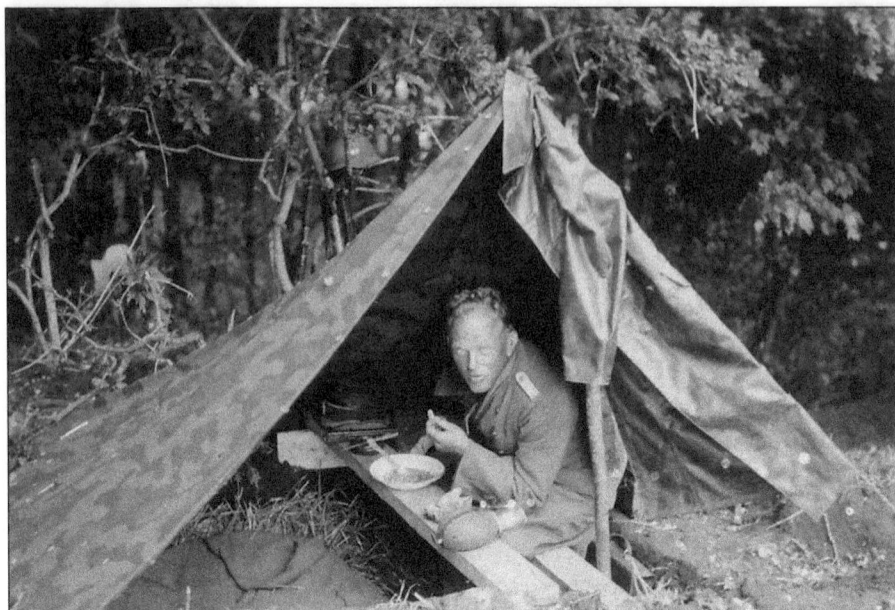

Beigel dines in his tent. Throughout the eastern campaign, Beigel never spent one night inside a building, instead preferring to camp outside. The ostensible reason was bugs and lice inside Russian huts.

Grimm was awarded the Iron Cross Second Class on 18 June and finally replaced the War Merit Cross ribbon in his buttonhole. Some of the battalion's officers, like Grimm, wore fabric sleeves over their shoulderboards to conceal their status as officers from enemy snipers.

During lunch on 18 June 1942, Grimm's entire company was assembled for an award ceremony. Grimm handed out ten Iron Crosses. From left: Grimm, Feldwebel Willi Platzer (leader of 3rd Platoon), Unteroffizier Fritz Rauschenbach (squad leader in 1st Platoon) and Unteroffizier Adam Pauli (squad leader in 2nd Platoon).

Corps commander General Heitz (left) and Hauptmann Beismann (middle) inspect and approve the 8-tonne bridge over the Oskol River, 2 July 1942. On the far right is Grimm.

Leutnant Beigel, wearing a field-made camouflage jacket, is ready for reconnaissance of crossing sites over the Don River, mid-July 1942.

The grave of Pionier Heinrich Erath in Orekhovskiy. His death on 30 July 1942 set a tragic train of events in motion that would haunt Grimm for the rest of his life.

In early August 1942, Soviet assaults were expected at any moment against the battalion's sector near Hill 243.3. Note the rifle, spare ammo clips, egg- and stick grenades. Gefreiter Rinck is on the right.

Feldwebel Platzer, sitting outside his foxhole in the Hill 243.3 sector, wears a Soviet amoeba-pattern camouflage suit used by combat engineers and snipers.

Hygiene standards slipped due to the lack of water. This is Unteroffizier Pauli and Unteroffizier Bub, the first man in the company to receive the Iron Cross First Class for his role in destroying a T-34 on 24 June 1942.

Lack of water also affected officers, but Beismann expected them to maintain higher standards. Grimm prepares to wash his dirty hands and face in a precious bucket of water brought forward by his orderly in his ex-British Morris car. Behind is a German half-track disembowelled by a direct hit.

This unassuming patch of woods was the scene of a nightmarish battle for 1st and 2nd Companies, mid-August 1942. The copse was traversed by two cuttings from bottom to top. A pioneer is in position in the foreground. Note the burnt grass, set on fire by the Soviet defenders one night when winds were favourable.

Amongst Grimm's prisoners were these tough officer candidates hastily taken from schools and thrown in the path of the German advance. The pioneers were impressed by the morale and bearing of these teenagers.

The Ostrovskiy bridge was captured by 16th Panzer Division on 16 August. The Bodensee Division took over the bridge three days later and Beismann's pioneers set to work improving and maintaining it.

Grimm and Vorherr look annoyed ("mildly worn out" wrote Grimm) after running the gauntlet of aimed artillery fire near Vertyachii, September 1942. Fortunately, Vorherr's Zündapp KS750 was up to the task.

Beismann (middle) personally pinned the Iron Cross First Class on Oberleutnant Beigel and Leutnant Hepp. The smile could not be wiped from Beigel's face, while Hepp seemed blasé throughout the entire ceremony.

On 18 September, the division's Catholic priest conducted 2nd Company's first field service about 1400 metres behind the front-line. Grimm kept an eye on the skies. Mid-service, two Soviet fighters zoomed overhead, collided and crashed to the ground only 150 metres away. Some perceived it as divine intervention.

Grimm posed for some final photos near Gorodishche on 13 October 1942, the day before the assault on Stalingrad's industrial district. This is Oberleutnant Staiger, Grimm and Gefreiter Nuoffer, Grimm's orderly. Except for helmets, both officers are fully equipped for battle, although their MP-40s are not loaded.

The task of clearing multi-storey buildings in the workers settlement was daunting. Considering that Soviet soldiers could be lurking behind any window, the actions of this soldier are foolhardy, 14 October 1942.

In cooperation with assault guns, Grimm's company methodically worked its way south along a railway embankment on 16 October, blotting out resistance nests with grenades and demolition charges.

While a comrade keeps watch, Gefreiter Rinck (left) takes a catnap near Hall 5 during a pause in the fighting on 23 October 1942. By this stage, he had already been wounded twice in the vicious urban fighting.

The battalion's paymaster, Oberzahlmeister Max Keppler, chats with Hauptmann Traub near Gorodishche, October 1942. A few months on operations has stripped all fat from Traub's once-portly frame

Richard and Charlotte Grimm celebrated their 50th wedding anniversary in 1989, eight months before his death. Prostate cancer ended his life on 24 May 1990, a week after his 77th birthday.

About the Author

Jason D. Mark has written ten books – several titles have been translated into Spanish, Swedish and Chinese – and over thirty articles. He runs the Leaping Horseman Books publishing house (www.leapinghorseman.com) and is the Editor-in-Chief of the quarterly journal *Kampfzone*.

Other Books by the Author

www.ingramcontent.com/pod-product-compliance
Lightning Source LLC
Chambersburg PA
CBHW061103220326
41599CB00024B/3900